TECHNOLOGY, SOCIETY AND MAN

RICHARD C. DORF

University of California, Davis

BOYD & FRASER PUBLISHING COMPANY

3627 Sacramento Street, San Francisco, California 94118

To my parents,
W. Carl and Marion F. Dorf,
and to my teachers at
the Bronx High School of Science, 1947–1951.

Richard C. Dorf: **TECHNOLOGY, SOCIETY AND MAN**

Library of Congress Catalog Card Number: 74-76445

ISBN: 0-87835-047-0

1 2 3 · 6 5 4

CONTENTS

iii

1

TECHNOLOGY AND SOCIETY

THE WESTERN WORLD, and increasingly the Eastern world, are technological worlds. Man lives with the tools he has developed; he depends upon them for his livelihood. *Technology* is science plus purpose. While science is the study of the laws of nature, technology is the practical application of those laws toward the achievement of some purpose or purposes. One may define technology as "the organization of knowledge for the achievement of practical purposes." A more expanded definition of the term is "a use of devices and systematic patterns of thought and activity to control physical and biological phenomena in order to serve man's desires with a minimum of effort and a maximum of efficiency."[1]

CONTEMPORARY INFLUENCES

The contemporary world is to a great extent influenced by technology. Major technological changes can set population shifts in motion, determine development patterns and create or solve pollution problems. For example, population throughout the world is shifting from rural to urban areas because farming technology allows fewer people to produce more food. Our complex energy systems have given us a wide array of conveniences, but at the cost of the degradation of the air, water and land.

Perhaps no change has had an influence as great as the general introduction of the automobile in the United States. With the introduction of the automobile, new industries were no longer limited to locating along a river or railroad line. People were able to work in the city and yet live far away from it. The scale and speed of technological change may well have outstripped the ability of our institutions to control and shape the human environment.

TECHNOLOGY AND CIVILIZATION

At least since the invention of the wheel or the lever, man has applied technology to overcome his physical limitations. In the future, by prudent use of improved technology, land can be made to grow more food, substitutes can be developed for scarce natural resources, and devices can be made for controlling pollution.

However, new technology can also create many new and often unanticipated problems. Automotive air pollution, persistent pesticides and nondegradable solid wastes, for example, are the consequences of technological innovations.

For good or ill the contemporary world is and will continue to be substantially shaped by technology. As Professor Elting Morison observed[2], "All earlier history has been determined by the fact that the capacity of man had always been limited to his own strength and that of the men and animals he could control. But, beginning with the nineteenth century, the situation had changed. His capacity is no longer so limited; man has now learned to manufacture power and with the manufacture of power a new epoch began."

Civilization is often thought to have started in the lower valleys of the Tigris and Euphrates, and its basic technology, which permitted development, was irrigation. Yet the very process of irrigation, through evaporation, left salt in the topsoil. The salt eventually destroyed the fertility of the land and the civilization based thereon. Again, modern civilization is based on technologies which may lead to its destruction.

In the middle of the twentieth century, two technological projects of the United States demonstrated to its citizens that technology could be harnessed to national goals. The two triumphs of American technology were produced by large government-funded enterprises. The first was the successful development of atomic weapons and the second was the conquering of interplanetary space and setting foot on the moon. Both were American conquests of "the impossible." But both gave individual Americans a new sense of powerlessness. Perhaps the most nearly comparable American achievement was the building of the railroad across the United States. Charles Francis Adams, Jr. called the transcontinental railroad a new "enormous, incalculable force ... exercising all sorts of influences—social, moral and political; precipitating upon us novel problems which demand immediate solution; banishing the old before the new is half matured to replace it; bringing the nations into close contact before yet the antipathies of race have begun to be eradicated; giving us a history full of changing fortunes and rich in dramatic episodes." The railroad, he stated, might be "the most tremendous and far-reaching engine of social change which has ever either blessed or cursed mankind."[3]

TECHNOLOGY IN THE UNITED STATES

The United States Government is the largest financial supporter of the development of new technologies. The amount expended for the fiscal year 1974 (July 1973 until June 31, 1974) for expenditures on research and development was $16,790 million, which is 6.25% of the total federal budget. Approximately $9,000 million of the total was spent for military research and development conducted by the Department of Defense and the Atomic Energy Commission. Another $3,000 million went to the National Aeronautics and Space Administration, leaving only $4,800 million for the development of new technologies for civilian needs.

The economic strength and national security of the United States rest on a technological foundation. Largely on the basis of the technological skills, industries and businesses in the United States have gained a predominant position in the world. This view, held by most citizens of the United States, is aptly summarized by the President's message of 1972. In his Message to the Congress on Science and Technology in March 1972, President Nixon said,

... innovation is essential to improving our economic productivity .... By fostering greater productivity, technological innovation can help us to expand our markets at home and abroad, strengthening old industries, creating new ones, and generally providing more jobs for the millions who will soon be entering the labor market.

• • •

This work is particularly important at a time when other countries are rapidly moving upward on the scientific and technological ladder, challenging us both in intellectual and in economic terms. Our international position in fields such as electronics, aircraft, steel, automobiles, and shipbuilding is not as strong as it once was. A better performance is essential to both the health of our domestic economy and our leadership position abroad.

If the United States is to meet these challenges, its technological resources must be utilized in a most effective manner with a minimum of undesirable side effects.

The rapid growth of a technology and its impact on a society can be adequately illustrated by the growth of the use of the passenger car. In Table 1–1 the number of autos, the population of the United States and the average number of persons per auto are shown. While population grew 45% during the 25 year period, the number of autos grew 214%. The automobile is now an integral part of American life.

This form of technological growth is not limited to the United States. For example, in Table 1–2 are listed several indicators of the use of technologies in

TABLE 1-1

The Growth of the Number of Autos and People in the United States.

Year	Passenger Cars	People	Persons per Auto
1947	30,718,852	144,126,000	4.69
1953	46,422,443	160,184,000	3.45
1959	59,561,726	177,830,000	2.99
1965	75,400,000	194,303,000	2.58
1972	96,397,000	208,837,000	2.17

Source: *Christian Science Monitor*, May 4, 1972.

TABLE 1-2

Indices of Technological Use in the World.

Indices (Per thousand persons)	U.S.A.	Ten European Nations of the Common Market	Soviet Russia	Japan
Autos	460	218	7	85
Television sets	399	231	127	214
Telephones	567	203	50	194
Energy in Kilowatt-hours used in industries	3.30	1.74	1.90	1.86

other parts of the world. It is clear that Europe and increasingly Japan and Russia are also becoming dependent upon technological devices in their countries.

RESPONSIBILITY AND TECHNOLOGY

Each citizen of the United States is a secondary party to every decision on the exploitation of technology. Many persons are unwilling and unknowing partners in each implementation of technology. We are all connected in the web of life, on which technology has a profound effect.

Often our traditional legal mechanisms for redressing civil wrongs are no longer as effective as they were when only two parties were involved. It is difficult to determine and place responsibility for technological consequences; many feel thwarted in their desires to allocate responsibility for the acts and consequences of technology.

Many individuals are frustrated by a world where the things they purchase are too complicated for them to comprehend or repair. Furthermore, they are not sure what performance they can expect from their purchases, and they are unhappy about the cost of repairing them. They never see the man who made what they buy and the item lacks the personal craftsmanship of the imagined past.

These sources of frustration are familiar, and they often do stem from technology. But many are a result of the frustration of rising expectations and our inability to satisfy them. Thus technology stands accused of failing human society, where perhaps much of the frustration results from the failure of our social institutions to use widely, and to distribute fairly, the benefits of technology.[4]

Society, in the form of government, intervenes on behalf of its citizens to attempt to obtain solutions to their problems. Yet safety and reliability are difficult to achieve at an optimum level. Safety is never absolute and the actions of the man must be accounted for. For example, after four decades of experience in improving the safety features of farm machinery, manufacturers are frustrated to find that owners of the safer equipment simply push it to a higher performance, accepting a constant level of hazard. A tractor redesigned to lessen the likelihood of tipping over is then driven on a steeper hillside, for instance.

One means of improvement, proposed by Lewis Branscomb, is to incorporate industrial standards in design.[5] It is proposed to place engineering standards on a performance basis.

RESPONSIBILITIES OF TECHNOLOGISTS

Gene Marine, the writer, has stated a stringent criticism of the acts of engineers in his recent book.[6] He believes that the engineer or technologist divides the world into that which is useful to man and that which is useless. He believes there is an "engineering mentality." This mentality, he states, is represented by an engineering approach to public questions, to planning, even to the correction of malfunctions resulting from earlier technical developments. It is an approach, he believes, which ignores the resulting side effects and never seeks a larger context for problems. For example, in the case of the construction of a bridge, he states:

The point is not that we ought not to build bridges; the point is that the Engineers—all of those who take the engineering approach: build the bridge and get the people and the cars from one side of the river to the other and to hell with the side effects—are shaping the nation unchecked, molding the land and murdering thousands of its inhabitants, raping America while the rest of us look the other way.

Perhaps this highly critical, unsympathetic view can be summarized by the following passages from Marine's book *America the Raped.**

*From *America the Raped: The Engineering Mentality and the Devastation of a Continent* by Gene Marine, Copyright Simon and Schuster, Inc., 1969. With permission.

Fifty years ago, more or less, Americans rose up in anger against the rape of their country.

The fight of a few dozen men, led by giants—Theodore Roosevelt, Gifford Pinchot, John Muir, John Wesley Powell—was against the uncaring lumbermen, who despoiled hill and valley and left eroding soil and sick rivers in their wake; against the unthinking farmers and stockmen, who replaced precious and fertile grasslands with thorn scrubs and dust bowls; against the mindless hunters who wiped out a hundred species and endangered a hundred more. The spoils of their victory are the national parks and forests, the wildlife refuges and wilderness areas, the national seashores and monuments, the soil conservation services, the hundreds of state parks and forests and preserved areas that followed them into existence.

Left when the battle had ended was a mushy purr-word—"conservation"—and a vague, persisting conviction that except for a few renegade lumber companies and mining firms, the rapine had ended. In fact, it has hardly begun.

But while a dozen groups have arisen to keep the old rapists in check—or at least to try to—the new rapists are loose upon the land. Theirs, still, are the vicious, violent, *laissez-faire* techniques of the turn of the century. They are not necessarily employed by lumber companies or mining companies or railroads; a lot of them work for you and me. They are the public servants who work for the Port of New York Authority or for the state highway commissions. They work for the United States Forest Service or the National Park Service. They are in the Army's Corps of Engineers and the Bureau of Reclamation and the Bureau of Public Roads. They are dedicated, single-minded men. And when they talk—which is as rarely as they can manage—theirs is the language of fanatics.

They are called Engineers.

They build bridges and dams and highways and causeways and flood-control projects. They *manage* things. They commit rape with bulldozers.

They are hard to fight off, because they must be fought off with words, and the weapons we have are inadequate. In New Jersey there is a fantastic land of wonders, still substantially as it was when the glaciers retreated three thousand years ago. It is called the Great Swamp. The Engineers of the Port of New York Authority want to put a jetport on it—an absurd and irreversible crime. But who needs a "swamp"?

The salt marshes of the Georgia coast have become an outstanding laboratory for the study of the bases of life; there, the University of Georgia Marine Station has learned much of how shrimp and other seafood depend on the unusual—and diminishing —estuarine conditions for their life. Yet Dr. Eugene Odum, the leading researcher in the field, reports that "we are often asked, 'Of what value is the salt marsh?' or 'What can be done with all that wasted land?'"

The Engineers know: build a dam, build a levee, build a wall, dredge, fill, *change.* The marsh grass will die, the phytoplankton will die, the algae will die—and thus the shrimp and the bass will die, but the Engineers don't care. What good is a salt marsh? Who needs a swamp?

LIMITATIONS OF TECHNOLOGY

Numerous examples of technological successes are evident in our daily lives. Perhaps the most glamorous and widely-hailed successes are in the fields of medicine and space technology. However, most notably, technology is criticized for its failure to solve the problems of transportation and to ameliorate the serious urban and environmental problems.

Indeed, if one examines a variety of national programs of the past decade which have been developed to attack major social problems, one is tempted to draw the conclusion that a technological solution to such problems is destined to be incomplete. In the area of education, for example, problems of *de facto* segregation yield only in part to technical solutions such as the use of computer aided instruction and educational television. Such can be ameliorated, slowly, with the proper application of technological approaches merged wisely with social and political awareness of the many dimensions of the issue.

Such an approach to the problem of health care has been recently proposed.[4] An illustration of the application of technology to health care follows.

From ... a model ... we can formulate a lengthy list of alternatives—specific governmental or national programs which effect a positive change in the health status of this country within the existing social, political, and economic constraints. Such a list would certainly include, for example, the following:

a) the design of an array of ambulances and emergency vehicles similar to that now in existence in the Soviet Union and ranging from general purpose to highly specialized;

b) the development of low-cost special-purpose minicomputers for hospital information systems;

c) the realization of a greatly expanded nationwide network of artificial-kidney centers to treat the 30 000 patients annually who are now unable to obtain help;

d) research directed toward malnutrition tests which can be administered early and at low cost to large numbers of people, particularly expectant mothers, where nutritional deficiencies seem to affect the child adversely.

Just these four possibilities point out the wide range of technological difficulties which can be anticipated for the complete list of alternatives. The programs a) and b) are straightforward from a technical standpoint—they require no new technology. Indeed, a) merely awaits some assurance of a suitable market. Program b) requires at least a modest study of the true needs of this aspect of the health care system, or perhaps even more, an agreement among hospital administrators and managers.

Item c) begins to introduce technological difficulties, or at least uncertainties, because of on-going research directed toward cost reduction, simplification, and portability of artificial kidney machines. Finally, the nutritional-test program demands a significant research effort and hence involves more uncertainty as to success.

In each of the hundred or so possible programs which can be listed for improving health care by technology, the description of the program must also include the model for the social and economic constraints within which the new technology must operate. For example, the ambulance-redesign effort (studied recently by both HEW and the National Academy of Engineering) is severely limited by the confusing multiplicity of responsibilities for ambulance operation in the typical city, the stringent financial constraints under which both municipal and entrepreneurial systems operate, the inadequate training programs for ambulance personnel, and a disarray of Local and State statutes governing operation.

Thus a model is to be developed where a list of alternatives has been generated, each complete with the social, political and economic portions of the model and the program plan. The actual implementation of any such program requires decisions on priorities. It is at this point that the political processes assume the decision-making responsibility. Unfortunately, the political processes are not always effectively assisted by the engineering and scientific communities in this matter of priorities. Many professional engineering or scientific societies have maintained a detachment from the political process, often abdicating a responbility to be involved in the decision process to ensure appropriate and intelligent uses of technologies that will serve national needs and priorities.

In setting goals for society and performance standards for our technologies, we must be judicious in timing our commitment of resources and effort. To commit too soon is to risk failure and frustration. To hesitate too long deprives society of needed resources and developments. The decision–making process must account for the fact that the impact of a new technology is not linear, but rather a complex set of mutually dependent matters.

PRUDENCE AND FORESIGHT

The late Paul Goodman wrote about technology and its liberating and practical benefits in the following article. Goodman discussed and urged prudence, *i.e.*, foresight, and simplification of the technical system. He recommended a resistance to apply every new technology or device such as the supersonic transport plane. Finally, he advised the less–developed countries to innovate and utilize technology, one would hope without a repetition of the mistakes of the already developed countries. Goodman had much to say both to the critic and to the advocate of technology.*[7]

For three hundred years, science and scientific technology had an unblemished and justified reputation as a wonderful adventure, pouring out practical benefits, and liberating the spirit from the errors of superstition and traditional faith. During this century they have finally been the only generally credited system of explanation and problem-solving. Yet in our generation they have come to seem to many, and to very many of the best of the young, as essentially inhuman, abstract, regimenting, hand-in-glove with Power, and even diabolical. Young people say that science is antilife, it is a Calvinist obsession, it has been a weapon of white Europe to subjugate colored races, and manifestly—in view of recent scientific technology—people who think that way become insane. With science, the other professions are discredited; and the academic "disciplines" are discredited.

The immediate reasons for this shattering reversal of values are fairly obvious.

Hitler's ovens and his other experiments in eugenics, the first atom bombs and their frenzied subsequent developments, the deterioration of the physical environment and the destruction of the biosphere, the catastrophes impending over the cities because of technological falures and psychological stress, the prospect of a brainwashed and drugged 1984. Innovations yield diminishing returns in enhancing life. And instead of rejoicing, there is now widespread conviction that beautiful advances in genetics, surgery, computers, rocketry, or atomic energy will surely only increase human woe.

In such a crisis, in my opinion, it will not be sufficient to ban the military from the universities; and it will not even be sufficient, as liberal statesmen and many of the big corporations envisage, to beat the swords into ploughshares and turn to solving problems of transportation, desalinization, urban renewal, garbage disposal, and cleaning up the air and water. If the present difficulty is religious and historical, it is necessary to alter the entire relationship of science, technology, and social needs both in men's minds and in fact. This involves changes in the organization of science, in scientific education, and in the kinds of men who make scientific decisions.

In spite of the fantasies of hippies, we are certainly going to continue to live in a technological world. The question is a different one: is that workable?

PRUDENCE

Whether or not it draws on new scientific research, technology is a branch of moral philosophy, not of science. It aims at prudent goods for the commonweal and to provide efficient means for these goods. At present, however, "scientific technology" occupies a bastard position in the universities, in funding, and in the public mind. It is half tied to the theoretical sciences and half treated as mere know-how for political and commercial purposes. It has no principles of its own. To remedy this—so Karl Jaspers in Europe and Robert Hutchins in America have urged—technology must have its proper place on the faculty as a learned profession important in modern society, along with medicine, law, the humanities, and natural philosophy, learning from them and having something to teach them. As a moral philosopher, a technician should be able to criticize the programs given him to implement. As a professional in a community of learned professionals, a technologist must have a different kind of training and develop a different character than we see at present among technicians and engineers. He should know something of the social sciences, law, the fine arts, and medicine, as well as relevant natural sciences.

Prudence is foresight, caution, utility. Thus it is up to the technologists, not to regulatory agencies of the government, to provide for safety and to think about remote effects. This is what Ralph Nader is saying and Rachel Carson used to ask. An important aspect of caution is flexibility, to avoid the pyramiding catastrophe that occurs when something goes wrong in interlocking technologies, as in urban power failures. Naturally, to take responsibility for such things often requires standing up to the front office and urban politicians, and technologists must organize themselves in order to have power to do it.

Often it is clear that a technology has been oversold, like the cars. Then even though the public, seduced by advertising, wants more, technolgists must balk, as any professional does when his client wants what isn't good for him. We are now repeating the same self-defeating congestion with the planes and airports: the more the technology is oversold, the less immediate utility it provides, the greater the costs, and the more damaging the remote effects. As this becomes evident, it is time for technologists to confer with sociologists and economists and ask deeper questions. Is so much travel necessary?

Are there ways to diminish it? Instead, the recent history of technology has consisted largely of a desperate effort to remedy situations caused by previous overapplication of technology.

Technologists should certainly have a say about simple waste, for even in an affluent society there are priorities—consider the supersonic transport, which has little to recommend it. But the moon shot has presented the more usual dilemma of authentic conflicting claims. I myself believe that space exploration is a great human adventure, with immense aesthetic and moral benefits, whatever the scientific or utilitarian uses. Yet it is amazing to me that the scientists and technologists involved have not spoken more insistently for international cooperation instead of a puerile race. But I have heard some say that except for this chauvinist competition, Congress would not vote any money at all.

Currently, perhaps the chief moral criterion of a philosophic technology is modesty, having a sense of the whole and not obtruding more than a particular function warrants. Immodesty is always a danger of free enterprise, but when the same disposition is financed by big corporations, technologists rush into production with neat solutions that swamp the environment. This applies to packaging products and disposing of garbage, to freeways that bulldoze neighborhoods, highrises that destroy landscape, wiping out a species for a passing fashion, strip mining, scrapping an expensive machine rather than making a minor repair, draining a watershed for irrigation because (as in Southern California) the cultivable land has been covered by asphalt. Given this disposition, it is not surprising that we defoliate a forest in order to expose a guerrilla and spray teargas from a helicopter on a crowded campus.

Since we are technologically overcommitted, a good general maxim in advanced countries at present is to innovate in order to simplify the technical system, but otherwise to innovate as sparingly as possible. Every advanced country is overtechnologized; past a certain point, the quality of life diminishes with new "improvements." Yet no country is rightly technologized, making efficient use of available techniques. There are ingenious devices for unimportant functions, stressful mazes for essential functions, and drastic dislocation when anything goes wrong, which happens with increasing frequency. To add to the complexity, the mass of people tend to become incompetent and dependent on repairmen—indeed, unrepairability except by experts has become a desideratum of industrial design.

When I speak of slowing down or cutting back, the issue is not whether research and making working models should be encouraged or not. They should be, in every direction, and given a blank check. The point is to resist the temptation to apply every new device without a second thought. But the big corporate organization of research and development makes prudence and modesty very difficult; it is necessary to get big contracts and rush into production in order to pay the salaries of the big team. Like other bureaucracies, technological organizations are run to maintain themselves but they are more dangerous because, in capitalist countries, they are in a competitive arena.

I mean simplification quite strictly, to simplify the *technical* system. I am unimpressed by the argument that what is technically more complicated is really economically or politically simpler, e.g., by complicating the packaging we improve the supermarkets; by throwing away the machine rather than repairing it, we give cheaper and faster service all around; or even by expanding the economy with trivial innovations, we increase employment, allay discontent, save on welfare. Such ideas may be profitable for private companies or political parties, but for society they have proved to be an accelerating rat race. The technical structure of the environment is too important to be a political or economic pawn; the effect on the quality of life is too disastrous; and the hidden social costs are not calculated, the auto graveyards, the torn-up streets, the longer miles of commuting, the advertising, the inflation, etc. As I pointed out in *People or Personnel*, a country with a fourth of our per capita income, like Ireland, is not necessarily less well off; in some respects it is much richer, in some respects a little poorer. If possible, it is better to solve political problems by political means. For instance, if teaching machines and audio-visual aids are indeed educative, well and good; but if they are used just to save money on teachers, then not good at all—nor do they save money.

Of course, the goals of right technology must come to terms with other values of society. I am not a technocrat. But the advantage of raising technology to be a responsible learned profession with its own principles is that it can have a voice in the debate and argue for *its* proper contribution to the community. Consider the important case of modular sizes in building, or prefabrication of a unit bathroom: these conflict with the short-run interests of manufacturers and craft unions, yet to deny them is technically an abomination. The usual recourse is for a government agency to set standards; such agencies accommodate to interests that have a strong voice, and at present technologists have no voice.

The crucial need for technological simplification, however, is not in the advanced countries—which can afford their clutter and probably deserve it—but in underdeveloped countries which must rapidly innovate in order to diminish disease, drudgery, and deepening starvation. They cannot afford to make mistakes. It is now widely conceded that the technological aid we have given to such areas according to our own high style—a style usually demanded by the native ruling groups—has done more harm than good. Even when, as frequently if not usually, aid has been benevolent, without strings attached, not military, and not dumping, it has nevertheless disrupted ways of life, fomented tribal wars, accelerated urbanization, decreased the food supply, gone wasted for lack of skills to use it, developed a do-nothing élite.

By contrast, a group of international scientists called Intermediate Technology argue that what is needed is techniques that use only native labor, resources, traditional customs, and teachable know-how, with the simple aim of remedying drudgery, disease, and hunger, so that people can then develop further in their own style. This avoids cultural imperialism. Such intermediate techniques may be quite primitive, on a level unknown among us for a couple of centuries, and yet they may pose extremely subtle problems, requiring exquisite scientific research and political and human understanding, to devise a very simple technology. Here is a reported case (which I trust I remember accurately): In Botswana, a very poor country, pasture was overgrazed, but the economy could be salvaged if the land were fenced. There was no local material for fencing, and imported fencing was prohibitively expensive. The solution was to find the formula and technique to make posts out of mud, and a pedagogic method to teach people how to do it.

In *The Two Cultures*, C.P. Snow berated the humanists for their irrelevance when two-thirds of mankind are starving and what is needed is science and technology. They have perhaps been irrelevant; but unless technology is itself more humanistic and philosophical, it is of no use. There is only one culture.

Finally, let me make a remark about amenity as a technical criterion. It is discouraging to see the concern about beautifying a highway and banning billboards, and about the cosmetic appearance of the cars, when there is no regard for the ugliness of bumper-to-bumper traffic and the suffering of the drivers. Or the concern for preserving an historical landmark while the neighborhood is torn up and the city has no shape. Without moral philosophy, people have nothing but sentiments.

EVOLUTION OF THE ROLE OF TECHNOLOGY

Professor Daniel Bell has coined the phrase "the post-industrial society" to describe the economic and social forces present in our technologically-oriented society. In his current book, Bell describes the new society as society of "new men" who are the scientists and engineers of the new intellectual technology.[8] He believes that the United States is evolving from a manufacturing economy into a service economy. More than half the work force has moved from the assembly line to service functions in health, education and the like.

Secondly, he observes that the post-industrial society is a knowledge society where knowledge provides power and status to the technologists. Bell also forecasts the decline of the corporation and the rise of the knowledge-based institution such as the research university. Bell says we are moving from individual choice to political choice. Finally, Bell sees a clash between society's need for experts such as engineers and the populist vision of equality and maximum participation.

One of the characteristics of technology is the subdivision of the necessary tasks. John K. Galbraith discusses technology and its characteristics in his important book *The New Industrial State.**[9]

Technology means the systematic application of scientific or other organized knowledge to practical tasks. Its most important consequence, at least for purposes of economics, is in forcing the division and subdivision of any such task into its component parts. Thus, and only thus, can organized knowledge be brought to bear on performance.

Specifically, there is no way that organized knowledge can be brought to bear on the production of an automobile as a whole or even on the manufacture of a body or chassis. It can only be applied if the task is so subdivided that it begins to be coterminous with some established area of scientific or engineering knowledge. Though metallurgical knowledge cannot be applied to the manufacture of the whole vehicle, it can be used in the design of the cooling system or the engine block. While knowledge of mechanical engineering cannot be brought to bear on the manufacture of the vehicle, it can be applied to the machining of the crankshaft. While chemistry cannot be applied to the composition of the car as a whole, it can be used to decide on the composition of the finish or trim.

Nor do matters stop here. Metallurgical knowledge is brought to bear not on steel but on the characteristics of special steels for particular functions, and chemistry not on paints or plastics but on particular molecular structures and their rearrangement as required.

Nearly all of the consequences of technology, and much of the shape of modern industry, derive from this need to divide and subdivide tasks and from the further need to bring knowledge to bear on these fractions and from the final need to combine the finished elements of the task into the finished product as a whole.

---

*From *The New Industrial State* by J.K. Galbraith, Houghton Mifflin Co. With permission of the author and publisher.

SOCIAL DISCONTINUITY CAUSED BY TECHNOLOGY

In a recent book, Professor Peter Drucker explores the discontinuities in technology and society among others.[10] The application of knowledge to the technology of work will result, in the author's view, in profound changes. Technology will cause a discontinuity in society. An excerpt follows*.

The greatest of the discontinuities around us is the changed position and power of knowledge.

Seven thousand years ago or more, man discovered skills. There were great artists before this discovery. There have never been painters greater than those prehistoric men who left behind the cave paintings in France and Spain or the rock paintings in the Sahara. But there were no skilled craftsmen in those days. Skill supplied the tools to make average people without towering genius capable of competent, predictable performance, and capable of advancing from generation to generation by organized systematic apprenticeship. Skill created the division of labor and therefore made possible economic performance. By the year 2000 B.C. or so, our ancestors in the irrigation civilizations of the Eastern Mediterranean had developed every single one of the basic social, political, and economic institutions of society, every single one of our occupations, and most of the tools man had at his disposal until two hundred years ago. The discovery of skill created civilization.

Now we are about to make another major move. We are beginning to apply knowledge to work. We are not yet much further ahead, I submit, than were those remote ancestors of ours who first made hunting a specialized occupation with its own special tools, and with a skill to be learned in long arduous apprenticeship. But even the first faltering steps we have taken have shown that applying knowledge to work is a big idea, and an exciting one. Its potential may be as great as was the potential of skill when first discovered. The development may take as long. But the impacts already are very great—and the changes they imply are tremendous, indeed.

As great and as profound as any of these impacts *of* knowledge are the impacts *on* knowledge. Above all the shift to knowledge as the foundation of work and performance imposes responsibility on the man of knowledge. How he accepts this responsibility and how he discharges it will largely determine the future of knowledge. It may even determine whether knowledge has a future.

The vision of a clash between science and its colleague technology and literary intellectuals, while not new, has been expounded by C.P. Snow in his book *The Two Cultures* and its sequel *A Second Look*.[11,12] Snow believes that a polarization is increasingly developing between the humanists (or literary intellectuals) and the scientists and engineers. The engineers are ignorant of Shakespeare and the humanists know little of the Second Law of Thermodymanics. This split in cultures, he states, is a loss to society.

The two cultures are marked by two contrasting approaches to understanding the world (the rational, analytical, logical method versus that of the intuition and the creative imagination). The two cultures gap is nothing new.

*From *The Age of Discontinuity* by Peter F. Drucker, Harper & Row, 1969. With the permission of the author and publisher.

Swift, Pope, Blake and later Yeats attacked science and technology for its effect on the area of the spirit and the damage to nature.

Snow believes a great misfortune in society today is the innocence of each culture of the other culture. He pleads for more knowledge of science and technology among the humanists and more knowledge of literature and the social sciences among scientists and engineers. As Matthew Arnold wrote in the ninteenth century, a genuine humanism is scientific; certainly today many would avow that a genuine technology would be humanistic.[13]

POPULAR FEAR OF TECHNOLOGY

Why do people fear technology? There are many reasons that people fear technology and its consequences, but some are predominant at this time.[5] Technology appears to many to have too much momentum; it is difficult to "control." Technology produces change too fast and without effective opportunities to debate its effects and the desirability of its introduction into use. This rapidity of change is effectively portrayed in Figure 1–1, which shows the acceleration of social change with technological innovations. The span of time (shown by a solid line in the figure) from discovery of a new technology to practical application is constantly decreasing. Thus, while it took 35 years from discovery of the vacuum tube to a practical radio, it requirrd only three years from discovery of the transistor to practical application. With increasing rapidity, knowledge is converted to useful products which produce social change. Hence the problem of anticipating social change becomes increasingly difficult.

Our society is increasingly accustomed to calling on technology to solve problems, whether they be problems of war, poverty or disease. For example,

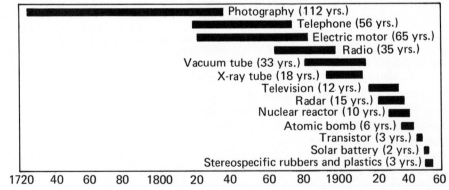

FIGURE 1-1   The increasing speed with which a technological innovation moves from discovery to practical application. From *World Facts and Trends*, published by the Center for Integrative Studies, School of Advanced Technology, State University of New York, Binghampton, New York, 1969.

until Vietnam, our national experience with technology in war was very successful. Now some realize that massive applications of technology can be ineffective, often counter-productive, in the absence of effective social goals.

POLITICAL SIGNIFICANCE OF TECHNOLOGY

In an incisive article, Wilbur Ferry argues that we have overcommitted ourselves in allegiance to technology. It enjoys too privileged a place on our political agenda. The first portion of Ferry's article follows.*[15]

MUST WE REWRITE THE CONSTITUTION TO CONTROL TECHNOLOGY?

Wilbur H. Ferry

I shall argue here the proposition that the regulation of technology is the most important intellectual and political task on the American agenda.

I do not say that technology *will* be regulated, only that it *should* be.

My thesis is unpopular. It rests on the growing evidence that technology is subtracting as much or more from the sum of human welfare as it is adding. We are substituting a technological environment for a natural environment. It is therefore desirable to ask whether we understand the conditions of the new as well as we do those of the old, and whether we are prepared to do what may be necessary to see that this new environment is made suitable to men.

Until now, industrial man has only marginally and with reluctance undertaken to direct his ingenuity to his own welfare. It is a possibility merely—not a probability—that he will become wise enough to commit himself fully to that goal. For today the infatuation with science and technology is bottomless.

Here is where all the trouble begins—in the American confidence that technology is ultimately the medicine for all ills. This infatuation may, indeed, be so profound as to undercut everything of an optimistic tone that follows. Technology is the American theology, promising salvation by material works.

I shall argue that technology is merely a collection of means, some of them praiseworthy, others contemptible and inhumane. There is a growing list of things we *can* do that we *must not* do. My view is that toxic and tonic potentialities are mingled in technology and that our most challenging task is to sort them out.

A few cautionary words are in order.

First, I am aware of the distinctions between science and technology but intend to disregard them because the boundary between science and technology is as dim and confused as that between China and India. Besides, it is impossible to speak of public regulation of technology while according the mother-lode, science, a privileged sanctuary. At the same time, it must be granted that the scientists have been more conscientious than the technologists in appraising their contributions and often warning the community of the consequences of scientific discovery.

Next, I shall use everyday examples. Some will therefore consider my examples

---

superficial. But it appears to me to be better to illustrate the case by situations about which there is considerable general knowledge. I shall rely on well known contemporary instances of technological development chiefly to show the contrast between their popular aspects, including popular ideas about control, and those less well known side effects that in the long run threaten to cancel out promised benefits.

The first point to be made is that technology can no longer be taken for granted. It must be thought about, not merely produced, celebrated, and accepted in all its manifestations as an irrepressible and essentially benign human phenomenon. The treason of the clerks can be observed in many forms, but there is no area in which intellectuals have been more remiss than in their failure to comprehend technology and assign it its proper place in humane society. With many honorable exceptions—I give special recognition to Lewis Mumford, who for forty years has been warning against the castration of spirit by technique—the attitude of the physical scientists may be summarized in advice once proffered to me, "Quit worrying about the new scientific-technical world and get with it!" And the disposition of the social scientists, when they notice technology at all, is to suggest ways of adjusting human beings to its requirements. Kenneth Keniston says in *The Uncommitted*: "We have developed complex institutions to assure (technology's) persistence and acceleration (and we) seldom seek to limit its effects."

We are here near the core of the issue. Technology is not just another historical development, taking its place with political parties, religious establishments, mass communications, household economy, and other chapters of the human story. Unlike the growth of those institutions, its growth has been quick and recent, attaining in many cases exponential velocities. Federal expenditure for research and development in 1940 was $74,000,000—less than 1 per cent of total government spending. In 1966 it was $16 billion—15 per cent of federal spending. This is not history in the old sense, but instant history. Technology has a career of its own, so far not much subject to the political guidance and restraints imposed on other enormously powerful institutions.

This is why technology must be classed as a mystery and why the lack of interest of the intellectuals must be condemned. A mystery is something not understood. Intellectuals are in charge of demystification. Public veneration is the lot of most mysteries, and technology is no exception. We can scarcely blame statesmen for bumbling and fumbling with this phenomenon, for no one has properly explained it to them. We can scarcely rebuke the public for its uncritical adoration, for it knows only what it is told, and most of the information comes now from the high priests and acolytes of technology's temples. They are enraptured by the pursuit of what they most often call truth, but what in fact is often obscene curiosity, as when much of a nation's technological quest is for larger and more vicious ways of killing—the situation today.

There is an analogy between the rise of modern economics and that of the new technology that one would have thought intellectuals would be especially eager to examine. Technological development today is in the enshrined position in political-economic theory that was accorded to economic development in the nineteenth and early twentieth centuries. Unguided and self-directed technology is the free market all over again. The arguments justifying *laissez faire* were little different from those justifying unrestrained technology. The arguments in both cases are either highly suspect or invalid. The free market dwindles in real importance, though the myth remains durable enough. But we know now that the economic machine needs to be managed if it is not to falter and behave eccentrically and needlessly injure people. That we have not yet conquered the political art of economic management only shows how arduous and thought-demanding a process it is, and why we should get after the equivalent task in technology at once.

Quite a lot of imaginative writing has been done about the world to come, whether that world develops from the technological tendencies already evident or is reconstructed after a nuclear war. This future-casting used to be known as Utopian writing. Utopias today are out of fashion, at least among novelists and poets, who are always the best guides to the future. With only two exceptions, the novels I have read tell of countries that no one here would care to live in for five minutes.

The conditions imagined are everywhere the same. High technology rules. Efficiency is the universal watchword. Everything works. All decisions are made rationally, with the rationality of the machines. Humans, poor folk, are the objects of the exercise, never the subjects. They are watched and manipulated, directed, and fitted in. The stubborn few in whom ancient juices of feeling and justice flow are exiled to Mars or to the moon. Those who know *how* are the ones who run things; a dictator who knows *all* reigns over all; and this dictator is not infrequently a machine, or—more properly—a system of procedures. I need go no further, for almost everyone is familiar with Orwell's *1984*.

Ferry then proceeds to illustrate, by example, cases of the malignant or undesirable capacities of technology. In his view the proposed SST and the proposed National Data Center are clear examples of undesirable technologies. He then proceeds with some prescriptions for the future. Specifically, Ferry recommends that the United States Constitution be rewritten in light of the immense power of technology.

One must nevertheless be grateful to those few members of the sanhedrin who keep pointing out the dangers as the nation turns doubtful corners. Dr. Murray Gell-Mann of the California Institute of Technology says that "society must give new direction to technology, diverting it from applications that yield higher productive efficiency and into areas that yield greater human satisfaction. ... " Carl Kaysen of the Institute for Advanced Studies at Princeton emphasizes that government institutions are no longer equal to the job of guiding the uses of technology.

Scientists and technologists are the indubitable agents of a new order. I wish to include the social scientists, for whose contributions to the technological puzzle I could find no space in this paper. Whether the political and social purpose of the nation ought to be set by these agents is the question. The answer to the question is no. We need to assign to their proper place the services of scientists and technologists. The sovereignty of the people must be reestablished. Rules must be written and regulations imposed. The writing must be done by statesmen and philosophers consciously intent on the general welfare, with the engineers and researchers summoned from their caves to help in the doing when they are needed.

How specifically to cope? How to regulate? Answers are beginning to filter through. Not many years ago it was considered regressive and ludditish even to suggest the need for control of technology. Now a general agreement is emerging that something must be done. But on what scale, and by whom?

E. J. Mishan, the British economist, calls for "amenity rights" to be vested in every person. He says, "Men [should] be invested by law with property rights in privacy, quiet, and clear air—simple things, but for many indispensable to the enjoyment of life." The burden would be on those offending against these amenities to drop or mend their practices, or pay damages to victims. Mishan's argument is scholarly and attractive, though scarcely spacious enough for the problems of a federal industrial state of the size of the United States. It does not seem likely that we can maintain our amenities by threat of tort suits against the manifold and mysterious agents, public and private, that are the "enemy."

The most comprehensive and thoughtful approach to the problem of regulation is that of U.S. Congressman Emilio Q. Daddario, chairman of the House Subcommittee on Science, Research, and Development. Representative Daddario starts with the necessity for "technological assessment," which he characterizes as urgent. It will amount to a persisting study of cause-effect relationships, alternatives, remedies. Representative Daddario does not speak of tonic and toxic, but of desirable, undesirable, and uncertain effects.

The subcommittee's study is only beginning, but it is based on some of the convictions that animated the writing of this article. Thus, the introduction to the first Congressional volume on technological assessment speaks of the dawning awareness of "the difficulties and dangers which applied science may carry in its genes" and of "the search for effective means to counter them."

It would be unfair to summarize the scope and method of this promising document in a sentence or two. I must leave it to those interested to look further into a first-rate beginning. It is too early to guess whether Congressman Daddario's group will come out where I do on this matter, but it seems unlikely. The subcommittee will probably come out for certain statutory additions to the present political organization as the proper way to turn back or harness technique's invading forces. There is ample precedent.

We can regard the panoply of administrative agencies and the corpus of administrative law as early efforts in this direction. They have not been very effective in directing technical development to the common good, although I do not wish to minimize the accomplishments of these agencies in other ways. Perhaps they have so far prevented technology from getting wholly out of hand. But it is very clear from examples like the communications satellites that our statutory means for containing technology are insufficient.

America is not so much an affluent as a technical society; this is the essence of the dilemma. The basic way to get at it, in my judgment, would be through a revision of the Constitution of the United States. If technology is indeed the main conundrum of American life, as the achieving of a more perfect union was the principal conundrum 175 years ago, it follows that the role and control of technology would have to be the chief preoccupation of the new founding fathers.

Up to now the attitude has been to keep hands off technological development until its effects are plainly menacing. Public authority usually has stepped in only after damage almost beyond repair has been done: in the form of ruined lakes, gummed-up rivers, spoilt cities and countrysides, armless and legless babies, psychic and physical damage to human beings beyond estimate. The measures that seem to me urgently needed to deal with the swiftly expanding repertoire of toxic technology go much further than I believe would be regarded as Constitutional.

What is required is not merely extensive police power to inhibit the technically disastrous, but legislative and administrative authority to *direct* technology in positive ways: the power to encourage as well as forbid, to slow down as well as speed up, to plan and initiate as well as to oversee developments that are now mainly determined by private forces for private advantage.

Others argue that I go too far in calling for wholesale revision of our basic charter. They may be right. Some of these critics believe that Constitutional amendment will do, and that what is needed is, in effect, reconsideration of the Bill of Rights, to see that it is stretched to cover the novel situations produced by technology. This is a persuasive approach, and I would be content as a starter to see how far Constitutional amendments might take us in protecting privacy and individual rights against the intrusions of technique.

But I also think that such an effort would soon disclose that technology is too vast, too pervasive to be dealt with in this way. The question is not only that of American rights, but of international relations as well, as Comsat illustrates. Technology is already tilting the fundamental relationships of government, and we are only in the early stages. A new and heavy factor has entered the old system of checks and balances. Thus, my perception of the situation is that the Constitution has become outdated by technical advance and deals awkwardly and insufficiently with technology's results.

Other critics tell me that we are sliding into anarchy, and that we must suffer through a historical period in which we will just "get over" our technological preoccupations. But I do not face the prospect of anarchy very readily.

So that my suggestion of fundamental Constitutional revision is not dismissed as merely a wild gasp of exasperation, I draw attention to the institutions dominating today's American scene which were not even dimly foreseen by the Founding Fathers. I refer to immense corporations and trade unions; media of communication that span continent and globe; political parties; a central government of stupendous size and worldshattering capabilities; and a very unJeffersonian kind of man at the center of it all.

It seems to me, in face of these novelties, that it is not necessarily madness to have a close look at our basic instrument in order to determine its ability to cope with these utterly new conditions, and especially with the overbearing novelty of technique. Technology touches the person and the common life more intimately and often than does any government, federal or local; yet it is against the aggrandizement of government that we are constantly warned. Technology's scope and penetration places in the hands of its administrators gigantic capabilities for arbitrary power. It was this kind of power the Founding Fathers sought to diffuse and attenuate.

Constitutional direction of technology would mean planning on a scale and scope that is hard now to imagine. Planning means taking account, insofar as possible, of the possibilities of technique for welfare. It means working toward an integrated system, a brand-new idea in this nation.

I recognize all the dangers in these suggestions. But leaving technology to its own devices or to the selfish attentions of particular groups is a far more hazardous course. For it must not be forgotten that the enormous proliferation of technology is today being planned by private hands that lack the legitimacy to affect the commonwealth in such profound measure.

The wholesale banning of certain techniques becomes absolutely necessary when technical development can no longer help but only harm the human condition. Scientists Jerome Wiesner and Herbert York exemplify this dictum in its most excruciating aspect when they say:

Both sides in the arms race are ... confronted by the dilemma of steadily increasing military power and steadily decreasing national security. It is our considered professional judgment that *this dilemma has no technical solution*. ... If the great powers continue to look for solutions in the area of science and technology only, the result will be to worsen the situation.

Though many lives are being wrecked, though the irrationality and human uselessness of much new technology is steadily becoming more evident, we are not yet over the edge. I close with Robert L. Heilbroner's estimate of the time available:

... the coming generation will be the last generation to seize control over technology before technology has irreversibly seized control over it. A generation is not much time, but it is *some* time. ...

LOSS OF POPULAR BELIEF

The present discontent with technology is clearly expressed by Barry Commoner, who says, "the age of innocent faith in science and technology may be over." Professor Rene Dubos says, " ... the technological and other practical applications of science have been oversold. In fact, the production of goods and the development of what is now called technological 'fixes' may not be the most valuable contribution that scientists make to society." Nevertheless, one must ask if the innocent faith in technology is *ever* justified. This confidence or faith in technology was partly based on the success with which it was possible to manufacture nuclear weapons in the four years between 1941 and 1945. In a similar manner, people ask, cannot technology make people healthy, prosperous and even wise?

But now, with the loss of faith, man may become more concerned about the failure of his former religion. He attacks the technocratic society, which he may have come to see as coldly rational and inhuman—a world of technology in which human ends are forgotten in the search for rational techniques and bureaucratic efficiency. In some sense man is dominated by technology, and some people contend that technology determines not only means, but also ends. Technology is the search for the one most efficient way, with no room for human choice or judgment about values.

SCIENCE, EDUCATION AND TECHNOLOGY

It is science that underlies technology and provides the fundamental knowledge for technological application. It is in this sense that science has become the contemporary ideology of the technologist. One view of the common man overwhelmed by a tidal wave of scientific and technological knowledge is shown in Figure 1–2.

FIGURE 1-2 One view of man overwhelmed by the tidal wave of scientific and technological knowledge. Courtesy Kaiser Aluminum News © 1966.

The *Oxford Dictionary* defines *liberal education* as "education fit for a gentleman." That is still an acceptable definition, but the idea of a gentleman has changed in this century. A century ago, when Europe awoke to the need for technological education, a gentleman belonged to the leisure class, which did not require any knowledge of science or technology. Modern gentlemen belong to the knowledge class; that is, their work requires knowledge of science and technology. A case can be made for including technology in the liberal education. Furthermore, it could become a unifying force between science and the humanities, because technology depends, in its proper use, on both. Technology is inseparable from men and their communities.

So the study of technology, its benefits and social consequences is the work of the humanist as well as the scientist and the engineer. For technology shapes our communities and is an integral portion of our lives. All men live with technology and should know its essence. A truly educated person must understand the technological forces present in the world and thus control them.

QUESTIONS FOR NOW AND THE FUTURE

In the following chapters there will be some answers to the myriad criticisms of technology as it is applied in the United States and the world. There will be analysis of the basis of the question: Is technology responsible for our social ills?

Some popular negative views are examined. Do machines deprive people of employment? Does technology rob people of privacy? Is technology autonomous and uncontrollable by man? Does technology dehumanize man? Is a bureaucratic state a natural consequence of a commitment to a technological society? These are questions we shall attempt to explore. The reader will, of course, formulate his own conceptual view of technology, society and man. Man the maker and user of tools is modern man. Is his future one of liberation?

CHAPTER 1 REFERENCES

1. R. Parkman, *The Cybernetic Society*, Pergamon Press, New York, 1972.
2. E.E. Morison, *Men, Machines and Modern Times*, M.I.T. Press, Cambridge, 1966.
3. D. Boorstin, *The Americans: The Democratic Experience*, Random House, New York, 1973.
4. J.G. Truxal, "Goals for Technology", *IEEE Transactions on Systems, Man and Cybernetics*, Vol. SMC–2, No. 5, Nov. 1972, pp. 595–598.
5. L.M. Branscomb, "Why People Fear Technology," *The Futurist*, Dec. 1971, p. 232.
6. G. Marine, *America the Raped: The Engineering Mentality and the Devastation of a Continent*, Simon and Schuster, New York, 1969.
7. P. Goodman, *New Reformation*, Random House, Inc., New York, 1969.
8. D. Bell, *The Coming of the Post-Industrial Society*, Basic Books, New York, 1973.
9. J.K. Galbraith, *The New Industrial State*, Second Edition, Houghton Mifflin Co., Boston, 1971.
10. P.F. Drucker, *The Age of Discontinuity*, Harper and Row, New York, 1969.
11. C.P. Snow, *The Two Cultures and The Scientific Revolution*, Cambridge University Press, New York, 1959.
12. C.P. Snow, *A Second Look*, Cambridge University Press, New York, 1969.
13. W.H. Davenport, *The One Culture*, Pergamon Press, New York, 1970.
14. I. Illich, *Tools for Conviviality*, Harper and Row, New York, 1969.
15. W.H. Ferry, "Must We Rewrite the Constitution to Control Technology?" *Saturday Review*, March 2, 1968, pp. 36–39.

CHAPTER 1 EXERCISES

1-1. One of the great triumphs of American technology was the conquest of interplanetary space. Man has set foot on the moon and has constructed a space station for further experiments. Does the United States space effort contribute to the common man's feeling of helplessness toward technology? Do you find in you a sense of awe and helplessness? Approximately \$3 billon is allocated to NASA each year. Is this allocation excessive?

1-2. Many critics find technology acceptable, but are unhappy with the distribution of the benefits among the citizens of the nation. Is this, if true, a failure of technology or of our social institutions?

1-3. Paul Goodman states that technology is a branch of moral philosophy, not of science. Do you view technology in this way? Review Goodman's article and discuss his concept of technological overcommitment and the need for simplification.

1-4. Drucker states that technology is causing a discontinuity in our society. Identify areas where you believe we are experiencing this discontinuity.

1-5. The concept of the two cultures is attributed to C.P. Snow. Is there a polarization on your campus between the humanists and the engineers? Describe the characteristics of this split, if any, on your campus or in your community. Should a true humanist have some knowledge of science and technology? Explore this dichotomy in our society.

1-6. The age of innocent faith in the benefits of technology may be over, if it ever did exist. What attitude should our society hold toward technology over the next decade?

1-7. As technology has progressed, many of its critics have pointed to its failures. Ivan Illich criticizes the technology of medicine in his book *Tools for Conviviality*; he asks why as medicine has become more advanced, the practice of medicine has become more institutionalized and monopolized and the cost multiplied beyond the ability to pay.[14] Those who are so ill that they will die are destined to spend their last days with plastic tubes and machines, deprived of loving care and contact from their families. He also asks why faster surface transportation has not made people free, but in his opinion has created distances between people and communities. Consider the effects of technology and compare the benefits of technology with its "dehumanizing" qualities.

1-8. Where do the children play? In an age of disavowal of God on the one hand and wide acceptance of engineering technology on the other, the engineer is finding it increasingly difficult to find the proper place for his technology in a humane world. Many modern songs of discontent express this quandary. The song "Where Do the Children Play" by Cat Stevens is illustrative of the mood of distrust of technology. Find several other modern and popular folk songs which express this concern with technology and explore the criticisms evident in the songs.

Where Do The Children Play?*
by Cat Stevens

Well I think it's fine building Jumbo
planes, or taking a ride on a cosmic
train, switch on summer from a slot
machine, yes get what you want to,
if you want, 'cause you can get
anything.

I know we've come a long way, we're
changing day to day, but tell me,
where d' th' ch'ldr'n play?

Well you roll on roads over fresh
green grass, for your lorry loads
pumping petrol gas, and you make them
long and you make them tough, but they
just go on and on, and it seems that
you can't get off.

Oh. I know we've come a long way,
we're changing day to day, but tell
me where d' th' ch'ldr'n play?

Well you've cracked the sky, scrapers
fill the air, but will you keep on
building higher 'til there's no more
room up there. Will you make us laugh,
will you make us cry, will you tell
us when to live, will you tell us when
to die.

I know we've come a long way, we're
changing day to day. But tell me,
where d' th' ch'ldr'n play?

2

ENGINEERING AS A DISCIPLINE

IN MANY WAYS the story of the development of new technologies to aid man is the story of civilization itself. Civilization has been described as the process whereby man strives to overcome the obstacles in his way and to become master of his environment. Persons who are professionally involved in the development of new technologies are usually called engineers. Engineers are constantly seeking to understand the laws of nature so as to modify the environment beneficially.

ENGINEERING DEFINED

One view of engineering is to describe it as the activity of problem solving under constraints. An engineer defines a problem, asks questions that determine its true nature, and ascertains whether the problem is over-constrained, so that no solution exists, or under-constrained, so that several exist. The engineer relaxes constraints selectively, seeking better solutions. Also, he studies the range of possible solutions, seeking the one which is optimal in as many ways as possible. Usually he studies the solutions using an iterative process which examines a cycle of steps toward a suitable design.

Engineering represents a variety of activities associated with technology, but we shall use a somewhat expanded definition as follows:

Engineering is the profession in which a knowledge of mathematical, natural and social sciences gained by study, experience and practice is applied with judgment to develop ways to utilize economically the materials and forces of nature for the benefit of mankind.

Thus, for example, the design and construction of a bridge such as the Golden Gate Bridge requires the knowledge of the sciences which an engineer would gain in college and in practice. This knowledge is coupled with good judgment to choose from alternative designs for a bridge, and to examine the economics of various approaches. The engineer can then utilize the materials such as steel and cement to construct a bridge for the benefit of man.

The ultimate objective of engineering work is the design and production of specific items, often called *hardware*. Engineers may design and build a ship, a radio station, a digital computer or a rocket. Each item is designed to meet a different set of requirements or objectives. The idea of design is to bring into existence something not already existing. Engineering design is thus distinguished from a reproduction of identical items, which is the manufacturing process.

THE ENGINEERING PROCESS

The essential features of the engineering process are:[1]

1. Identification of a feasible and worthwhile technical objective and definition of this objective in quantitative terms.
2. Conception of a design which, in principle, meets the objective. This step involves the synthesis of knowledge and experience.
3. Quantitative analysis of the design concept to fix the necessary characteristics of each part or component and to identify the unresolved problems —usually problems of materials, of component performance, and of the interrelationship of components.
4. Exploratory research and component tests to find solutions of the unresolved problems.
5. Concepts for the design of those components which are not already developed and available.
6. Re-analysis of the design concept to compare the predicted characteristics with those specified.
7. Detailed instructions for fabrication, assembly and test.
8. Production or construction.
9. Operational use, maintenance, field service engineering.

Usually, if possible, it is desirable to satisfy the technical objective (of step 1) with a best or optimal design. The design objective of a bridge might call for a certain number of traffic lanes and a specified loaded strength, and with a minimal cost. Then the best design is one that meets all the steps of the process, one through nine, and which results in satisfying the objective with minimal cost. Of course, the engineer must recognize that minimal construction cost may not yield a minimal maintanance cost, and both must be accounted for in the design objective. The engineering solution is the optimum solution, the most desirable end result taking into account many factors. It may be the cheapest for a given performance, the most reliable for a given weight, the simplest for a given safety or the most efficient for a given cost.

EFFICIENCY

For the engineer a most important quality is efficiency, which is defined as the ratio of the output to input and written as:

$$\eta = \frac{\text{output}}{\text{input}}$$

The ratio may be expressed in terms of energy, materials, money, time or persons. Commonly, energy or money is used in this indicator. A highly efficient device yields a large output for a given input. An example of a high-efficiency device is an electric generator in a power plant. An example of a low-efficiency device is a passenger automobile, which ordinarily has an efficiency of approximately 0.15 (or 15%). (An auto provides only 15% of the energy stored in the gasoline as energy output at the tires of the car.)

ENGINEERING AND SCIENCE: MEASUREMENT

So the engineer is called upon to conceive an original solution to a problem or need, and to predict its performance, cost and efficiency. Then he is required to construct the device. The engineer uses the sciences and materials to build desirable products. The engineer is concerned with design and construction while the scientist is concerned with developing the fundamental knowledge of nature. The engineer must know science, but he is not necessarily a scientist, although many engineers are involved in pure science as well as their engineering work.

Engineering is built on the use of quantitative data, as is graphically shown in Figure 2–1. Thus, measurement of important data is often carried out as part of an engineering task. Often the engineer utilizes empirical approaches to the solution of a problem.

FIGURE 2-1   A graphic representation of an engineer's use of quantifiable data and
the measurement of important variables. Courtesy Kaiser Aluminum
News © 1966.

ECONOMY, SOCIAL RESPONSIBILITY AND HISTORY

Engineering is concerned with the proper use of money, a capital resource, as well as all other resources. Engineering requires the economical use of energy and materials so that there is a minimum of waste and a maximum of efficiency. It is also the safe application of the forces and materials of nature. Safety is an important quality and must be incorporated in all objectives.

The engineer of the earliest civilizations was an educated man, very likely to be a priest or physician. For example, Imhotep (*circa* 2500 b.c.), the builder of the first great pyramid of Egypt, was also the court physician.

As the engineering function became more elaborate it emerged in early Europe as that of the *architekton* or *architectus* of Greece or Rome, the predecessor of both the modern architect and civil engineer. He was concerned with building bridges, aqueducts, harbors and roads. The Roman engineer was called an *ingeniator* or builder of ingenious devices after 200 a.d.; he was highly concerned with building useful, practical devices such as roads and aqueducts. ("Engineering," therefore, is not a noun associated with the word "engine" but rather with "ingenuity.") The ingenious man with a call to social responsibility has been the type of man holding the qualifications of the engineer since Roman times. Such was Thomas Telford, a Scottish engineer who had a vision of road and canal construction to relieve the poverty of the highlands of Scotland in the early 1800s. So also was American civil engineer Loammi Baldwin, who resigned in 1822 from the Union Coal Company over a dispute concerning the role of an engineer with respect to safety.[2] Engineers have a responsibility to speak out on issues concerning their work, and they have attempted to do so for many centuries. They must do so today. Recently Ralph Nader reaffirmed the responsibility for the employed engineer in his talk "The Determinants of Professional Slavery."[3] He called for the scientist and engineer to provide information about new technologies to the public through the media and the courts.

Engineering as a calling is a profession rather than a trade and the engineer must assume personal responsibility for his work. This is true whether the individual is working for a government agency, a large corporation or as an individual consultant. Taking responsibility may mean speaking out if an organization is heading in an indefensible direction; an engineer may have to refuse to cooperate in work if it is not in the public interest.

TECHNOLOGICAL CHANGE AND THE ENGINEER

Engineering is intimately involved with the process of technological change, whereby new methods, new machines, new organizational techniques and new products are (1) discovered and developed to the market stage; (2) introduced in the fields of both production and consumption; and (3) improved, refined and replaced as they become obsolete. Engineering is active in each of the three stages. A simple model of the nature of the engineering process is shown in Figure 2–2. Sources of new energy, new information and knowledge and new materials lead to new activities and results. When society has a need for new tools, processes, methods and organizations, new technologies are formulated and devised. The actual results of the process are new capabilities and new products. In addition, as side effects of the process, new problems may result.

PHILOSOPHY OF ENGINEERING

Although the philosophy of science was already well developed in antiquity, the philosophy of engineering is not fully developed or articulated even today. Despite the influence of engineering upon society and philosophy, no one philosopher has defined the philosophy of engineering. An article by Henry Greber has attempted to delineate the foundation philosophy of engineering. This article follows:[4]*

Historically, engineering evolved from craft; it was based upon experience and had little to do with science. Just as the horse-drawn buggy was replaced by the automobile, so was the buggy maker replaced by the modern engineer. Similarly, the ancient wine maker was replaced by the contemporary chemical engineer who still relies—but to a decreasing extent—on the know-how of the ancient wine maker. Modern engineering applies practical experiences even though they may not be understood scientifically. The quantitative knowledge of involved physical phenomena is important, but if it is not available, it can be replaced by suitable empirical approximations for engineering purposes.

The Greeks had a word for it

The close connection between engineering and craftsmanship is in conformance with the Greek concept of *techne*—practical knowledge used for the purposeful formation of matter to serve man's needs. The Greek word *logos* means science. These are the two roots of the word "technology." A striking feature of the ancient Greek civilization was the separation of theoretical science (theoria) from practical knowledge (techne). The latter was the unwritten skill passed from the craftsman of one generation to the apprentice of the next. *Theoria* was meant only for the delight of the intellect in contemplation.

The essence of engineering—an historical overview

An essential aspect of the Hellenic civilization was that the ancient Greeks held themselves to be above manual labor, which was performed for them by slaves. Not only did the Greeks disdain manual work, but also they were ashamed of being interested in it. Archimedes apologized for his inventions in his writings. He claimed to have made them only for amusement.

Plato considered the use of geometry for land measurement to be an undue vulgarization of science. Although glass was known in ancient Greece, and used for many household items, the Greeks never devised a lens. Consequently, they never invented a

*Reprinted with the permission of the author, Henry Greber, and the Institute of Electrical and Electronic Engineers. Published in *The IEEE Spectrum*, October, 1966.

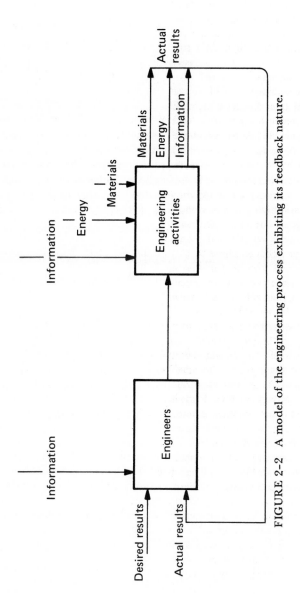

FIGURE 2-2 A model of the engineering process exhibiting its feedback nature.

microscope or a telescope. Why did they not develop these devices that are so essential to the studies of biology and astronomy? The answer is that in ancient Greece, science was in the hands of philosophers who were too proud to learn anything from nonphilosophers —let alone from slaves making glassware.

Thus, through their contempt for manual labor, the Greeks of antiquity divorced themselves from reality. The absence of an immediate contact with the physical surrounding—and particularly the almost total lack of experimentation—is the principal weakness of Greek classical science. This may have been the prime reason for the eventual downfall of the ancient Greek civilization.

Leading questions—partial answers. What is the essence of engineering? What is its purpose, and the means and methods leading to it? Such considerations are usually initiated by consulting a dictionary in which engineering is described as "an applied science that utilizes materials and physical forces for the production of structures, machines, equipment, and manufactured items. All these serve for the satisfaction of human needs." Although this description may well explain the *meaning* of the word "engineering," it is insufficient in expressing the essence of engineering. This essence is not easy to describe.

In a report, presented in 1951 at an engineering conference in The Hague by a task-force committee that had worked on it for two years, the definition of engineering is expressed in 300 words—much too long to be quoted here. A short definition, such as "engineering is the art and science by which matter and energy are used for the benefit of mankind," is far too general to reveal the essence of engineering. In lieu of this, the writer would suggest another definition:

Engineering is the skill of utilizing knowledge, acquired from science and from practical experience, to produce devices for the use of mankind.

The products of engineering, although they may serve for the *use* of mankind, are not necessarily for its benefit. For example, engineering in ancient Rome was a military craft. The word "engineer" is derived from the Latin *ingeniator*, which means the one who invents *ingenia*—war machines. Even today, products such as alcoholic liquor, tobacco, and some drugs, manufactured or processed by engineering skills and techniques, are of questionable benefit to mankind.

The merger of 'techne' and 'theoria.' Today, the separation between practical and theoretical knowledge is only formal. There is a continuous shortage of available theoretical knowledge that is badly needed in engineering. Many engineering problems must be solved in spite of the fact that the theoretical knowledge for their solution is not yet at hand.

Even though the modern concept of engineering has been with us for a few centuries, it is still in the process of formation and fusion. Some branches of engineering, such as electronics and aeronautics, have existed only for a few decades. Older branches of engineering are drastically changed as new scientific discoveries are made. It is, therefore, not surprising that no precise definition of engineering can be given today, and many different definitions and redefinitions can be expected in the future. Nevertheless, the essence of engineering—its reason for being—will remain unchanged.

The goals and the methods. The goal of engineering is to provide goods and services to mankind; its method in pursuing this goal is to apply scientific or any other available, verified knowledge. The goal of engineering is economical; its method is scientific. The goal and the method clearly place engineering in a position between economics and science.

The method of engineering

Problem solving. There is a characteristic procedure in the solution of engineering problems that can be broken down into six cardinal steps:

1. Recognition of the economical need.
2. Formulation of the problem.
3. Breaking down the problem into concepts that will suggest a solution.
4. Finding elements for the solution.
5. Synthesizing the solution.
6. Simplifying and optimizing the solution.

Neither deductive nor inductive. The engineering method is neither deductive nor inductive; it starts from facts known either to science or from practical experience, and these facts are applied for the solution of an essentially economic problem. Technological devices are not derived from this knowledge; rather, they are invented so that the knowledge can be applied.

There is often more than one scientific principle that can be used for the solution of a given engineering problem, and there is also more than one engineering problem that can be solved by the application of a given scientific principle. Thus there can be *more than one answer* to an engineering question. Conversely, there are many engineering questions to which the same answer can be given.

The direct solution. Some solutions are more direct than others. The most direct solution, which is the simplest (and usually the best), can be found by an incisive formulation and reformulation of the problem. This can be done only by finding its underlying economics.

By way of illustration, consider the problem of a river crossing. Here, the primary objective is not to build a bridge, but to get to the other side of the river. This can be done in many ways: by ferry, cableway, underwater tube, helicopter, etc. Which means do we choose? This depends upon the economics of the problem, because getting to the other side of the river is not the final objective: there must be a *reason* for being on the other side of the river. Perhaps this reason is to extract ore from a mine, to farm land, or to build a suburb. It is only when the ultimate economic objective is known that it is possible to find a direct solution to the problem, provided such a solution exists at a given level of science.

Criteria for problem solving. Although there is a standard approach to the solution of engineering problems, the solutions themselves must fulfill a set of standard requirements that includes safety; reliability; longterm economy; a minimum of labor; practicality in ease of manufacture, installation, operation, and maintenance; and esthetic design, insofar as possible.

In the majority of cases, all of these criteria cannot be satisfied at the same time. Therefore, the final engineering solution has to be a compromise.

The safety requirement dictates the need for the thorough checking of any engineering work and the continual testing of components during construction and manufacture. The primary reason for this is that an engineering mistake may endanger life and property. A secondary reason for caution is that an engineering work is predicated upon many assumptions, some of which may be conflicting. Thus the compatibility of these assumptions, and their influence upon the end result must be constantly investigated. It follows, therefore, that to compromise some conflicting assumptions and to optimize intermediate results, reiteration is indispensable.

The engineering assumption is an inductive step, generalizing features that are found in many similar devices. The experienced engineer is expected to be able to make correct assumptions. In fact, this ability is what makes the difference between the experienced and the inexperienced engineer. But correct assumptions also must be optimized and coordinated to suit the project at hand, so that even the experienced engineer must check and recheck his computations and designs as he proceeds in his work. As a measure of the advancement of science, however, the engineer of the future will undoubtedly depend less upon his assumptions; yet, he will never eliminate them altogether.

Making assumptions is not unique to engineering. In science, assumptions (or hypotheses) are formed first. When these hypotheses become experientially verified, the hypotheses become scientific theories.

Synthesis and analysis. Engineering effort is best described by the words *designing* and *inventing*. The difference between them is quantitative rather than qualitative. The terms imply "putting parts together so that they can work together"—or, more precisely, that natural forces can work through these parts to satisfy human needs. The engineering method is synthetic; scientific work is usually analytic, but it can be also synthetic.

In conclusion, although there is certainly a pattern in solving engineering problems in which specific methods are used more often than others, *there are no unique features that distinguish the engineering method from methods used in other branches of human activities.* The scientific method is used in engineering as far as possible, but it cannot be used far enough. In general, the engineering method is far less rigorous than the scientific.

Engineering education

The work of an engineer, at least in its methodology, resembles that of the physician and the attorney. In these professions, a body of generalized theories is applied to particular cases. But although there is relatively little discussion concerning the education of physicians and lawyers, engineering education seems to be the subject of continual, worldwide discussion. The global consensus is that engineers are not as useful as they might be—or even as they should be.

The elements of engineering. Since engineering is the application of scientific and practical knowledge to economic ends, it consists of three elements:

1. Engineering know-how, or technology.
2. Scientific knowledge.
3. Knowledge of economics and of business conditions.

It is only by possessing these three kinds of knowledge that the engineer can fulfill his function. Technology is the engineering science proper. Within it, machines, apparatus, and equipment and its components are subject to scientific classification and investigation. It is interesting to note in technology that the scientific method used for the investigation of natural phenomena is applied to study man-made means for harnessing natural phenomena.

The 'weak spots.' It has been alleged that the engineer is not sufficiently equipped with a knowledge of science and economics, nor even with the proper degree of technological knowledge. Much of the latter remains to be acquired by him on his first jobs.

Ideally, the future engineer should learn in school that which he will be asked to do in his actual work. The contemporary engineer cannot talk with competence on science since he is not a scientist; neither can he talk with authority on economics since he is not an economist. As already stated, since his knowledge of technology depends largely on his job experience, there can be but one conclusion: *Engineering education must be upgraded.*

The upgrading process. All trades and professions, including engineering, are involved in a continuous process of upgrading. The urgency of improving engineering education is based on the following premises:

1. The explosive growth of science.
2. The rapid growth of technology.
3. The increasing complexity of economics.
4. The desire of industry to reduce the long apprenticeship period of the engineer.

'Creative engineering.' It is surprising that creative engineering, which encompasses the devising of new design concepts and inventions, is not taught at all as a part of the regular curricula in engineering schools. A few courses on this subject are conducted, on a highly experimental basis, in some colleges, but these are the exceptions.

From the preceding analysis, it can be seen that imaginative synthesis is the very essence of engineering. As Herbert Hoover, the engineer President of the United States, put it: "Engineering without imagination sinks into a trade." Thus the lack of creative engineering in the curricula of the majority of our colleges and universities seems to do little to upgrade the engineering profession. Further, the lack of sufficient knowledge in technology, science, and economics often prevents engineers from assuming positions as captains of industry.

Status of engineering in the modern society

The purpose of engineering is to satisfy the economic needs of society; it is not the engineer's responsibility to decide whether or not these needs have to be satisfied. If the engineer had the final say as to whose needs should be satisfied and to what extent, and what should and should not be done, he would be practically in control of society. This would contradict the principle of democracy, according to which the control of society belongs to society itself. It does not mean, however, that the engineer is not responsible for what he is doing, for there are moral values that cannot be violated. The engineer must respect these moral values and be responsible for his work, just as the lawyer and the physician are responsible for theirs. On the other hand, the engineer cannot be solely responsible for all the sociological aspects of technology since he shares this responsibility with society in general.

Social position of the engineer. Engineers have not always been satisfied with their position in society, and, in the early 1930s, the "technocracy" movement was originated. The movement's originators, Thorstein Veblen, Howard Scott, and others, believed that because engineers and scientists contributed so much to the creation of modern civilization, society should be controlled by them. The concept of technocracy rested on the idea that the majority of social problems are those of production and distribution of goods and services. Therefore, engineers and scientists, by virtue of their education and experience, should be capable of handling such problems.

It was the thesis of the technocrats that the control of society by engineers and scientists should be based on the natural laws that govern human society, and they believed these laws could be derived from scientifically accurate measurements of social phenomena.

Although there can be little doubt that engineers and scientists have better judgment concerning some problems than those in other professions, there can also be little doubt that the idea of technocracy contradicts the concept of democracy. In a democracy, no man—no matter what his qualifications may be—can have a larger share in the control of society than that of another man.

The status of the engineer in modern society is not as high as it could be. He is not generally a leader of industry. As far as the application of scientific discoveries, such as atomic energy, is concerned, the engineer is often overshadowed by the physicist. Grand-scale engineering schemes are usually initiated by business leaders who are essentially trained in economics. To assume a leading position in industry, the engineer must have the comprehensive educational background, the skills, and the experience for such a position. Again, this can be achieved by means of a substantial upgrading of engineering education.

The future of engineering

It is certain that a considerable upgrading of the engineering profession is imminent. A powerful factor that will contribute to this upgrading is the widespread use of computers. They will free the engineer from his routine chores and make him available for more creative work.

As the role of technology grows in our contemporary era, there will be more and more creative work to be done. The history of engineering clearly demonstrates that enormous use has been made of comparatively small scientific breakthroughs.

The 'outer limits.' Even the most optimistic scientist will agree that there are limits to man's scientific penetration into nature's citadel. There is no hope that man will ever know everything that he would like to know. Nevertheless, this limitation does not bar man's way to mastery over certain aspects of nature. Man's mastery over nature, through knowledge, does not imply that he will be able to run it according to his wish. This is clearly impossible. By thoroughly utilizing all of the scientific knowledge he can obtain, however, man will free himself from many of the bondages of natural forces; to that extent he will be independent of nature.

Within a limited scope, a bright forecast

Even within the limited scope of man's mastery, fantastic achievements can be attained. In time, agriculture will cease to be a source of food, which will be chemically synthesized. The recycling of biological material and waste products, and their conversion into food, will become feasible. Chemical raw materials and water will be extracted from the sea and from the atmosphere to supplement automated, deep-level mining. Land and sea transportation will be almost completely superseded by air transportation.

By altering the climate and, to some extent, the topography of this planet, man will be able to modify his environment to suit his needs. He will travel to other planets, change their orbits, and possibly create new, artificial planets. He will be able to modify existing forms and create new forms of matter. He may even succeed in creating new forms of life. Natural laws will be the only limitation on his possibilities.

By becoming a limited master of nature, man will also become a master of his own nature. Highly developed technology will be able to satisfy all the needs of mankind. When this Nirvana is reached, the level of human happiness will be proportional to the level of engineering.

Engineers are obligated by the nature of their profession to *understand* what they do and its influences as well as to *explain* it to others. One mark of a mature profession is consciousness of its own history. Another equally important sign of a mature profession is conscious dedication to an explicit ideal goal.[5] Medical men are dedicated, for example, to the liberation of mankind from the ills of the body. Engineers are dedicated to the liberation of mankind from the limitations of the physical world. Thus the engineer is leading man in the conquest and utilization of matter, energy, time and space. Engineering has flourished in the context of the assumption that the physical world was created for a good purpose and that it is to be used and treasured for the necessary ground of man's life. As Professor Lynn White, Jr. has observed:[5]

Growing exponentially, Western technology has now led to the globalizing of human experience and the smashing of the physical barriers between peoples. This is the prerequisite to breaking through the other barriers between them. Whatever the incidental problems, it is a prime spiritual achievement. In the industrial nations, technology has likewise led to an increase in the standard of living and of education which has broken the old functional division between educated rulers and uneducated workers. Engineering today is pushing over the geographical and social fences which have prevented mankind from unifying its total experience and thus discovering itself not as classes or tribes but as humanity.

Engineers are the chief revolutionaries of our time. Their implicit ideology is a compound of compassion for those suffering from physical want, combined with a Promethean rebellion against all bonds, even bonds to this terrestrial ball. Engineers are arch-enemies of all who, because of their fortunate position, resist the surge of the mass of mankind towards a new order of plenty, of mobility, and of personal freedom. Within the societies which have consolidated about the Marxist and the Western democratic revolutions, engineers' activities are the chief threat to surviving privilege.

Without deliberate intent, but by the nature of their activity, engineers have largely destroyed the contemporary validity of the older aristocratic humanism which was a cultural weapon in the hands of the ruling class. When engineers in greater numbers come to know explicitly what they are doing, when they recognize their dedication, they can join with alert humanists to shape a new humanism which will speak for and to a global democratic culture.

Engineering as a profession has always differed from the professions of law and medicine in that attorneys and physicians are usually self-employed, while engineers have typically been employed by corporations or government agencies. Since the last half of the nineteenth century most engineers have worked for large bureaucratic entities. The careers of such well-known engineers as Herbert Hoover, Charles Steinmetz and Thomas Edison were atypical because they were individual and glamorous. The engineer's knowledge and ability are task-oriented, and our society manages tasks with organizations. Professional competence is not an end in itself, but an important part of an organized enterprise. For example, one great engineering achievement of the last decade, that of the Appollo space program, was undertaken by the vast government organization known as the National Aeronautics and Space Administration (NASA).

As Edwin Layton has observed in *The Revolt of the Engineers*, twentieth century professional engineering societies do not closely resemble their medical and legal counterparts, since the engineering societies have never completely controlled access to the profession or effectively disciplined their members' activities.[6] Furthermore, from a public-policy viewpoint, they have never acted with autonomy as as bloc distinct from business. These characteristics are the result of the fact that engineers are largely employed by government and business.[6] In a sense the majority of engineers are essentially employees of organizations and ultimately of society, and they place their defined role to a great extent within the needs and power of the organizations they serve.

Physicians and attorneys are usually self-employed and autonomous and to an extent hold self-defined roles. Currently our society is criticizing the independent professions for a role which is both too extensive and too much self-defined, perhaps even self-serving, while at the same time it is also criticizing engineers for a role too extensively defined by corporations. Engineers need to respond with greater responsibility to society, to humane values and to human life. This is a challenge to the profession.

QUALITIES AND CRITERIA FOR ENGINEERS

Because an engineer works in organizations, he often works in groups. In groups an engineer finds that the ability to communicate well with others and to work harmoniously with other people are desirable characteristics. Another characteristic of the engineering profession is that an engineer must often be

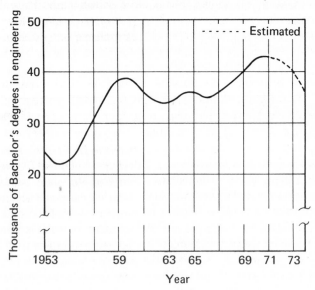

FIGURE 2-3   The number of Bachelor's degrees awarded each year in United States engineering colleges.

willing to make decisions on the basis of incomplete information. This requires courage and maturity, which are often gained by experience.

The standard criteria of a profession are knowledge, autonomy, obligation and commitment. Generally speaking, engineers possess all these qualities to a great extent, with autonomy their weakest characteristic. Usually the members of a profession have a certain unity of educational experience and work activity. These characteristics are more commonly held by members of the professions of law and medicine than that of engineering. Engineers may attend over 180 accredited programs of instruction; often these programs differ in curricula and process. There are many (over 25) professional societies within the engineering profession, rather than one professional society as is the case in law and medicine. Nevertheless, the engineering profession has offered a great opportunity to many individuals to contribute to their society with a significant commitment to service, an obligation to ethical practice and a responsibility to maintain a currency of knowledge. Leadership is a quality often found among persons with such commitment; it is not surprising, therefore, that over 25% of the top 1,000 executives of the 600 largest United States firms hold degrees in engineering.[7]

PREPARATION OF ENGINEERS

One important difference between an occupation and a profession is that the subject of a profession has sufficient technical content to require formal educational preparation. An engineer normally is the graduate of a four-year college or university with a degree in one of the disciplines of engineering. In many specialties of engineering, a master's degree is required for full status in the field. During the decade of the 1960s, the number of engineering graduates grew by 3% per year while the United States population grew by 1% per year.

As society has established new tasks for the engineering profession, the demand for new engineering graduates has fluctuated, resulting in a surplus or deficit of graduates. While there was a surplus of graduates in 1971, there is a deficit predicted for 1975. The number of bachelor's degrees in engineering awarded in United States colleges is shown in Figure 2–3.

There are currently approximately 1,200,000 people employed as engineers, of which 800,000 are engineering college graduates. Thus one third of those who hold the title of engineer are not graduates, but may have several years of college and many years of experience which have permitted them to qualify for their position in the view of their employers. Approximately 40,000 bachelor's degrees are awarded in engineering each year. This number may drop off below 40,000 for several years in the mid-1970s because a surplus of graduates in 1969 to 1971 caused a drop in the number of new students choosing to enroll in engineering subsequently.

There were 15,152 master's degrees and 2,921 doctor's degrees in engineering awarded in 1968, contrasted with 7,159 master's and 783 doctor's degrees in engineering awarded in 1960. It is clear that advanced education has become a most important factor in the profession of engineering.

Nevertheless, engineering education has dropped dramatically over the past 20 years as a proportion of all college education. Thirteen per cent of all male graduates were engineering graduates in 1953, and 16.5% of male graduates were engineering graduates in 1960. In 1973 only 8% of the male graduates were engineering graduates. Thus engineering graduates are a small percentage of college graduates and will probably stay at 8% of the male graduates for the remainder of this decade with a normal graduating class of 37,000 to 40,000 engineers each year.

Until recently, fewer than 1% of the engineering graduates each year were women. There is now a tendency for the percentage to increase. Nevertheless, it will take many years before women engineers comprise a significant percentage of the practicing engineers.

CHAPTER 2 REFERENCES

1. J.R. Whinnery, *The World of Engineering*, McGraw-Hill Book Company, New York, 1965.
2. J.B. Rae, "The Engineer in Society: Past and Present," American Society for Engineering Education, Annual Meeting, June 1971, Annapolis, Maryland.
3. "No Protection for Outspoken Scientists, says Nader," *Physics Today*, July 1973, pp.77–78.
4. H. Greber, "The Philosophy of Engineering," *IEEE Spectrum*, Oct. 1966, pp. 112–115.
5. L. White, Jr., "Engineers and the Making of a New Humanism," *Engineering Education*, Jan. 1967, pp. 375–376.
6. E.T. Layton, Jr., *The Revolt of the Engineers: Social Responsibility and the American Engineering Profession*, Case Western Reserve University Press, Cleveland, 1971.
7. R. Perrucci and J.E. Gerstl, *Profession Without Community: Engineers in American Society*, Random House, New York, 1969.

CHAPTER 2 EXERCISES

2–1. There are many ways to define "engineering." Prepare your own definition and explore its relation to those provided in the chapter.

2–2. Select a recent engineering project such as the design and construction of the BART transit system or the Golden Gate Bridge, and identify and name the essential features of the engineering process that contributed to the project.

2–3. As an example of the term *efficiency*, calculate the efficiency of your bicycle, refrigerator or other appliance.

2–4. Engineers have often spoken out on their responsibilities and concerns for new engineering developments. Nevertheless, they have often been accused of neglecting this responsibility. Explore some recent subjects of dispute, such as the SST, the Alaskan pipeline or the missile defense system, and determine if engineers have met their responsibility in the case you select.

2–5. Gene Marine, in his book *America the Raped* (see Chapter 1), criticizes what he calls "the engineering mentality." This approach, in his view, ignores the side effects of technology and never seeks a larger context for a problem. His example is the building of a bridge. Another might be the construction of a highway right through a residential neighborhood of a city.

Can you identify other examples?

Do you agree that there is an engineering mentality? If so, where does an engineer acquire it? In college or on the job?

2–6. Many engineering professional societies are considering a more direct involvement in the political decision-making process in order to insure that a more informed use of technology occurs. Would you recommend that professional engineering societies engage in what may be called lobbying? What are your reasons?

2–7. Engineers are ideally dedicated to the liberation of mankind from the limitations of the physical world. Explore the ways that man has been liberated from his limitations. As liberators, engineers may also be described as revolutionaries. Identify a revolutionary change resulting from a technological development.

2–8. Engineering as a profession often differs in many ways from the professions of law and medicine. Interview a practicing lawyer, physician and engineer and identify those characteristics they hold in common concerning their professions as well as those that differentiate them.

2–9. Women have traditionally comprised fewer than 1% of an engineering graduate class. The social and professional factors leading to this very low percentage are slowly disappearing. Interview several women engineering students and practicing engineers and determine what are the social and professional factors discouraging a greater influx of women into the profession.

3

TECHNOLOGY AND HISTORY

TECHNOLOGY INFLUENCES SOCIETY and therefore history. In the history of the United States, technology has played a central part and the United States of America might be said to be a country with a technological foundation. First came the opportunistic phase, with the pioneers, adventurers and entrepreneurs seizing opportunities as they found them at hand. The second phase was that of invention, or at least the application of technology. Then came the development of methods of mass-manufacturing and marketing. Perhaps a new phase is now developing with a focus on consumerism and equality.

INVENTION

As an example of the phase of invention, Boorstin tells of Samuel Kier, who inherited salt mines in western Pennsylvania in the 1840s.[1] There was so much seepage of oil into the mines that they were rendered useless. So Kier bottled the oil and sold it as medicine (for $1.00 per bottle) before it was discovered that the oil was better for lamps than for aches. Then Edwin Drake, seizing upon the idea of oil as fuel, began drilling for oil. An early oil drilling site is shown in Figure 3-1.

FIGURE 3-1 An early oil drilling site, Signal Hill, California. The rush to exploit new-found resources led to over-drilling and waste of finite petroleum reserves. Courtesy of Shell Oil Company.

Another example of an opportunistic invention is in the beef cattle business. After the Civil War, beef cattle were raised in the Southwest and shipped all over the country. Gustavus F. Swift developed the idea of the refrigerated freight car, which led to the use of centralized slaughterhouses in Chicago. In these centralized slaughterhouses, moving production lines for slaughtering and dressing the beef were developed to feed into the freight cars.

Trolley lines made it possible for large central department stores to serve a wide area. In 1898, Rural Free Delivery further centralized the nation by delivering mail and newspapers to all areas. Such unglamorous developments as the flat-bottomed paper bag and the machine-made cardboard box made significant changes in retailing of food and merchandising.

The city was made possible by the elevated railway, the trolley and the vertical railway—the elevator, invented by Otis. With elevators, tall buildings could be built to house people and offices.

Thomas Alva Edison was a systematic inventor. In 1876 Edison moved a crew of assistants into a 30-by-100 foot building in Menlo Park, New Jersey with the idea of turning out inventions "just as regularly and as intentionally as a factory could turn out any other product … a minor invention every ten days and a big thing every six months or so." By an "invention" Edison meant a marketable product. However, Edison was a social inventor; that is, he responded to the needs of society.

Before he was 21, Edison was granted a patent for a telegraphic vote-recording machine. By 1874 he had invented his quadruplex telegraph, which allowed two messages to be sent in each direction.

Americans also invented the jet engine, the gyrocompass, power steering, Bakelite, the helicopter, the Kodakchrome process, the zipper and the self-winding wristwatch, among others.[1]

When the Edison Electric Light Company was formed in 1878, electric arc lamps were used infrequently for lighting where they were permitted and were safe. Gas lights were used as the primary source of home and street lighting, but Edison saw a future for electric lighting: an electrical system with a central station and a wire network to every house. Edison moved from the arc lamp to the filament or incandescent lamp with his first trial device, and he applied for a patent on November 1, 1879. Edison announced that Menlo Park would be illuminated by the new lights on New Year's Eve, 1880. Many came to see his system of 40 incandescent bulbs all lit from an simple electric dynamo. Edison then established the first central electric station, the Pearl Street Station in New York, serving 85 customers on the first day, Sept. 4, 1882. The first Niagara Falls electric station is shown in Figure 3–2. Edison was representative of the inventive engineer who helped to shape the country. He combined technical talents with the ability to recognize a social need and to market a device to fill the need.

PROTECTION THROUGH PATENTS

The English system of patents, adopted in the United States, was of great assistance toward the encouragement of technological innovations. The English Statute of Monopolies of 1623, restricting the issue of patents to the first inventor of a new technique, became the model for most subsequent patent systems. When the Constitution of the United States was written in 1787, inventions were protected by Article I, Section 8, which is a charge to Congress: "To promote the progress of science and the useful arts by securing for limited times to authors and inventors the exclusive right to their respective writings and discoveries."[2] An invention is essentially an engineering achievement which is legally patent-able. Other engineering achievements, while equally laudable, may not be patentable because they are not unique or attributable to one person.

The development of new technologies may occur when conditions are ripe for it. (See Figure 3–3.) In other cases, an invention requires an act of individual inspiration. One can argue that the late eighteenth century provided the materials, experience and need that spawned the steam engine. Similarly, the

FIGURE 3-2   Generators for the first Niagara Falls, New York electric power station
in 1896. Courtesy of Niagara Mohawk Company.

airplane had to await the existence of a suitably light power plant. When it was
built, the Wright brothers achieved the first powered flight, shown in Figure 3–4.

TRAINING OF ENGINEERS

The first recognition of the need for systematic training of engineers came in
France with the founding of the École des Ponts et Chaussees (School of Bridges
and Roads) in 1747. The school taught civil and military engineering, and one of
its graduates founded the École Polytechnique in 1794 as a major school of
engineering. The École Polytechnique had such famed graduates as Carnot and
Eiffel. During the nineteenth century, polytechnic institutes were founded on the
French model in Europe and the United States.

In Britain, however, engineers were the product of after-work-hours trade
schools and apprenticeship programs. Not until the 1840s was engineering
offered at the Universities of London and Glasgow. Nevertheless, Britain led the
way in the formation of professional engineering societies with the Institute of
Civil Engineers in 1828.

FIGURE 3-3   An early Baker electric car. During the 1890s the electric automobile
enjoyed a greater popularity than the gasoline car. The weakness of the
electric car was that it could not go fast or far without running down
its battery—a shortcoming of "electrics" even today. Courtesy of Motor
Vehicles Manufacturers Association.

The first engineering school in the United States was the United States
Military Academy, founded in 1802; its curriculum emphasized military engineering. Rensselaer Polytechnic Institute, of Troy, New York was founded in
1824 on the European model. Engineering colleges were introduced at Yale
(1850) and Harvard (1847) and new polytechnics were founded, including the
Massachusetts Institute of Technology in 1861. Congress provided a stimulus to
engineering education with the Morrill Land-Grant College Act of 1862, granting
land to the states for the support of "colleges of the agriculture and the mechanic
arts."

The first engineering professional societies in the United States were the
American Society of Civil Engineers (1852), the American Institute of Mining
Engineers (1871), the American Society of Mechanical Engineers (1880) and the
American Society of Engineering Education (1893).

FIGURE 3-4   The Wright brothers' first flight at Kitty Hawk on December 17, 1903.
Courtesy of the Smithsonian Institution.

GROWTH OF INDUSTRIAL TECHNOLOGY

The introduction of new technologies in the United States can be illustrated by the rise in the number of manufacturing establishments, shown in Table 3-1. The number of manufacturing installations grew by a factor of 3.6, while the number of manufacturing workers grew by a factor of 3.1 during the last 40 years of the nineteenth century. The horsepower of machinery per worker used in these establishments grew by a factor of 50% during this period, allowing each worker to produce more.

The purpose of machines was to expand the capability of the worker and to do new things men could not do by themselves. For example, if operators were required today to accomplish by manual methods all the automatic telephone switching, half of the population would be required to work as operators.

Machines and new technologies influence the course of events, but can they be said to determine history? The cotton gin, invented by Eli Whitney in 1793, lowered the price of cotton in the United States and made it very competitive with cotton from other parts of the world. The economic advantages of the invention caused a dependency of the Southeastern United States on the plantation system and the institution of slavery. By the middle of the nineteenth

TABLE 3-1.

The rise of manufacturing establishments during the last 40 years of the nineteenth century in the United States.

	1859	1869	1879	1889	1899
Number of Manufacturing Establishments (thousands)	140	252	254	354	510
Number of Manufacturing Workers (thousands)	1311	2054	2733	4129	5098
Value added (millions of dollars)	854	1395	1973	4102	5475
Installed Horsepower (thousands)	1600	2346	3664	6308	9633
Capital (millions of dollars)	—	—	2718	5697	8663
Capital per Worker	—	—	$995	$1380	$1699
Horsepower per Worker	122	114	134	153	189
Value added per Worker	652	679	722	994	1074

century, some historians claim, slave labor had proved itself technologically obsolete. The movement toward the abolition of slavery cost us a civil war, since the South sought to maintain a separate economic and social system based in part on the institution of slavery.

Robert Heilbroner, the noted author, has written an article entitled "Do Machines Make History?" This article considers this very important question; it follows in its entirety.

Do Machines Make History?*

ROBERT L. HEILBRONER

The hand-mill gives you society with the feudal lord; the steam-mill, society with the industrial capitalist.

MARX, *The Poverty of Philosophy*

That machines make history in some sense—that the level of technology has a direct bearing on the human drama—is of course obvious. That they do not make all of history, however that word be defined, is equally clear. The challenge, then, is to see if one can say something systematic about the matter, to see whether one can order the problem so that it becomes intellectually manageable.

To do so calls at the very beginning for a careful specification of our task. There are a

*From *Technology and Culture* by Robert L. Heilbroner, The University of Chicago Press, Vol. 8, No. 3, 1967. With the permission of the author and publisher.

number of important ways in which machines make history that will not concern us here. For example, one can study the impact of technology on the *political* course of history, evidenced most strikingly by the central role played by the technology of war. Or one can study the effect of machines on the *social* attitudes that underlie historical evolution: one thinks of the effect of radio or television on political behavior. Or one can study technology as one of the factors shaping the changeful content of life from one epoch to another: when we speak of "life" in the Middle Ages or today we define an existence much of whose texture and substance is intimately connected with the prevailing technological order.

None of these problems will form the focus of this essay. Instead, I propose to examine the impact of technology on history in another area—an area defined by the famous quotation from Marx that stands beneath our title. The question we are interested in, then, concerns the effect of technology in determining the nature of the *socioeconomic order*. In its simplest terms the question is: did medieval technology bring about feudalism? Is industrial technology the necessary and sufficient condition for capitalism? Or, by extension, will the technology of the computer and the atom constitute the ineluctable cause of a new social order?

Even in this restricted sense, our inquiry promises to be broad and sprawling. Hence, I shall not try to attack it head-on, but to examine it in two stages:

1. If we make the assumption that the hand-mill does "give" us feudalism and the steam-mill capitalism, this places technological change in the position of a prime mover of social history. Can we then explain the "laws of motion" of technology itself? Or to put the question less grandly, can we explain why technology evolves in the sequence it does?

2. Again, taking the Marxian paradigm at face value, exactly what do we mean when we assert that the hand-mill "gives us" society with the feudal lord? Precisely how does the mode of production affect the superstructure of social relationships?

These questions will enable us to test the empirical content—or at least to see if there *is* an empirical content—in the idea of technological determinism. I do not think it will come as a surprise if I announce now that we will find *some* content, and a great deal of missing evidence, in our investigation. What will remain then will be to see if we can place the salvageable elements of the theory in historical perspective—to see, in a word, if we can explain technological determinism historically as well as explain history by technological determinism.

I

We begin with a very difficult question hardly rendered easier by the fact that there exist, to the best of my knowledge, no empirical studies on which to base our speculations. It is the question of whether there is a fixed sequence to technological development and therefore a necessitous path over which technologically developing societies must travel.

I believe there is such a sequence—that the steam-mill follows the hand-mill not by chance but because it is the next "stage" in a technical conquest of nature that follows one and only one grand avenue of advance. To put it differently, I believe that it is impossible to proceed to the age of the steam-mill until one has passed through the age of the hand-mill, and that in turn one cannot move to the age of the hydroelectric plant before one has mastered the steam-mill, nor to the nuclear power age until one has lived through that of electricity.

Before I attempt to justify so sweeping an assertion, let me make a few reservations.

To begin with, I am fully conscious that not all societies are interested in developing a technology of production or in channeling to it the same quota of social energy. I am very much aware of the different pressures that different societies exert on the direction in which technology unfolds. Lastly, I am not unmindful of the difference between the discovery of a given machine and its application as technology—for example, the invention of a steam engine (the aeolipile) by Hero of Alexandria long before its incorporation into a steam-mill. All these problems, to which we will return in our last section, refer however to the way in which technology makes its peace with the social, political, and economic institutions of the society in which appears. They do not directly affect the contention that there exists determinate sequence of productive technology for those societies that are interested in originating and applying such a technology.

What evidence do we have for such a view? I would put forward three suggestive pieces of evidence:

1. *The Simultaneity of Invention*

The phenomenon of simultaneous-discovery is well known. From our view, it argues that the process of discovery takes place along a well-defined frontier of knowledge rather than in grab-bag fashion. Admittedly, the concept of "simultaneity" is impressionistic, but the related phenomenon of technological "clustering" again suggests that technical evolution follows a sequential and determinate rather that random course.

2. *The Absence of Technological Leaps*

All inventions and innovations, by definition, represent an advance of the art beyond existing base lines. Yet, most advances, particularly in retrospect, appear essentially incremental, evolutionary. If nature makes no sudden leaps, neither, it would appear, does technology. To make my point by exaggeration, we do do not find experiments in electricity in the year 1500, or attempts to extract power from the atom in the year 1700. On the whole, the development of the technology of production presents a fairly smooth and continuous profile rather than one of jagged peaks and discontinuities.

3. *The Predictability of Technology*

There is a long history of technological prediction, some of it ludicrous and some not. What is interesting is that the development of technical progress has always seemed *intrinsically* predictable. This does not mean that we can lay down future timetables of technical discovery, nor does it rule out the possibility of surprises. Yet I venture to state that many scientists would be willing to make *general* predictions as to the nature of technological capability twenty-five or even fifty years ahead. This too suggests that technology follows a developmental sequence rather than arriving in a more chancy fashion.

I am aware, needless to say, that these bits of evidence do not constitute anything like a "proof" of my hypothesis. At best they establish the grounds on which a prima facie case of plausibility may be rested. But I should like now to strengthen these grounds by suggesting two deeper-seated reasons why technology *should* display a "structured" history.

The first of these is that a major constraint always operates on the technological capacity of an age, the constraint of its accumulated stock of available knowledge. The application of this knowledge may lag behind its reach; the technology of the hand-mill, for example, was by no means at the frontier of medieval technical knowledge, but technical realization can hardly precede what men generally know (although experiment may incrementally advance both technology and knowledge concurrently). Particularly from the mid-nineteenth century to the present do we sense the loosening constraints on technology stemming from successively yielding barriers of scientific knowledge —loosening constraints that result in the successive arrival of the electrical, chemical, aeronautical, electronic, nuclear, and space stages of technology.

The gradual expansion of knowledge is not, however, the only orderbestowing constraint on the development of technology. A second controlling factor is the material competence of the age, its level of technical expertise. To make a steam engine, for example, requires not only some knowledge of the elastic properties of steam but the ability to cast iron cylinders of considerable dimensions with tolerable accuracy. It is one thing to produce a single steam-machine as an expensive toy, such as the machine depicted by Hero, and another to produce a machine that will produce power economically and effectively. The difficulties experienced by Watt and Boulton in achieving a fit of piston to cylinder illustrate the problems of creating a technology, in contrast with a single machine.

Yet until a metal-working technology was established—indeed, until an embryonic machine-tool industry had taken root—an industrial technology was impossible to create. Furthermore, the competence required to create such a technology does not reside alone in the ability or inability to make a particular machine (one thinks of Babbage's ill-fated calculator as an example of a machine born too soon), but in the ability of many industries to change their products or processes to "fit" a change in one key product or process.

This necessary requirement of technological congruence gives us an additional cause of sequencing. For the ability of many industries to co-operate in producing the equipment needed for a "higher" stage of technology depends not alone on knowledge or sheer skill but on the division of labor and the specialization of industry. And this in turn hinges to a considerable degree on the sheer size of the stock of capital itself. Thus the slow and painful accumulation of capital, from which springs the gradual diversification of industrial function, becomes an independent regulator of the reach of technical capability.

In making this general case for a determinate pattern of technological evolution—at least insofar as that technology is concerned with production—I do not want to claim too much. I am well aware that reasoning about technical sequences is easily faulted as *post hoc ergo propter hoc*. Hence, let me leave this phase of my inquiry by suggesting no more than that the idea of a roughly ordered progression of productive technology seems logical enough to warrant further empirical investigation. To put it as concretely as possible, I do not think it is just by happenstance that the steam-mill follows, and does not precede, the hand-mill, nor is it mere fantasy in our own day when we speak of the coming of the automatic factory. In the future as in the past, the development of the technology of production seems bounded by the constraints of knowledge and capability and thus, in principle at least, open to prediction as a determinable force of the historic process.

II

The second proposition to be investigated is no less difficult than the first. It relates, we will recall, to the explicit statement that a given technology imposes certain social and political characteristics upon the society in which it is found. Is it true that, as Marx wrote in *The German Ideology*, "A certain mode of production, or industrial stage, is always combined with a certain mode of cooperation, or social stage," or as he put it in the sentence immediately preceding our hand-mill, steam-mill paradigm, "In acquiring new productive forces men change their mode of production, and in changing their mode of production they change their way of living—they change all their social relations"?

As before, we must set aside for the moment certain "cultural" aspects of the question. But if we restrict ourselves to the functional relationships directly connected with the process of production itself, I think we can indeed state that the technology of a society imposes a determinate pattern of social relations on that society.

We can, as a matter of fact, distinguish at least two such modes of influence:

1. The Composition of the Labor Force

In order to function, a given technology must be attended by a labor force of a particular kind. Thus, the hand-mill (if we may take this as referring to late medieval technology in general) required a work force composed of skilled or semiskilled craftsmen, who were free to practice their occupations at home or in a small atelier, at times and seasons that varied considerably. By way of contrast, the steam-mill—that is, the technology of the nineteenth century—required a work force composed of semiskilled or unskilled operatives who could work only at the factory site and only at the strict time schedule enforced by turning the machinery on or off. Again, the technology of the electronic age has steadily required a higher proportion of skilled attendants; and the coming technology of automation will still further change the needed mix of skills and the locale of work, and may as well drastically lessen the requirements of labor time itself.

2. The Hierarchical Organization of Work

Different technological apparatuses not only require different labor forces but different orders of supervision and co-ordination. The internal organization of the eighteenth-century handicraft unit, with its typical man-master relationship, presents a social configuration of a wholly different kind from that of the nineteenth-century factory with its men-manager confrontation, and this in turn differs from the internal social structure of the continuous-flow, semi-automated plant of the present. As the intricacy of the production process increases, a much more complex system of internal controls is required to maintain the system in working order.

Does this add up to the proposition that the steam-mill gives us society with the industrial capitalist? Certainly the class characteristics of a particular society are strongly implied in its functional organization. Yet it would seem wise to be very cautious before relating political effects exclusively to functional economic causes. The Soviet Union, for example, proclaims itself to be a socialist society although its technical base resembles that of old-fashioned capitalism. Had Marx written that the steam-mill gives you society with the industrial *manager*, he would have been closer to the truth.

What is less easy to decide is the degree to which the technological infrastructure is responsible for some of the sociological features of society. Is anomie, for instance, a disease of capitalism or of all industrial societies? Is the organization man a creature of monopoly capital or of all bureaucratic industry wherever found? These questions tempt us to look into the problem of the impact of technology on the existential quality of life, an area we have ruled out of bounds for this paper. Suffice it to say that superficial evidence seems to imply that the similar technologies of Russia and America are indeed giving rise to similar social phenomena of this sort.

As with the first portion of our inquiry, it seems advisable to end this section on a note of caution. There is a danger, in discussing the structure of the labor force or the nature of intrafirm organization, of assigning the sole causal efficacy to the visible presence of machinery and of overlooking the invisible influence of other factors at work. Gilfillan, for instance, writes, "engineers have committed such blunders as saying the typewriter brought women to work in offices, and with the typesetting machine made possible the great modern newspaper, forgetting that in Japan there are women office workers and great modern newspapers getting practically no help from typewriters and typesetting machines." In addition, even where technology seems unquestionably to play the critical role, an independent "social" element unavoidably enters the scene in the *design* of technology, which must take into account such facts as the level of education of the work force or its relative price. In this way the machine will reflect, as much as mould, the social relationships of work.

These caveats urge us to practice what William James called a "soft determinism" with regard to the influence of the machine on social relations. Nevertheless, I would say that our cautions qualify rather than invalidate the thesis that the prevailing level of technology imposes itself powerfully on the structural organization of the productive side of society. A foreknowledge of the shape of the technical core of society fifty years hence may not allow us to describe the political attributes of that society, and may perhaps only hint at its sociological character, but assuredly it presents us with a profile of requirements, both in labor skills and in supervisory needs, that differ considerably from those of today. We cannot say whether the society of the computer will give us the latter-day capitalist or the commissar, but it seems beyond question that it will give us the technician and the bureaucrat.

III

Frequently, during our efforts thus far to demonstrate what is valid and useful in the concept of technological determinism, we have been forced to defer certain aspects of the problem until later. It is time now to turn up the rug and to examine what has been swept under it. Let us try to systematize our qualifications and objections to the basic Marxian paradigm:

1. *Technological Progress Is Itself a Social Activity*

A theory of technological determinism must contend with the fact that the very activity of invention and innovation is an attribute of some societies and not of others. The Kalahari bushmen or the tribesmen of New Guinea, for instance, have persisted in a neolithic technology to the present day; the Arabs reached a high degree of technical proficiency in the past and have since suffered a decline; the classical Chinese developed technical expertise in some fields while unaccountably neglecting it in the area of production. What factors serve to encourage or discourage this technical thrust is a problem about which we know extremely little at the present moment.

2. *The Course of Technological Advance Is Responsive to Social Direction*

Whether technology advances in the area of war, the arts, agriculture, or industry depends in part on the rewards, inducements, and incentives offered by society. In this way the direction of technological advance is partially the result of social policy. For example, the system of interchangeable parts, first introduced into France and then independently into England failed to take root in either country for lack of government interest or market stimulus. Its success in America is attributable mainly to government support and to its appeal in a society without guild traditions and with high labor costs.

The general *level* of technology may follow an independently determined sequential path, but its areas of application certainly reflect social influences.

3. *Technological Change Must Be Compatible with Existing Social Conditions*

An advance in technology not only must be congruent with the surrounding technology but must also be compatible with the existing economic and other institutions of society. For example, labor-saving machinery will not find ready acceptance in a society where labor is abundant and cheap as a factor of production. Nor would a mass production technique recommend itself to a society that did not have a mass market. Indeed, the presence of slave labor seems generally to inhibit the use of machinery and the presence of expensive labor to accelerate it.

These reflections on the social forces bearing on technical progress tempt us to throw aside the whole notion of technological determinism as false or misleading. Yet, to relegate technology from an undeserved position of *primum mobile* in history to that of a mediating factor, both acted upon by and acting on the body of society, is not to write off its influence but only to specify its mode of operation with greater precision. Similarly, to admit we understand very little of the cultural factors that give rise to technology does not depreciate its role but focuses our attention on that period of history when technology is clearly a major historic force, namely Western society since 1700.

IV

What is the mediating role played by technology within modern Western society? When we ask this much more modest question, the interaction of society and technology begins to clarify itself for us:

1. *The Rise of Capitalism Provided a Major Stimulus for the Development of a Technology of Production*

Not until the emergence of a market system organized around the principle of private property did there also emerge an institution capable of systematically guiding the inventive and innovative abilities of society to the problem of facilitating production. Hence the environment of the eighteenth and nineteenth centuries provided both a novel and an extremely effective encouragement for the development of an *industrial* technology. In addition, the slowly opening political and social framework of late mercantilist society gave rise to social aspirations for which the new technology offered the best chance of realization. It was not only the steam-mill that gave us the industrial capitalist but the rising inventor-manufacturer who gave us the steam-mill.

2. *The Expansion of Technology within the Market System Took on a New "Automatic" Aspect*

Under the burgeoning market system not alone the initiation of technical improvement but its subsequent adoption and repercussion through the economy was largely governed by market considerations. As a result, both the rise and the proliferation of technology assumed the attributes of an impersonal diffuse "force" bearing on social and economic life. This was all the more pronounced because the political control needed to buffer its disruptive consequences was seriously inhibited by the prevailing laissez-faire ideology.

3. *The Rise of Science Gave a New Impetus to Technology*

The period of early capitalism roughly coincided with and provided a congenial setting for the development of an independent source of technological encouragement —the rise of the self-conscious activity of science. The steady expansion of scientific research, dedicated to the exploration of nature's secrets and to their harnessing for social use, provided an increasingly important stimulus for technological advance from the middle of the nineteenth century. Indeed, as the twentieth century has progressed, science has become a major historical force in its own right and is now the indispensable precondition for an effective technology.

. . .

It is for these reasons that technology takes on a special significance in the context of capitalism—or, for that matter, of a socialism based on maximizing production or minimizing costs. For in these societies, both the continuous appearance of technical advance and its diffusion throughout the society assume the attributes of autonomous process, "mysteriously" generated by society and thrust upon its members in a manner as indifferent as it is imperious. This is why, I think, the problem of technological determinism—of how machines make history—comes to us with such insistence despite the ease with which we can disprove its more extreme contentions.

Technological determinism is thus peculiarly a problem of a certain historic epoch —specifically that of high capitalism and low socialism—*in which the forces of technical change have been unleased, but when the agencies for the control or guidance of technology are still rudimentary.*

The point has relevance for the future. The surrender of society to the free play of market forces is now on the wane, but its subservience to the impetus of the scientific ethos is on the rise. The prospect before us is assuredly that of an undiminished and very likely accelerated pace of technical change. From what we can foretell about the direction of this technological advance and the structural alterations it implies, the pressures in the future will be toward a society marked by a much greater degree of organization and deliberate control. What other political, social, and existential changes the age of the computer will also bring we do not know. What seems certain, however, is that the problem of technolgical determinism—that is, of the impact of machines on history—will remain germane until there is forged a degree of public control over technology far greater than anything that now exists.

CHAPTER 3 REFERENCES

1. D.J. Boorstin, *The Americans: The Democratic Experience*, Random House, New York, 1973.
2. J.B. Rae, "The Invention of Invention," from *Technology in Western Civilization*, Vol. I, Edited by M. Kranzberg and C.W. Pursell, Oxford University Press, 1967, pp. 325–336.
3. R.L. Heilbroner, "Do Machines Make History?" *Technology and Culture*, Vol. 8, No. 3, 1967, pp. 335–345.
4. "A Channel Tunnel Finds Firm Support," *Business Week*, Sept.22, 1973, pp. 36–38.
5. W. Ley, *Engineers' Dreams*, The Viking Press, New York, 1954, Chapter 1.

6. D.S.L. Cardwell, *Turning Points in Western Technology*, Science History Publications (Neale Watson Academic Press), New York, 1972.
7. M. Kranzberg and W.H. Davenport, *Technology and Culture*, Schocken Books, New York, 1972.

CHAPTER 3 EXERCISES

3-1. A tunnel under the English Channel connecting France and Great Britain was first proposed in 1751. The British and French governments have announced plans to begin construction of a tunnel in 1975. The five-year project is estimated to cost $2.1 billon.[4] A new $290 millon railroad is also planned to link London with Cheriton so that eventually a London-to-Paris train trip would take only 3 hours and 40 minutes. Two consortiums of private companies, the British Tunnel Company and the Societé Française du Tunnel sur la Manche, will finance and supervise the tunnel construction. Explore the history of the plans for a tunnel under the English Channel. Discuss the technological developments that have occurred which have permitted the now-planned construction.[5] Also, discuss the political impediments to such a tunnel and the changes that permit its construction today.

3-2. Building in natural stone probably started with the wall and ceremonial buildings encircling the pyramid of King Zoser at Saggarah in 2,600 B.C. The King's vizier Imhotep was the first engineer and architect known to us by name. Explore the engineering work of Imhotep and the builders of the pyramids of Egypt. What tools and machines were used to construct these buildings, some of which are over 4,000 years old?

3-3. The recording and knowledge of the passage of time by using clocks is critical to an industrialized nation. The first water clocks were invented in Alexandria. Explore the development of water clocks during the period 300 to 600 B.C. and their influence on the civilization of that time.

3-4. The sources of the industrial revolution are the steam engine, the factory system, the spinning jenny and the railroad, among others. In fact, the industrial revolution was an age of continued, but rapid, technological evolution in which earlier developments were fulfilled. Examine the tendencies and developments of the periods preceeding the closing decades of the eighteenth century. Why did the industrial revolution begin in Britain?

3-5. In his article Professor Heilbroner explores the question, do machines make history? Is industrial technology the necessary and sufficient condition for capitalism? Consider these questions in light of Professor Heilbroner's article and provide a critique of his reasoning.

3-6. As technology takes on a special significance in the context of capitalism, in Heilbroner's view, explore the special role of technology in a modern socialist state such as Russia or East Germany.

4

ECONOMICS, PRODUCTIVITY AND TECHNOLOGY

THE CONTRIBUTION OF TECHNOLOGY to economic growth is a large and significant one. Technology is important to the soundness of the economy, international trade and the affluence of the citizens of a country. Technology is a means whereby the goods available can be invested so that more people may share in the gains of the society. Thus, lower-income groups can achieve their goals without having to reduce the shares of others.

GROWTH OF PRODUCTIVITY AND TECHNOLOGY IN THE UNITED STATES

Productivity measures the effectiveness of the way we combine methods, materials and labor to produce goods or services. In the United States the productivity of the worker has increased over the past 70 years. The output per man-hour increased by 1% per year during the period 1869–1889; 1.6% per year during the period 1889–1919, and about 2.5% per year since then. The real income, in goods produced per person, increased fivefold from 1890 to the present. At the same time, hours worked per week have dropped by a third, from over 62 to 40. These changes were brought about by increased use of capital, energy, management systems and new technologies. Technology was used to increase the efficiency of manufacturing processes in order to increase productivity.

The increase in productivity per man-hour of work is shown in Figure 4–1. The productivity increased by a factor of four during the period 1910–1970.

An example of the growth of technology in the United States can be seen in the growth of the use of electrical energy. In Figure 4–2 the population of the United States, and the electrical energy produced, are plotted for the period 1900–1970. Using the estimated growth of these two indices, the curves have been extended to the year 2000; the dotted lines indicate that the projections are only estimates or forecasts. These curves are plotted on a graph with a logarithmic vertical scale. Thus, while the curves are straight lines in this figure, the growth is exponential.

Recall that a dependent variable, y, is an exponential function of an independent variable x, when represented by the relationship

$$y = e^{mx} \qquad\qquad (4\text{--}1)$$

where $e = 2.71828$, the base of the natural logarithm. Taking the natural logarithm of this relationship, we have

$$\ln y = mx \qquad\qquad (4\text{--}2)$$

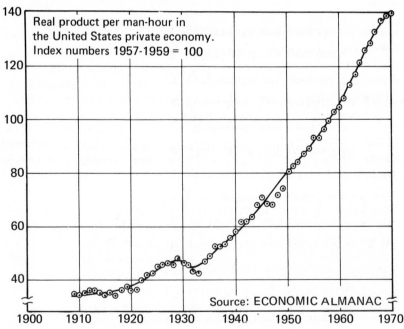

FIGURE 4-1   The increase in United States productivity per man-hour of work during the period 1900-1970.

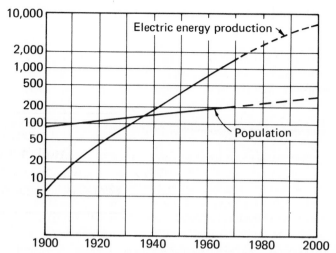

FIGURE 4-2 Growth of the United States population (in millions) and of electric
energy production (in billions of kilowatt-hours) over the period
1900-1970, with projected growth for the period 1970-2000. Note
that the vertical scale is logarithmic.

If the vertical scale of a chart is the logarithm of y and the horizontal scale is
linear in x, we obtain a straight line with a slope of m for the exponential
relationship of Equation 4-1. The growth of the population in the United States
and the growth of electric energy production in the United States are both
essentially exponential in nature.

Agriculture, which in 1890 employed 42% of the United States workers, now
has only 5%. To a large degree, this change can be attributed to increasing use of
technology on the farms. Industry and mining together then had 33% and now
have 39%. The service industries then took 25% of the employed workers, but
now account for 60%. This significant change in work patterns is shown in Figure
4-3. The reduction in the average work week is shown in Figure 4-4. Technology
is used extensively to increase the productivity of agriculture. Farm employment
has continually dropped while output per acre has risen, as shown in Figure 4-5.
This productivity increase has been made possible by means of technological aids
such as tractors and fertilizers. The persons supported per farm worker rose from
10.7 in 1940 to 34.5 in 1970.

RATES OF GROWTH

Productivity gains slowed down in some years in the United States. For
example, they were only approximately 1% per year for 1969 through 1971.
However, the productivity growth of 1973 exceeded 3%. Rates of productivity

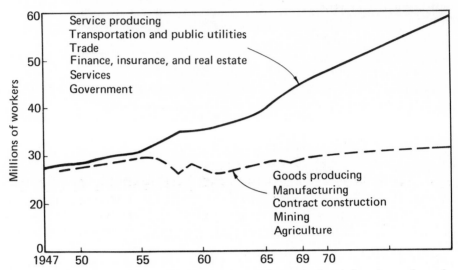

FIGURE 4-3   The number of workers in goods–producing industries and service industries. Source: Bureau of Labor Statistics.

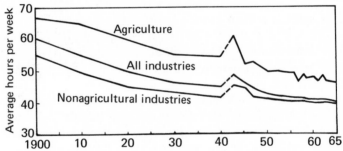

FIGURE 4-4   The annual average work week in the United States during the period 1900–1965. Source: Bureau of Labor Statistics.

have grown most rapidly, in recent years, in communications, public utilities and farming with an average productivity growth of 5% during the period 1948–1969.

Japan has shown what productivity growth is possible for the less developed portions of the world. Japan experienced a yearly gain of 14% in productivity per worker during the period 1965–1970. A growth through the use of machinery and technology for the less-developed nations of at least 4% per year is not considered over-ambitious. The growth per capita in these nations will depend on the increase in population. If the population growth is controlled or the population decreases, then the per capita income could improve significantly for these nations. The productivity of several European and the United States is shown in Figure 4–6.[2]

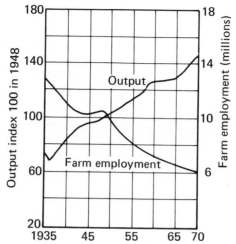

FIGURE 4-5. Farm employment and output during the period 1935-1970.

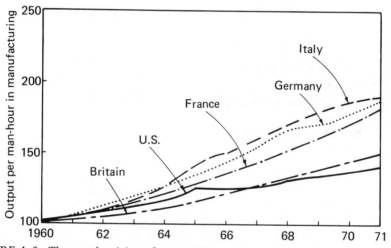

FIGURE 4-6 The productivity of several European nations and the United States during the period 1960-1971.

TECHNOLOGY INFLUENCES GROWTH

Technology permits industries to grow by providing new devices and improved efficiencies and output. By increasing the performance of a device continually, or with the introduction of new devices, new performance standards are achieved. For example, mechanical power output has climbed four orders of

magnitude (a factor of 10^4) since the start of the industrial revolution. The growth
in power performance is illustrated in Figure 4–7 for the period 1700–1970. Note
that the vertical scale is logarithmic. The growth of power of the internal–
combustion engine, for example, was exponential for many years. The trend in
the performance of illuminating devices is shown in Figure 4–8. The trend is
essentially a straight line on the logarithmic coordinates. This means that
performance grew exponentially over the past 100 years. The performance of
computers has grown exponentially, as shown in Figure 4–9. In all these cases,
the improved performance of the devices permitted new applications and uses of
the devices.

The application of new technologies is equally as rapid as the development
of innovations. The use of three major new technological devices is shown in
Figure 4–10 for the period 1920–1960. For example, the continuous milling
machine was placed into use by 60% of the major industrial firms only five years
after its introduction in 1945.

The growth of society in the United States, Western Europe and Japan runs
on the fuel of new technology. Productivity gains and the application of new
technologies lead to widely available consumer products, readily available energy
to serve us and convenient transportation available for our use. Technology is
used to assist in the abolition of poverty, and establishment of new mass transit
and programs for health care and education as well as to assist in the elimination
of pollution. It is only with increasingly efficient and productive devices that
these goals can be realized. If increased output is desired for a greater proportion
of the population, this must be achieved with more efficient devices and more
productive systems in order not to overtax our resources, be they capital, natural
or human.

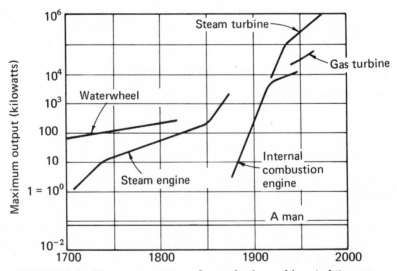

FIGURE 4-7   The power output of some basic machines in kilowatts.

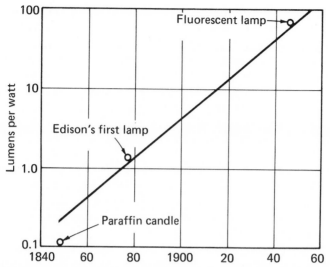

FIGURE 4-8 The trend in the performance of illuminating devices.

The affluence of the United States can be represented in part by the growth of the use of technological amenities as shown in Table 4–1. It is the general availability of such devices that leads to an affluent society.

GROWTH MEASURED BY RESEARCH FUNDING

Technological progress is maintained, to a large extent, by the expenditures for research and development by government and industry. While research and development expenditures do not directly lead to new products and devices in

Table 4-1

Affluence: the growth of the use of technological amenities in the United States in *per capita* figures.

Items in Use	Initial Value		Final Value		Per Cent Increase
Automobiles	.208	(1940)	.416	(1968)	100%
Telephones	.165	(1940)	.540	(1968)	227
Automatic Home Heating Units	.042	(1946)	.133	(1960)	217
Refrigerators	.145	(1946)	.277	(1960)	91
Clothes dryers	.0013	(1949)	.053	(1960)	4,000
Percent of Households with Air Conditioners	.02%	(1948)	13.6%	(1960)	6,700

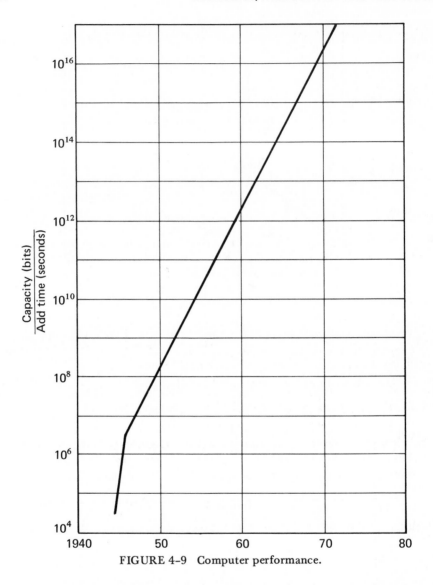

FIGURE 4-9   Computer performance.

every case, they contribute to new approaches and often to new technologies.

In the United States the research and development budget is 6½% of the federal budget. The federal expenditures for research and development expressed as the percent of the total federal budget expenditures are shown in Figure 4–11. During the period 1950–1957 a number of events led to increased spending for research and development, particularly in the military sector. The Korean War and the Cold War, along with the expenditures for the fission and fusion bombs,

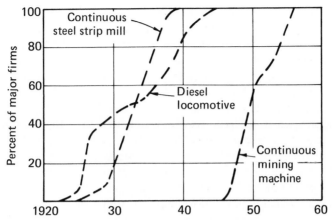

FIGURE 4-10 The trends in the use of new devices by major firms for three important devices. The vertical axis shows the percent of major firms using
a given device for a specific year.

were predominant in that period. Following the launching of the Soviet
spacecraft Sputnik, the United States increased its expenditures for space
research and development during the period 1958–1965. Funding as a percent of
total spending reached a peak in 1965 and has dropped off since then. As yet the
United States has not provided increased research and development for domestic
and civilian needs. As military and space needs have decreased in priority, new
funds have *not* been allocated for new domestic programs.

The United States and Great Britain both spend about 2.75% of their GNP
on research and development. In 1972 the expenditures were $30.1 billion in the
United States and $3.6 billion in the United Kingdom. In both countries the

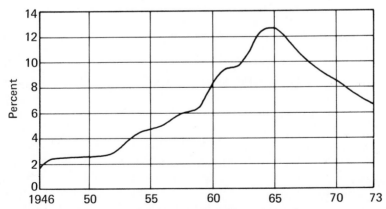

FIGURE 4-11 Federal expenditures for research and development expressed as a
percent of the total federal budget outlays during the period 1946–
1973.

government financed about 50% of the total expenditures for research and development. One important difference between the two countries is that in Britain the government carries out 25% of the nation's research and development in its own laboratories, while the corresponding figure is only 15% for the United States. Traditionally the United States government has preferred to contract with industry and universities for research and development activities.

Every industrial firm must do research and development to maintain its competitive position in its field. As the obsolesence life—that is, the time from the introduction of a new product to its replacement—becomes very short, it is necessary to spend a rapidly-increasing percentage of the firm's profit to maintain its present position. In Table 4–2 the relationship between obsolescence life and the percent of profit expended on research and development is given.[1] The actual expenditures are lower than theoretically required expenditures, because firms cannot afford to spend a very high percentage of profit on research and development. Actual expenditures are 36% for the aircraft industry, where the obsolescence life is only five years or so, to 12% in the electronics industry, where the obsolescence life is approximately 16 years. In the United States, industrial research and development expenditures normally are about 10% of net income (before taxes). Another 10% to 20% could be expended on design and tooling of new plants for new products.

Table 4-2
Obsolescence life and profit required to expend on research and development to maintain a firm's position.

Obsolescence Life (Years)	5	10	20	50	100
Per Cent of Profit needed to expend on research and development	85	70	55	40	17

New technology leads to new products, and more efficient plants and devices lead to more widely–available goods. It is these advantages of technology that make its use as an economic and social instrument so attractive to all nations. Of course, all benefits come with associated costs. The social and environmental costs of the uses of technology are significant and must be accounted for by the engineer. In the following chapters some of the benefits and costs of technology will be considered further.

CHAPTER 4 REFERENCES

1. "Too Much Research?" *European Scientific Notes*, ESN–21–11, Nov. 30, 1967.

2. C.E. Evanson, "Automation: A Tool for Productivity," Mechanical Engineering, Oct., 1973, pp. 24–28.

CHAPTER 4 EXERCISES

4–1. The number of telephones in use in the United States has grown exponentially since 1940. In 1940 there were 20 million telephones in use, while there are over 130 million now in use. Determine the statistics on the growth of the number of telephones in use in your city over the past 25 years and the cost to the user per telephone during that period.

Determine the obsolescence life of a telephone handset installed in your community. Also, determine what percentage of profit the nationwide telephone company invests in research and development.

4–2. The United States invests about 7% of its GNP in building new plants and in purchasing and installing new equipment in its manufacturing plants. Japan, Britain and Germany invest 22%, 11% and 16% respectively. Because of its vastly larger GNP and population, the United States invests, in absolute terms, the largest amount, $60 billion, while Japan invests $40 billion in new plants and equipment. How much do you believe a highly technological nation such as the United States should invest in upgrading its plants and manufacturing facilities? How important is it to have the latest up—to—date equipment?

4–3. Visit a local industry and obtain its productivity data for the last ten years. Plot a graph for productivity of the firm. Is it an exponential growth? What is the average annual growth rate? Attempt to determine what permitted the firm to increase its productivity over the period studied.

4–4. Visit a local industry that has been in operation for over 25 years (preferably 50 years). Obtain data on the normal work week for its workers during the past 50 years. Plot a graph for this data and determine its average annual percentage change.

4–5. Choose an electrical appliance or useful aid in the home and attempt to determine the growth of its use during the past 20 years. Compare the growth of its use with that of other amenities as provided in Table 4–1. As affluence increases in our nation, can we expect the number of items in use per capita to increase?

4–6. Research and development expenditures vary from industry to industry. Obtain such expenditure figures from two or three local industries over the past decade and compare the effect of differing expenditures.

5

TECHNOLOGY AND TRADE

IN 1972 PRESIDENT NIXON sent to Congress the first Presidential message devoted solely to science and technology. In it he promised to use technology in a "strong new effort" to raise productivity and restore a favorable trade balance. In the view of many, technology holds the key to favorable trade balances for the United States, because this country has held a lead in advanced technology for some 25 years. Nevertheless, as Figure 5–1 shows, the surplus in the trade of technological devices and products is shrinking.

HOW TECHNOLOGY AFFECTS INTERNATIONAL TRADE

The overall trade deficit for the United States in 1972 for all commercial transactions was the highest in history—$8 billion. One economist, Michael Boretsky, of the United States Commerce Department, believes that only by improving United States high technology will the country achieve a sound trade surplus position for the long term.[1] Many agree with Boretsky. The thesis is that the United States lags in new product development, in contrast to Japan and West Germany, in the fields of chemicals, automobiles and electronics. Boretsky's theory is that advanced technology plays a very large part in import-

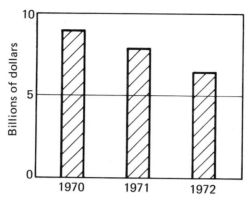

FIGURE 5-1   The United States trade surplus in technological devices and products.

export standings. It can make some industries produce more goods cheaply, thus making their prices more competitive. However, technological knowledge, in terms of quality and scope, is a strong determinant of international trade posture.

The United States has a unique capability to build Boeing 747's, and this capability has resulted in a strong trade surplus in airplanes. However, Japan has challenged the United States in electronics and automobiles in recent years. A recent publication of the Department of Commerce appears to confirm the fact that other industrialized nations are increasing their proportion of the advanced knowledge and development of new technology.[2,3] The share of all United States patents awarded to inventors who live outside the United States has climbed from 17% in 1961 to 29% in 1971. Four countries emerge as the leading recipients of United States patents: Germany, Britain, France and Japan. In three broad areas of textile weaving, foreign inventors obtained 70% of all United States patents that were issued in the period 1969–1971. In the field of metal shaping by rollers, 59% of the patents have been going to foreign inventors, and in devising tunneling methods foreign inventors have been obtaining 69% of the patents. Foreign countries lead in patents in the fields of polyamide resins similar to nylon, tracked air-cushion vehicles and the magnetohydrodynamic generation of electric power.

One measure of the technological development of a nation is its factor productivity, which is defined as the amount of output generated per unit of captial and labor input. The comparison of the level of technological development of several countries is shown in Figure 5–2 for the period of the mid-1960s.[8] In recent years the technological leadership of the United States has been seriously challenged by Japan and several European nations.

While there are no precise measures of technological developments, it is generally accepted that the Soviet Union lags behind the United States, Japan and Western Europe. This fact is somewhat surprising considering the emphasis on technology in the U.S.S.R. This result is in contrast to the percentage of Soviet

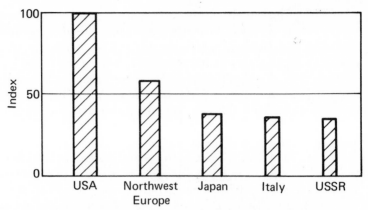

FIGURE 5-2   The levels of technological development of several developed countries
for the mid-1960s. Index USA = 100. The index is the gross national
product produced from each combined unit of capital and labor
employed.

GNP devoted to investment—higher in the U.S.S.R. than in the United
States—33% and 17% respectively in 1971.[8] Labor productivity in Soviet
industry and agriculture were a fraction of the United States level—41% and 11%
respectively in 1971. This fact may be due, in part, to the heavy investment in
"defense" in the Soviet Union. Nevertheless, the Ninth Five-Year Plan calls for
the modernization of the civilian economy and the improvement of the quality of
living.

SHARING THE LEAD IN TRADE AND TECHNOLOGY

The United States is not a sole leader in world trade or in technology.
Rather, the field is shared in a reasonable distribution among the European
Common Market, the United States, Russia and Japan, among others. For
example, in 1970 the European Common Market had a total of $45 billion,
passing the United States level of $43 billion for the first time. The United States
is falling behind in growth of productivity and new investment in plant. In 1970
the United States showed a trade deficit with Japan in the area of technological
goods of more than $1 billion. During the past decade a complacency has existed
in the field of technology with respect to our trade partners. The United States
has invested a relatively lower percentage in research and development for
civilian products and a relatively higher percentage for military technology for
over a decade. This preoccupation is now returning its fruits.
Even in the aircraft field, which is supported in part by military expenditures
for research and development, the commanding position of the United States has

eroded. About 80% of the aircraft in airlines of non-communist nations were built in this country. Now, however, challenges are being mounted by Japan and Europe, and even the Russian supersonic TU-144 may penetrate the world market. American dominance in aircraft may not hold in the future.

In his book *The American Challenge*, J. J. Servan-Schreiber states that if present tendencies continue, the third industrial power in the world after America and Russia could be, not Europe, but American industry in Europe.[6] Servan-Schreiber finds the American art of organization in development, production and marketing a challenge to Europe. The penetration of United States firms such as I.B.M. and General Motors and the amount of United States capital invested in Europe, over $15 billion, is a measure of the challenge to Europe. Innovation and technical knowledge are the powerful forces accompanying the capital to challenge the industries of Europe. As an example, Servan-Schreiber examines the field of computers in Europe and the dominance of American firms in the European market. Nevertheless, Europe has responded to this challenge with great strength in many fields of technology, including atomic energy and aviation. In Servan-Schreiber's view, Sweden and Japan are examples of nations with a growing technological base which yet remain unAmericanized. With a new resurgence of management and organizational innovations and technologies, Europe will remain an independent and important region of the world.

Two recent developments have affected the technological relationship between the United States and Japan. In 1969, for example, Japan exported new technology in the form of licenses, processes and other means at a value of $34 million while importing $314 million. The United States in that year exported $1.8 billion and imported only $194 million. Clearly, Japan imported many technologies. Coupled with its labor market and incentive to corporations, the importation of technology has resulted in a large increase in the growth rate of the Japanese gross national product.

INTERNATIONAL FUNDING FOR TECHNOLOGY

The industrial nations of the world have taken steps to encourage industrial technology, while the United States has not been as active in the civilian field. For example, in Canada companies receive grants equal to 25% of their capital expenses on research and development. In Britain companies can write off as much as 100% of their investments in new production facilities in the year they are made. In Germany, companies receive special tax write-offs of up to 50% of corporate research and development expenses, plus direct subsidies. The German, French and British aircraft industries are all subsidized to facilitate research and development. Many advocate that the United States should consider using research and development tax incentives, government development funding for new technologies in the civil sector, accelerated depreciation allowances encouraging new plant construction and other means.[4] Of course, many others oppose tax incentives for research and development or the construction of new plant as tax giveaways to business.

The United States spends only about 1.5% of its gross national product on research and development for non-military products, while Germany spends 2.6% and Japan 2.0%. The United States has invested heavily in military and space research and development and trained a large percentage of its engineers to serve in these fields. This unfortunate fact is now resulting in the increasing inability of the United States to compete in the trade marketplace with a corresponding surplus.

ENERGY

The problem of competition in international trade will become a particularly difficult issue if the United States becomes increasingly dependent on the Middle East for importing of petroleum for energy sources over the next ten years. Some have advocated that exports of agricultural products will enable the United States to balance the high cost, perhaps $2 billion per year by 1980, of imported oil.

The correlation between the use of energy per person and the gross national product per person is shown in Figure 5–3. As nations become more industrialized, they move up to the right along the trend line. Industrial nations are all obviously dependent upon the use of energy to improve their condition by powering their factories and transportation systems. This will be a problem for most industrialized nations as the importation of fuels grows over the next decade. Petroleum will, of necessity, be imported from the Middle East, South America and Indonesia. This need for energy, so vital to the industrialized countries, will tie them in new ways to less-developed countries which hold the natural resources of fuel.

TECHNOLOGY AND WORLD SOCIETY

Dr. Glenn Seaborg noted recently, "Today we must think of the New World in terms of the entire world, as a community of mankind whose future lies in pursuing the belief that knowledge—universally obtained, widely shared, and wisely applied—is the key to the viability of the human race and the earth that supports it."[5] Science and technology are to be put to work on an international scale. The sharing of such knowledge is a large and important enterprise. Each year there are more than 100 major international meetings of scientific and engineering societies. The world shares its technical knowledge in some 100,000 journals published in about 60 languages.

The earth is a vast system of forces: its oceans, its atmosphere and its lands

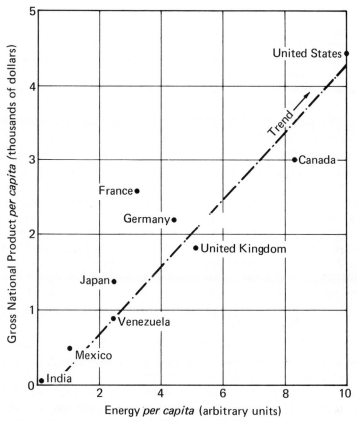

FIGURE 5-3 The energy per capita, plotted against the gross national product per capita for several countries.*

with which all men work to supply the world's food and resources for life. Science and technology wisely directed and supported will assume a central role in the development of individual nations and our international system of exchanging resources, products and services in the trade system.

It should be possible for the developing nations to avoid the errors of the past that are now obvious in the industrialized nations without significantly restricting their aspirations. World leadership has also been provided by the United Nations, specifically by the Food and Agricultural Organization, the United Nations Industrial Development Organization, the Organization for Economic Cooperation and Development, the International Atomic Energy Agency and other United Nations organizations. Critical to the development of nations are the problems of food supply and population growth. In many of the areas of the world there are indications that population is stabilized and that new

*Source: Adapted from *Problems of Our Physical Environment* by J. Priest, Addison–Wesley, 1973.

programs of family planning can regulate the birth rate, although this is not true everywhere.

It is estimated that more than two-thirds of the people of the developing nations depend directly upon agriculture for employment. A question of critical importance to nations as they industrialize is whether they will experience a shift of population from rural to urban areas (and with this shift, problems of urbanization). Mexico is currently experiencing such a movement to Mexico City. With urban populations now totaling 600 million, it is estimated they could increase to three billion by the end of the century. Such an increase could be overwhelming to the developing nations. Can newly-industrialized nations rapidly advance without the dislocations that Europe and the United States have experienced?

Perhaps these nations can move toward rural industrialization with small and light industry that will provide high employment and low environmental impact. In this case rural electrification and transportation could play an important part in the development of the industries and the potential for trade. Much can be done with technology to bring education to the rural areas of the developing nations. Currently, satellite and television technologies are being planned for the distribution of education to remote and rural areas.

The world is a closely-connected set of countries with shared weather, resources and food. With a shared technological base, this network of nations will be even more closely related. The interchange of technical knowledge is as important to the world as is the interchange of commodities and products. The world production of commodities over the period 1938–1968 is shown in Table 5-1. The world is increasingly based on technology, its products and the use of natural resources. Only with a rational, mutually advantageous process of shared technology and the associated standard of living can the people of "spaceship earth" live harmoniously.

Table 5-1

World production of commodities*

	1938	1948	1950	1960	1965	1968
Electric Energy (10^9 kwh)	459	810	957	2304	3377	4192
Petroleum (million tons)	250	467	523	1054	1511	1924
Natural Gas (10^9 cubic meters)	78	189	196	468	705	891
Aluminum (million tons)	0.6	1.3	1.5	4.5	6.2	8
Cement (million tons)	85	103	133	317	434	514

*Source: *Statistical Yearbook*, United Nations.

CHAPTER 5 REFERENCES

1. D. Shapley, "Technology and the Trade Crisis: Salvation Through a New Policy?" *Science*, March 2, 1973, pp. 881–3.
2. "Where the Action Is," "Science and the Citizen," *Scientific American*, July 1973, pp. 83–85.
3. *Technology Assessment and Forecast*, First Report of the Office of Technology Assessment and Forecast, United States Department of Commerce, May, 1973.
4. K.G. Harr, Jr. "Technology and Trade," *Vital Speeches of the Day*, Feb. 15, 1972, pp. 267–70.
5. G.T. Seaborg, "Science, Technology and Development: A New World Outlook," *Science*, Vol. 181, No. 4094, July 6, 1973, pp. 13–19.
6. J.J. Servan-Schreiber, *The American Challenge*, Avon Books, Inc., New York, 1969.
7. C.L. Hogan, "The Expanding Semiconductor World," *Vital Speeches of the Day*, Nov. 1, 1973, pp. 51–54.
8. J.P. Hardt and G.D. Holliday, "U.S.-Soviet Commercial Relations: The Interplay of Economics, Technology Transfer and Diplomacy," United States Government Printing Office, June 10, 1973.

CHAPTER 5 EXERCISES

5–1. The United States has been the leader in the world in the development of new technologies. Four technologies that initiated whole new industries are the computer, television, jet aircraft and xerography. These new industries led to favorable trade balances.

   Consider the effects which new uses of lasers may have on trade. Laser communication systems may be able to carry hundreds of thousands of telephone calls on a single laser light beam, for example.

5–2. The first satellite communication systems are now available for intercontinental communication. They can be used for transmitting television signals or radio signals. What is the potential for United States trade in the field of satellite communications?

5–3. The ability to build large ship canals led to the construction and opening of the Suez Canal by a French company in 1869. This canal has had a significant influence on world trade and diplomacy for over 100 years. Review the design and construction of the Canal and the subsequent effect on trade routes.

5–4. The world merchant fleet has increased from 105 million gross tons in 1956 to 268 million gross tons in 1972. The rate of growth of shipbuilding through the world is about 8% per year. This rate is indicative of the increased trade between nations. From 1950 to 1969, the cargo tonnages carried in worldwide trade increased four-fold. Historically, trade has been based on trading raw materials from underdeveloped countries against goods manufactured in the developed countries. Consider the effects that export-trading of oil, iron ore and other raw materials has on the development of an unindustrialized nation. How can trade help to raise the standard of living of the developed and underdeveloped nations?

5-5. The worldwide sales of electronic semiconductors has grown over the past ten years. In 1958 the total worldwide consumption of semiconductors was about $250 million, with about $20 million outside the United States. In 1973, sales worldwide were about $3.5 *billion*, divided equally between the United States and the rest of the world. The United States is the primary exporter of semiconductors, which are used in autos, cameras, electronic calculators, entertainment systems and computers. The semiconductor industry has shown a positive balance of trade for the United States over the past decade, and in 1973 a positive balance of $350 million was reached.[7] Examine the effect of new semiconductor uses on the sales of semiconductors around the world. As an example, consider the recent boom in low-priced electronic calculators.

5-6. The Soviet Union is interested in importing United States technological knowledge and equipment in the following areas: (1) large-scale petroleum and natural gas extraction; (2) management control facilities utilizing computers; (3) mass production machinery output such as that of trucks and cars; and (4) animal husbandry in agriculture.[8] Explore the potential effects which the wide introduction of these technologies would have on Soviet society.

5-7. The United States is currently exploring the joint development of Siberian natural gas resources with Soviet Russia. The gas made available would then, in part, be imported to the United States. These gas projects might require a United States investment of about $10 billion, largely for pipelines and tankers. How good an investment is Soviet energy exploitation? What are the potential effects of trade deficits due to these gas imports? What are the political benefits of the close ties resulting from such a large mutually-developed project?

6

AGRICULTURE AND TECHNOLOGY

AGRICULTURAL TECHNOLOGY IN THE UNITED STATES is highly developed. As we observe in Figure 6–1, farm productivity has increased by a factor of six while farm employment has dropped by a factor of three during the period 1950–1972.[1] American farms today utilize the world's most advanced equipment, and a typical farm has the appearance of the industrial enterprise it actually is. Agriculture may be the major growth industry of the 1970s. The demand for additional output is evident from the people of the industrialized nations as well as the less industrialized nations. The massive Russian purchase of grain from the United States in 1973 is an example of the large potential demand. In addition, there is growing world demand for wheat and for livestock feed grains.

Agriculture for many countries remains a protected area of trade today. For example, Japan and the European Common Market have protectionist agricultural policies (although these may be modified in the future by the demands of their own consumers).

INCREASING FARM OUTPUT

American farms are operating closer to capacity than ever before. In 1973, only 19 million acres were withheld from production, in contrast with 61 million

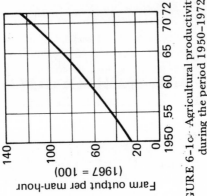

FIGURE 6-1c· Agricultural productivity during the period 1950–1972.

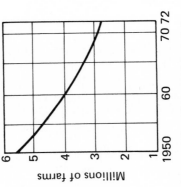

FIGURE 6-1b The number of farms in the United States during the period 1950–1972.

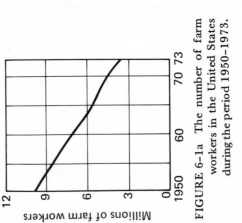

FIGURE 6-1a The number of farm workers in the United States during the period 1950–1973.

acres in 1972. It has been estimated that another 150 to 265 million acres could be added to agricultural use in the United States if the economics warranted this expansion.[1] Farm productivity has risen nearly 500% since 1945 and 27% in the last five years alone. Productivity increases are brought about by improved seed, fertilizers, pesticides, animal breeding and machinery.

The size of the average United States farm has grown from 191 acres to 394 acres since 1945; about 95% of American farms are owned by families. A well-equipped farm operated by two persons will normally achieve an optimal level of efficiency.

American farms have increased their exports, as shown in Figure 6–2, from $6 billion in 1965 to $11 billion in 1973. If this level of trade could be maintained, American farms would be working at record levels of production and efficiency. This level of trade would have a significant positive effect on the balance of trade between the United States and the rest of the world.

THE GREEN REVOLUTION

The high-yield crops of recent years are often called a portent of "the Green Revolution." To achieve such a revolution, new strains of plants—products of agricultural research and development—have been developed. Attempts are being made to breed strains of plants that will thrive in a variety of conditions throughout the world. This is particularly true of the cereal grasses which supply three-fourths of all of man's food. Agriculturalists are also developing a computerized data system to match fertilizer and herbicide application to soil

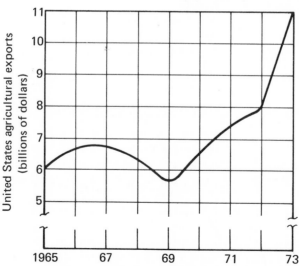

FIGURE 6-2   United States agricultural exports for the period 1965-1973.

properties for rational, efficient use and to prevent pollution.[2] It is an unfortunate fact that as much as a quarter of the world's food never reaches the consumer because it is destroyed in transit or in the market by insects, rodents, bacteria and mold. We must create and apply new technologies, particularly in the developing nations, to preserve and store grains and other produce effectively. Development of methods for the preservation of food, its transportation and distribution will help to alleviate the maldistribution of the world's food.

The Green Revolution can be described as the application of biological engineering to the development of edible plants. Agriculture, which was until recently viewed as a sector of human activity independent from industry, is now more clearly viewed as an area of technological attention. Some have said that the new agricultural revolution is bringing far-reaching changes to every segment of society. The agricultural revolution in the less developed countries may be the new technology that brings change as the steam engine did in the Industrial Revolution in Europe. According to Lester Brown, India's wheat crop increased 50% in the period 1965–1969 through the use of high-yield seeds.[3]

Mexico imported half of its wheat for many years. But using high-yield grain, Mexico became self-sufficient during the period 1944–1956, with the aid of an extensive wheat research program. In the Philippines, research developed two new strains of rice which transformed the country's rice production from a deficit to a surplus within three years. This has happened with and corn rice in Africa and the Far East. Japan has increased its production of rice per acre by a factor of two over the past 70 years by intensive use of fertilizers and land.

AMERICAN FARM FUTURES

Of course, the agricultural revolution is not limited to less developed nations. Predictions for the United States include:[4]

1. Wheat harvests of 300 bushels per acre compared with the present average of 30 bushels.
2. Corn outputs of 500 bushels per acre, compared with the present average of 75 bushels.

These increases, among others, may be made possible by new technological innovations such as fields covered with plastic domes in order to maintain environments. Perhaps we shall soon have remote-control harvesters that will pick and grade, pack and freeze, and transport the food to wholesale depots. The increase in corn and wheat yields per acre in the United States are shown in Figure 6–3a and 6–3b.

If any characteristic of American agriculture can be singled out, it is mechanization. The American farmer distinguished himself early by his willingness to adopt new tools and techniques for farming. Innovations such as the

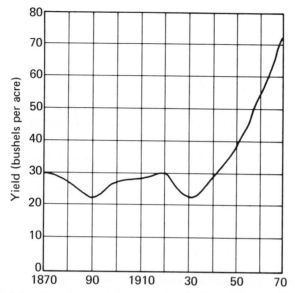

FIGURE 6-3a   Corn harvested per acre in the United States, 1870–1970.

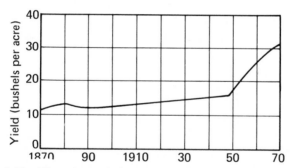

FIGURE 6-3b   Wheat harvested per acre in the United States, 1870–1970.

tractor, the steel plow, the reaper and the cotton harvester became part of the accepted agricultural practice by the turn of the twentieth century.[5]

The present ratio of farm workers to persons not on farms is about 1 to 48. Great social change has occurred as numbers of unemployed, untrained farm laborers have migrated to cities. Farmers process little of their own products because they are dependent upon the food processing, wholesaling and retailing industries. They also depend upon a multitude of other industries to supply machinery, fertilizers, pesticides, improved crop varieties and other supplies. For every farm worker, it is estimated there are two farm-support workers. About 20% of the United States work force and industries are involved in the food industries.

Power machinery has been used for agriculture for many years. The steam engine saw farm service in the United States in the early nineteenth century for threshers, sawmills and grist mills. During the latter half of the century, a portable engine was used for self-propelled farm vechicles. During the 1880s most traction engines were eight to 20 horsepower. By 1920 engines of 30 to 40 horsepower were common. With the use of the internal combustion engine instead of the steam engine, tractor size decreased although tractor power increased. The number of farm tractors in use with internal combustion engines has grown to approximately 5,000,000 in the United States, as shown in Figure 6–4. With the use of power machinery, the farmer is able to increase the yield of his crops. In Figure 6–5 the yield per unit of land (hectare) for an available power in horsepower per unit of land is shown. (Note that both scales are logarithmic.) The slope of the trend line is 0.7; that is, the vertical rise is a factor of 700 for a horizontal increase of 1,000. Therefore, if a country increases its available horsepower per hectare by a factor of ten, it can expect to increase its crop yield by a factor of seven. Or alternatively, if India desired to double its crop yield, it could, theoretically, do so by increasing its available horsepower per hectare by a factor of three by increasing the number of tractors and farm machines.

The number of tractor horsepower on American farms increased from five million in 1920 to 93 million by 1950 and is well over 200 million today.[9] Tractors on American farms consume about eight billion gallons of gasoline a year; some call modern agriculture "farming with petroleum." (A 1906 farm tractor is shown in Figure 6–6.) One calculation of the total energy consumed in

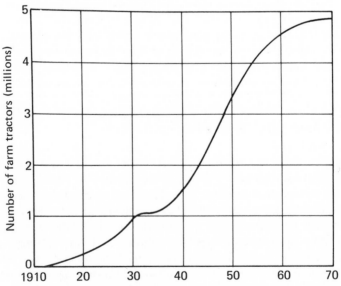

FIGURE 6-4 The number of farm tractors with internal combustion engines in the United States during the period 1910–1970.

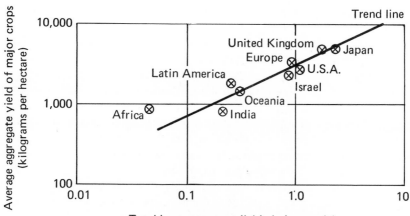

FIGURE 6-5 The average yield of major crops (kilograms) per unit of land (hectare) versus the total horsepower per unit of land available (horsepower per hectare) for several countries for 1967.

farming in the United States arrives at a figure equivalent to 30 billion gallons of gasoline per year.[12] In caloric value, the total energy used for agriculture is five times as much as the energy Americans ultimately get from the food they eat. Of course, we must recognize the fact that some American agricultural products are exported and many are used for fibers and other non-food products. In other words, we can say that it takes several calories of energy invested in machines, fertilizers and other inputs to yield one calorie of food energy for Americans.

One consequence of large mechanized farms and the capital investment necessary to operate them is the growth of large farms as a central agricultural supply source. The largest 223,000 farms in the United States, only 7.6% of the total number of farms, produce about 50% of American agricultural products.

FACTORS AND LIMITS IN PRODUCTIVITY

There is a limit in the ability to increase the yield per acre by simply increasing available machinery power. Also, one must realize that by use of tractors and machines, the farmer is using a fossil-fuel subsidy to the land. As long as fossil fuels such as gasoline are inexpensive relative to the return in crop yield, the use of farm machinery will continue to be attractive.

The main requirements beyond land itself are the technological factors of machinery, nutrients and water. The supply of plant nutrients can be increased with a favorable return. Fertilizers are made from non-renewable materials and the manufacturing technology is complex. Nitrogen products, potash products and phosphate products are used extensively for fertilizers in the United States.

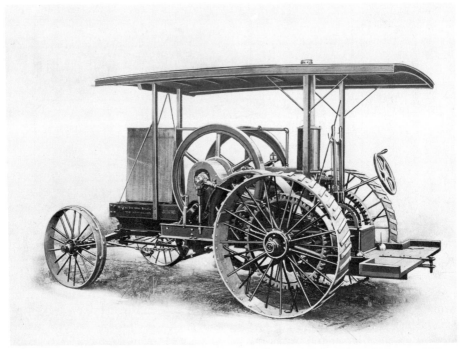

FIGURE 6-6 A 1906 model tractor. The tractor had high rear-wheel drive for maximum clearance under the rear axle and smaller narrow front wheels
designed to run between row crops. Courtesy of International Harvester Company.

Fertilizer production has been growing in the United States over the past few
years and it is a challenge to the chemical industry to meet the demand.[6] Figure
6-7 shows the growth of fertilizer use in the United States and Canada.[6]
Fertilizer technology is one of the foundations of modern agriculture.

Another technological aid to farmers is the availability of insecticides to
ameliorate the effects of pests that damage crops. Also, herbicides and fungicides
are used to control the effects of plant fungi. The growth in the production of all
three of these aids is shown in Figure 6–8. There exists a large dependency in the
United States on these aids and some side effects of their use have been detected
in our environment. It is well known that DDT ($C_{14}H_9Cl_5$), a pesticide, is a very
effective eliminator of pests but also has several undesirable characteristics.

Rachael Carson, in her book *Silent Spring*, pointed out the effects on our
environment caused by the use of DDT in vast quantities.[7] Carson and her
colleagues did not campaign against the use of *all* chemical pesticides, but rather
against the use of persistent pesticides—that is, those with strong residues over a
relatively long period of time.[13] One undesirable characteristic of DDT is its
stability. It does not tend to break down in the environment, but rather it is

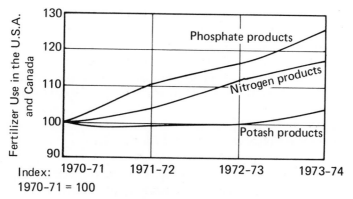

FIGURE 6-7   Fertilizer use in the United States and Canada. Each "year" is July 1 of
one year to June 30 of the next year.

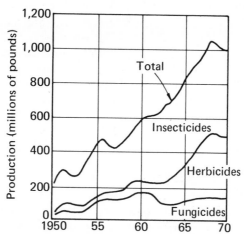

FIGURE 6-8   United States production of synthetic organic insecticides, herbicides
and fungicides.

recycled through food chains, accumulating in the tissues of those animals at the
top of the food chain. The pesticides of the group known as chlorinated
hydrocarbons tend to be cumulative in the body and to have deadly effects on
many animal species. In recent years the massive, indiscriminate use of persistent
pesticides with unknown undesirable side-effects has been reduced markedly.

Water is one of the several basic supports of agriculture. Most of it is
evaporated by the sun and the wind, thereby cooling the plants. Inadequate
water, even for a few days, can curtail the yield of a crop. The stable supply of
water can increase yields and is economically valuable. Thus, water supplied by
irrigation in dry areas is an important technological approach necessitating dams,
canals and pump systems.

In addition to fertilizers, irrigation and farm machinery, international business is considering new methods of procuring food or producing synthetic food. The oceans may become a major source of food with an increase in the fish catch, the use of fish meal as a protein supplement, and aquaculture—the raising and harvesting of sea animals and plants for food sources. Also, proteins may be produced from petroleum and natural gas and various vegetable sources.

EFFECTS OF TECHNOLOGICAL ADVANCES

Most farm technology has been beneficial to man, providing him with food necessary for a reasonable standard of living in most industrialized countries. The challenge is to achieve a reasonable and equitable distribution of food throughout the world. Farm machinery has been of assistance to the farmer; it allows him to increase his productivity. Driving a power cultivator, for intance, is less wearisome and faster than using a hoe.

The use of machinery has reduced the number of people employed on farms, although not all such reductions in farm labor were necessarily disadvantageous. While machinery reduces the number of harvest workers required, no one is sure that spending 10 hours a day, day after day in the hot sun was a worthwhile experience. Rather, the workers displaced by the mechanical harvester need to be offered new jobs and new training. During the past two decades, 240,000 farm jobs have been eliminated each year. This dislocation in the labor market is a challenge to our industries, our cities and our government.

MALTHUSIAN CONCLUSIONS

"Famine seems to be the last, most dreadful resource of nature. The power of the population is so superior to the power of the earth to provide subsistence ... that premature death must in some shape or other visit the human race." Thus did Thomas Malthus foretell the fate of the world's hungry in 1798. However, the race between population growth and food production is not yet over. World food production in 1972 increased 25% from a decade earlier while population increased 20% during that decade. Yet food production and availability is not evenly distributed, and many developing nations had their food production gains wiped out by population gains. World production of wheat fell 5% to 307.5 million tons in the year 1972–1973, mainly because of Soviet crop failures. Rice production also fell 5% in 1972–1973 because of drought in much of Asia. Also, many marine biologists now feel that the global catch of edible fish is near the maximum sustainable level. In recent years the harvest of fish from the sea have grown 5% per year.

Modern agriculture is highly energy-intensive. It has been estimated, for example, that about 2.9 million kilocalories of energy were used to raise an acre of corn in 1970—the equivalent of 80 gallons (2.5 barrels) of gasoline per acre. [12]

An estimated 330 million acres were planted in crops in 1970 in the United States. With about 200 million people in the United States, this averages about 1.65 acres per person. Since about 20% of our crops are exported, the estimated acreage planted for domestic consumption of the product is then 1.32 *per capita*. In terms of fuel per person for agriculture this is an equivalent of 106 gallons of gasoline per person. This figure can be compared with the figure of about 350 gallons of gasoline used per person per year for automobile travel.

Currently, for corn for example, a return of 2.8 kilocalories is achieved for a fuel input of 1 kilocalorie. As prices of fossil fuels increase, this ratio may be uneconomical and some return to human labor or smaller machines may occur.

During the year 1972–1973, the Soviet Union purchased 30 million tons of grain, much of it from the United States. It is believed that Russia will import at least ten million tons per year for several years. As nations increase their affluence, they increase the level of their diet to include milk and meat, which requires more grain.* India harvested only 95 million tons of grain in 1972–1973 against 108 million tons two years before. The drop was caused by drought conditions.

So the world has experienced a great increase in available food through the massive use of agricultural technologies. Nevertheless, the race between food supply and population continues. The race is far from secured, and a vast challenge to technology still exists. Man again depends upon technology to assist him in meeting his needs.

CHAPTER 6 REFERENCES

1. "Agriculture: Biggest Growth Industry in the United States," *Business Week*, April 28, 1973, pp. 62–66.
2. G.T. Seaborg, "Science, Technology and Development: A New World Outlook," *Science*, Vol. 181, No. 4094, July 6, 1973, pp. 13–19.
3. L. Brown, "The Optimist Outlook for World Food Production, *The Futurist*, August 1969, pp. 89–93.
4. C. Susskind, *Understanding Technology*, Johns Hopkins University Press, Baltimore, Maryland, 1973.
5. S.B. Hilliard, "The Dynamics of Power: Recent Trends in Mechanization on the American Farm," *Technology and Culture*, Vol. 13, No. 1, January 1972, pp. 1–14.

*Processing grain through cattle to produce milk and meat is at best extremely inefficient. No doubt future generations will be forced to make great use of milk and meat substitutes, made from soybeans and grains—an ironic twist of fate, perhaps.

6. "Shrinking Supplies Presage a Pinch," *Chemical Week*, Aug. 8, 1973, p. 13.

7. R. Carson, *Silent Spring*, Houghton Mifflin Co., Boston, 1962.

8. R. Keatley, "Once Again, Fears of World Food Shortage Arise; Affluence Is One Problem; Is Real Crisis at Hand?" *Wall Street Journal*, July 26, 1973, p. 28.

9. W. Cloud, "After the Green Revolution," *The Sciences*, October 1973, pp. 6–12.

10. "Riding the Farm Boom," *Business Week*, Oct. 27, 1973, pp. 74–78.

11. G.W. Thomas, "Can the World Afford to Feed Itself?" *Vital Speeches of the Day*, Oct. 15, 1973, pp. 28–32.

12. D. Pimental, *et. al.* "Food Production and the Energy Crisis," *Science*, Nov. 2, 1973, pp. 443–449.

13. F. Graham, Jr., *Since Silent Spring*, Houghton Mifflin Co., Boston, 1970.

14. *Environmental Quality*, 1972, Council on Environmental Quality, United States Government Printing Office, 1973.

15. C. Lerza, "The New Food Chain," *Environmental Action*, March 2, 1974, pp. 7–10.

CHAPTER 6 EXERCISES

6–1. In the United States the average horsepower of tractors has jumped from 46 in 1960 to 82 in 1973.[10] Over half the sales of new tractors are for machines with over 80 horsepower, and several companies sell tractors at 150 hp. and 200 hp. Within ten years, some companies are expect to sell a 500 hp. tractor. Productivity using tractors and other machines may be a key to feeding the United States and the world. It now takes only seven man-hours to produce 100 bushels of corn, while it required 108 man-hours in the late 1930s. Will increased tractor (and other machine) size result in increased productivity and decreased costs, or is agriculture working toward a point of diminishing returns?

6–2. The first mixed chemical fertilizers manufactured commercially in the United States were sold in Baltimore in 1849. The annual production of fertilizers in the United States had grown to 2 million tons by 1900. It has been estimated that the increased use of fertilizers was responsible for 55% of the increase in productivity per crop acre from 1940 to 1955. Are there more possibilities of more intensive use of fertilzers in the United States? Are all fertilizers beneficial, or are there disadvantages to their use?

6–3. One of the major mechanical improvements in agriculture during the nineteenth century was the reaper invented by the American Cyrus McCormick (1809–84). Prior to mechanized harvesting, the harvesting process required more labor than any other agricultural task. McCormick's reaper, patented in 1834, utilized the knife-and-cutter-bar principle and was pulled by a team of horses. If possible, obtain a drawing or an explanation of the way McCormick's reaper bound the sheaves of grain. Discuss the influence of the reaper on the growth of American agriculture. Consider the modern automatic tomato harvester and its influence on modern labor needs in agriculture.

6-4. Efforts to control agricultural pests have been instituted since early agricultural times. The use of poisons to eliminate the pests has been prevalent for over a century. The discovery of DDT in 1939 opened a new era in the warfare against insects. Examine the beneficial aspects of the use of DDT and the different side-effects. Should DDT be used in the United States?

6-5. Agriculture is heavily dependent upon technology in the United States. For example, over 2,500 pounds of water are used in the processes that eventually produce one pound of bread. Over 100 tons of water are associated with the process of producing one pound of beef at the supermarket. The United States consumes 115 pounds of beef annually per person now, while it consumed only 48 pounds per person in 1930. The beef industry is surely one large user of water.

Agriculture is also the largest single user of petroleum products in the United States.[11] Through fuel and water subsidies, total food production has increased many fold. However, the agricultural process is rather inefficient. Examine American diets, fuel economies and water use and suggest approaches to the provision of adequate and desirable diets at more efficient levels of operation.

6-6. Farming uses more petroleum than any other single industry. For example, for United States corn production, fuel consumption for all machinery rose from an estimated 15 gallons per acre in 1945 to about 22 gallons per acre in 1970.[12] Discuss the wisdom of the heavy use of petroleum for farm machinery, especially in light of the fuel oil crisis of the 1970s.

6-7. Some alternatives to the high use of energy from fossil fuels and electricity in agriculture exist. Increasing the input of farm labor is one possibility. For example, the application of a herbicide to corn by hand sprayer uses only 1/60 as much energy as that of a tractor application. As fuel costs rise, the cost of hand application will again become economical. Find other examples of the principle.

6-8. Manure from cows, cattle and hogs can be used as a substitute for chemical fertilizer. Discuss this alternative and others you can devise. What conditions are necessary to encourage alternative means to the high utilization of energy per acre in farming?

6-9. Pesticides such as DDT and chlordane are very effective in the elimination of unwanted agricultural pests. DDT is very effective in use against the mosquito and has probably saved many millions of people from immediate death from disease, staravation, or both. There are three main arguments against clorinated hydrocarbons used as pesticides:

1) They are universal poisons.
2) They degrade slowly.
3) They are fat-soluble and they accumulate in animal tissue.

In light of these risky characteristics would you recommend their use in underdeveloped nations where starvation is a very real possibility? What other means of pest control would you recommend?

6–10. What are the deterrents to reducing the energy-intensiveness of the American agricultural system? Would people be willing to live with a system with a subsidy ratio equal to 1? What do you expect will happen as petroleum prices rise over the next decade, forcing food prices up? Will the American and European people be willing to give up the beef products they enjoy? What forms of agricultural technology can help to reduce the average subsidy ratio?

7

MAN, THE MACHINE AND AUTOMATION

The machine provides a powerful aid to man and multiplies his power many times. A machine may be defined as an arrangement of fixed and moving parts for doing work and applying power. The machine, an amplifier of human power, utilizes the energy stored in fuels and permits man to exceed the limits of his bounded energies and power. In 1868, as the first transcontinental railroad was nearing completion, Charles Adams, Jr. predicted an impending transformation of the American experience as follows:[7]

Here is an enormous, an incalculable force ... let loose suddenly upon mankind; exercising all sorts of influences, social, moral, and political; precipitating upon us novel problems which demand immediate solution; banishing the old, before the new is half matured to replace it; bringing the nations into close contact before yet the antipathies of race have begun to be eradicated; giving us a history full of changing fortunes and rich in dramatic episodes. Yet, with the curious hardness of a material age, we rarely regard this new power otherwise than as a money-getting and time-saving machine ... not many of those ... who fondly believe they control it, ever stop to think of it as ... the most tremendous and far-reaching engine of social change which has ever either blessed or cursed mankind. ... Perhaps if the existing community would take now and then the trouble to pass in review the changes it has already witnessed it would be less astounded at the revolutions which continually do and continually must flash before it; perhaps also it might with more grace accept the inevitable, and cease from useless attempts at making a wholly new world conform itself to the rules and theories of a bygone civilization.

So it is again repeated with the automobile, the automatic assembly line and the robot. Man envisions a new engine which will either bless or curse mankind.

MACHINES AS MIXED BLESSINGS

Man sees the machine as a challenge to his abilities, his personhood and his livelihood. Perhaps this view is summarized by the lyrics of the song "The Workin' Machine" by Jimmy Sherman.

The Workin' Machine*

I was working real hard on my job one day,
When my boss came on the scene.
He said, "Son, go in an' get yo' pay–
An' make way—for th' workin' machine!"

I rolled my eyes, I sho' was sore,
My boss he sho' was mean.
He said, "Son, you know I don't need you no more.
You're fired! Cause I've got a machine."

Well, I got another job, that followin' day,
A-workin' harder than you've ever seen.
'Til I heard my boss in a loud voice say,
"Look out!" It was another machine.

Now I'm workin' like a slave, an' doin' real fine;
I'm gettin' these floors so clean.
An' I'm hopin' to keep this job of mine
Away! From the workin' machine.

The machine offers new power and efficiency to the benefit of some and to the immediate loss of others. The machine liberates the human spirit by removing drudgery, but only if the individual man is freed by its aid, not simply put out of a job. The machine produces abundance and is orderly. On the other hand, it also produces awe and fear in man. The machine is a *means*, although many people feel it becomes an end—the guiding force of society, in fact.

Many people feel negative toward machines. They believe that the machine kills the personal self and individuality and turns man into a part of the machine. Some observers state that machines necessitate a factory system which places man in a new serfdom to the machines. Arts and crafts are threatened, many believe, by machines and mass production. In mass production, the machines' rhythm is followed rather than natural human body rhythm, with an attendant

*Reprinted with permission of the author.

loss. The machine will replace thought, muscle and emotion. The machine, some feel, has separated man from nature and undermined his own nature.[2]

With computers, machines may soon make decisions for us. This may be one reason why so many negative statements about machines appear. A graphic illustration of man's interaction with his machines (at least in fantasy) is shown in Figure 7-1. In this picture, from the film *Modern Times*, Charlie Chaplin exemplifies man caught up in one of his machines.

As Carl Sandburg expressed this idea in his poem "The Hammer," published in 1910, [11]*

I have seen
The old gods go
And the new gods come.

Day by day
And year by year
The idols fall
And the idols rise.

Today
I worship the hammer.

MAN'S PSYCHE REACTS TO MACHINES

Man often identifies with a machine and finds it part of his total human fulfillment. Antoine de Saint-Exupéry, in *Wind, Sand and Stars*, states: "It is not with metal that the pilot is in contact. Contrary to the vulgar illusion, it is thanks to the metal, and by virtue of it, that the pilot rediscovers nature."[16] The pilot and his airplane, the rider and his motorcycle or the driver and his automobile all speak of the self-realization they achieve with their machines. The dependency on a mechanized instrument for self-identity reveals a dependence on technology which many would fear and others would deny. Yet what percentage of pilots, motorcyclists or auto drivers can otherwise achieve a good self-image? Clearly, when cars built to move at 100 m.p.h. are bought by urban dwellers one must ask for what purpose?

Perhaps an auto reveals its owner's chosen identity to others. One author has said that the relationship between man and his automobile is a love-hate relationship.[3] But one wonders if this is the result of man having to make his machine take the role of a person. This tendency is called anthropomorphism; that is, the attributing of human form or qualities to non-human things.

*From COMPLETE POEMS by Carl Sandburg, copyright, 1950, by Carl Sandburg. Reprinted by permission of Harcourt Brace Javanovich, Inc.

FIGURE 7-1   Charlie Chaplin caught up in his machine in the film *Modern Times*.
Courtesy of the Museum of Modern Art/Film Stills Archive.

THREE VIEWS OF MAN-MACHINE RELATIONSHIPS

We may consider several possible relationships: man gaining a fresh contact with nature through the machine; man as a slave of the machine; and man as a collaborator with the machine in destruction. The author Saint–Exupéry believed that the machine was a great asset to men. He was a pilot of single-engine propeller planes in the 1920s. He spoke as a romanticist of the machine.[4, 16]

The second image, which states that man is a slave of the machine, is found in many popular books such as Karel Čapek's *R.U.R.* In Čapek's book, in which the word *Robot* was coined (to mean artificial workers or mechanical men) the robots eventually become the masters of men. In *2001: A Space Odyssey* the machine which assists the astronauts in their journey to outer space, the computer HAL, attempts to take over the flight and to eliminate the astronauts. The one surviving astronaut is able to overcome the computer; he dismantles its brain.[6]

The third image is that of the machine as a collaborator in destruction. It is evident in the popular novels and movies about the spy James Bond. Bond relies on an extensive array of technical devices to be a potent adversary of evil. This is the same diamatic-technological basis that permits the members of the television series team of *Mission Impossible* to overcome their evil adversaries. This rise of technology as part of men, as extension of his power and ability, is extensive in these novels and television shows, which are popularly received. Bond possesses especially-equipped automobiles, magnetic watches, personal jet flight devices and the like. He is the technological (super) man of the 1970s.

And yet technology and the machine do provide man with abundance and the ability to overcome the drudgery of day-to-day work. In the song "John Henry," the coal miner laments that he is replaced with the steam-driven hammer machine. Yet who believes that man was destined to spend 10 hours per day, six days per week in the mines or tunnels digging with a shovel or pick (which are simple machines themselves)?

THE ASSEMBLY LINE

The mass production system known as industrialism is dependent upon many forms of machines. Industrialization is dependent upon transportation systems such as the railroad to deliver its products. All machines require the expenditure of energy, and energy in some form must be available to fuel the machines. New sources of energy, such as coal and oil, made new systems of transportation and communication available, thus facilitating the industrial revolution.

With machines, energy, communication and transportation, man was ready for the assembly line and for mass production. The purpose of the assembly line was to produce the maximum number of identical copies of the same product most economically. Assembly line is production by flow, and to keep the line moving at a constant rate requires breaking-down and splitting of tasks. This division of tasks demands precise timing and management of the small tasks. Thus the worker on the line has no possibility of satisfying his own taste or approach to the task, but rather must follow a prescribed order or process. Delay and variability in product are the two factors to be avoided in the operation of an assembly line.

The dream of Henry Ford was to make a new and better family car which everyone could afford and which would last a long time. To achieve mass production, Ford required a standard vehicle. It was the Model T. In 1908 Ford developed his assembly line, which was eventually to turn out millions of Model T's. A 1909 Model T Ford is shown in Figure 7–2. The ideal of the continuous production line was a non-stop flow from raw material to finished product.

A single magneto coil for the car's ignition required one individual 20 minutes to assemble. In 1913 Ford divided the operation among 29 men along a moving belt; in a year the construction time per magneto was reduced by assembly line techniques from 12 ½ hours to 1 ½ hours.[1] A photo of an assembly

FIGURE 7-2   A 1909 Model T Ford. Courtesy of Automobile Manufacturers Association.

line for making automobile magnetos is shown in Figure 7–3. This moving assembly line was begun in 1913.

Actually, the assembly line was not Ford's invention, but was adopted from the overhead trolley concept used earlier by the meat packers in the midwest. Frederick W. Taylor was the first man systematically to investigate the principles of scientific management of the technology of mass production. An American, Taylor wrote in 1911, "the man has been first; in the future the system must be first."[7] Taylor was interested in efficiency and economy. An engineering graduate of Stevens Institute of Technology in 1883, Taylor pursued a career of invention and secured a hundred patents including one for the largest steam hammer built in the United States.[1] Taylor's most important work was toward organizing factories for efficient work.

Taylor's prescription for a proper study of the division of the tasks was as follows:[1]*

First. Find, say 10 to 15 different men (preferably in as many separate establishments and different parts of the country) who are especially skillful in doing the particular work to be analyzed.

*From *The Americans: The Democratic Experience*, by Daniel J. Boorstin. Copyright © 1973 by Daniel J. Boorstin. Reprinted by permission of Random House, Inc.

FIGURE 7-3 One of the first applications of the moving assembly line was this magneto assembly operation at the Ford Motor Company Highland Park plant in 1913. Magnetos were pushed from one workman to the next, reducing production time by about one-half. By applying the same principle to total car assembly, Ford speeded up production until a finished Model T Ford came off the assembly line every 10 seconds. Courtesy of Educational Affairs Department, Ford Motor Company.

Second. Study the exact series of elementary operations or motions which each of these men uses in doing the work which is being investigated, as well as the implements each man uses.

Third. Study with a stop watch the time required to make each of these elementary movements and then select the quickest way of doing each element of work.

Fourth. Eliminate all false movements, slow movements, and useless movements.

Fifth. After doing away with all unnecessary movements, collect into one series the quickest and best movements, as well as the best implements.

This new method, involving that series of motions which can be made quickest and best, is then substituted in place of the 10 or 15 inferior series which were formerly in use. This approach is called time and motion study. Time and motion engineers became known as "efficiency experts," a not-entirely complimentary title. (The book *Cheaper By The Dozen* by Galbraith gives an amusing description of one such engineer.)

Items to be produced on an assembly line are designed and selected for production according to how quickly and economically they can be produced. The continuous–flow technology of the production line led to mass production at economical costs. It is this approach that permitted industrialized nations to raise the standard of living of their citizens.

WORKERS ON THE ASSEMBLY LINE

The feelings that the workers have about their work remain important; the worker is not just a part of the assembly line, although many workers do feel they are just that. In a famous experiment at Western Electric Company in Hawthorne, New Jersey in the 1930s it was revealed that no physical improvement in the production technique could overcome the feelings the workers had for their tasks. The experimenter, Elton Mays, juggled many physical variables such as the light in the workroom and the methods used in assembling telephones. Regardless of what the experimenters changed, the productivity rose. Finally Mays realized that the increases were caused by the experiment itself, and the show of interest in the workers and the *espirit* it gave them as a working team.[1]*

Ford startled the nation with a very high $5.00 daily wage for his assembly line workers in January, 1914. He explained it as profit sharing and an efficiency incentive. Ford built only Model T's, and said in 1911, "Any customer can have a car painted any color he wants so long as it is black." This one-model, one-color Model T was mass-produced so that by 1927 15 million Model T's had been produced. The Model T was inexpensive (as low as $265 at one time) and durable. Mass-produced, standardized parts were readily available. For example, a front fender cost $6.00.

THE LINE TAKES OVER

It was Alfred P. Sloan, Jr. who led General Motors Corporation toward the concept of an annual model of the automobile in order to keep new buyers interested. The idea was to create novelties and new additions to come. Sloan thus initiated a new challenge to the assembly line approach, since the line had to be retooled and reorganized each year as the annual models were changed.

In the 1960s a new production machine was added to the array of assembly line machines: the *industrial robot*. The industrial robot is an automatic

---

*The halo of good feeling that comes from being noticed is now a widely-recognized psychological phenomenon, usually called The Hawthorne Effect.

manipulator preprogrammed to perform a given task; it retains this program in its electronic memory. The robot, typically, has a pivoted mechanical arm with a hand-like gripper at one end as shown in Figure 7–4. Robots are used in dull and dangerous tasks such as die casting and forging. The tireless robot can work 24 hours every day without a break. There are several thousand robots of this kind in use in factories. A robot (often called an industrial manipulator) may operate in point-to-point tasks or continuous-motion tasks. A robot arm can be able to move vertically and horizontally and to rotate. The grip can move like a hand and wrist. Most robots use powerful hydraulic arms; they have electronic computers for program memory. The value of sales of industrial robots was about $15 million in 1972. It is growing at 30% per year.[8] Several hundred industrial robots are used in Japan and Europe also.

General Motors uses 26 industrial robots to put about 390 spot welds on its

FIGURE 7–4   A commercial robot called Versatran. The arrows show the permissible directions of movement for the machine. This robot can move up to 150 pounds with high speed. It can be used for loading or unloading, die casting, stamping, coating or spraying, among others. Courtesy of Versatran Automation Division, AMF Inc.

Vega automobile at its Lordstown, Ohio plant. The automatic line, with the robot welders, is shown in Figure 7–5. The price of a robot runs from $3,000 to $30,000 at present.[8] Currently work is proceeding on the development of a computer-controlled assembly machine with tactile and visual sensors. The grip carries touch sensors, and a television camera is used for a visual sensor. A robot arm controlled by a computer is shown in Figure 7–6.

AUTOMATION JUSTIFIED BY COSTS, EFFICIENCY

Automation of industry has been encouraged by the low cost of fossil–fuel energy. In 1973 only 3.6% of the producer's price was accounted for directly by energy. The ratio of production workers' wages to the cost of electricity increased steadily by 225% from 1951 to 1969. During that time the wholesale price index for electrical machinery increased by 50%. Therefore, it was greatly to the advantage of the producer to automate, using electrical machinery to replace human labor. As fuel costs increase in the late 1970's, this trend may be slowed down somewhat.

FIGURE 7-5   General Motors uses 26 industrial robots to put about 390 spot welds on its Vega automobile at its Lordstown, Ohio plant. Courtesy of Unimation, Inc.

FIGURE 7-6   A robot arm with a mechanical grip controlled by a computer. The television camera is the eye for the robot. The robot arm can rotate at the elbow and wrist and can grip with the two pincers. Courtesy of the Stanford University Artificial Intelligence Project.

As Lewis Mumford observed: "the clock, not the steam engine, is the key machine of the modern idnsutrial age."[9] The clock ensures an even flow of power and material. It permits standardization and automatic action. Through accurate timing the automated assembly line has been perfected. The gain in mechanical efficiency through coordination and close articulation of the day's events is a significant factor in the world of the machine. Man himself is required to accommodate to his machine, the clock.

Automated plants are often the modern answer to labor shortages or high labor costs. The Zenith Corporation has established a plant, which cost $5 million, and which has resulted in cutting in half the total labor involved in building a television set. The output per worker has also increased approximately 20 to 30 times greater than before. One Zenith automated plant inserts 155,000 components *per hour* into a variety of television circuits, enough to produce 7,000 television sets per day. The system uses small digital computers for 17 automatic sequencers and 31 insertion machines. To change from one circuit to another, the computer program is simply altered. The sequencer, on the automatic production line, is the contemporary clock.[10]

MAN–MACHINE RELATIONSHIPS IN LITERATURE

Poets have often summarized our concern over the advent of the machine.
W.B. Yeats (1865–1939) wrote the famous epigram:

Locke sank into a swoon;
The garden died;
God took the spinning jenny
Out of his side.

Yeats presents John Locke, the English philosopher of the seventeenth century,
as excessively oriented to the image of the garden—the natural, divinely-ordered
universe, a sort of Garden of Paradise. Eighteenth-century man, according to
Yeats, received an extension of himself in the form of a spinning machine
endowed with sexual significance.*[12]

Man and his relationship with his machines are often viewed as showing
man as a victim of the machine in its inexorable path. John Steinbeck, in *The
Grapes of Wrath*, depicts the tractors and their robot-like drivers, which "clean"
the land as the enemy of both man and nature. The tractors are pictured as
devices which rape the fields and which crush the houses, fill the wells, and drive
off the tenant farmers in the course of consolidating small farms into larger ones
during the Depression.[3]

In his novel *Player Piano*, Kurt Vonnegut tells of a society of the future
which is machine-dominated.[14] The persons working in the automated factories
hold positions assigned as a result of computer classification of their talents and
abilities. As jobs are eliminated, the characters in this novel find themselves
eliminated from the work force.

The relationship between man and machine has not always been one of
mutual benefit. Social ills such as the abuse of child labor in factories, suggested
in Figure 7–7, have often surrounded the introduction of new mechanical
processes. Children were suffering impaired health if not outright hazards to life
and limb well into the first decade of the present century in the United States.

THE FINAL QUESTION

The question remains: is the machine a servant of man, or his master? Will
man control the use of the machine or will it overwhelm him?

Perhaps the computer is the best illustration of a modern machine which
raises the spectre of slavery of man to a machine. The computer is a machine
capable of acting on its incoming data at a pace which man cannot comprehend.
We may not know—until it is too late—when to turn it off.

Perhaps this fear of the only half-understood machine may be illustrated by
the tale "The Monkey's Paw" by W.W. Jacobs, in which the Sergeant Major

*The spinning jenny was a basic part of the mechanized weaving system commonly associated with
the beginning of the Industrial Revolution in England.

FIGURE 7-7   This ten year old spinner in a North Carolina mill was photographed by Lewis Hine in 1909 for Hine's appalling report on child labor in cotton mills. Courtesy George Eastman House.

brings back from India a talisman which has the power to grant each of three people three wishes. Of the first recipient we are told only that his third wish is for death. The Sergeant Major, the second person whose wishes are granted, finds his experiences too terrible to relate. His friend, who receives the talisman, wishes first for £200. Shortly thereafter, an official of the factory in which his son works comes to tell him that his son has been killed in the machinery and that, without any admission of responsibility, the company is sending him as consolation the sum of £200. His next wish is that his son should come back, and the ghost knocks at the door. His third wish is that the ghost should go away.

Will man wish that his machines would go away, or will he learn to master them? Perhaps it can be expected that such a powerful concept as that of a machine has the ability to set mankind free from physical toil—or to enslave him to less-than-human subservience.

CHAPTER 7 REFERENCES

1. D. Boorstin, *The Americans: The Democratic Experience*, Random House, Inc., New York, 1973.
2. W.H. Davenport, *The One Culture*, Pergamon Press, New York, 1960.

3. D. Jewell, *Man and Motor: The Twentieth Century Love Affair*, Macmillan, New York, 1972.
4. A. De Saint-Exupéry, *Night Flight*, The Century Co., New York, 1932.
5. K. Čapek, *Rossom's Universal Robots*, English version by P. Selver and N. Playfair, Doubleday, Page and Co., New York, 1923.
6. A.C. Clarke, *2001: A Space Odyssey*, Signet Books, New York, 1968.
7. F.W. Taylor, *The Principles of Scientific Management*, Harper Bros., New York, 1911.
8. A. Rosenblatt, "Robots Handling More Jobs on Industrial Assembly Lines," *Electronics*, July 19, 1973, pp. 93–104.
9. L. Mumford, "The Monastery and the Clock" in *Technics and Civilization*, Harcourt, Brace and World, Inc., New York, 1934.
10 "Bringing TV Assembly Back to the United States," *Business Week*, Aug. 18, 1973, pp. 41–42.
11. C. Sandburg, "The Hammer," from *Complete Poems*, Harcourt, Brace and World, New York, 1956.
12. M. McLuhan, *Understanding Media*, McGraw-Hill Book Co., New York, 1964.
13. J. Steinbeck, *The Grapes of Wrath*, The Viking Press, Inc., New York, 1939.
14. K. Vonnegut, Jr., *Player Piano*, Charles Scribners and Sons, New York, 1952.
15. H.F. Williamson, "Mass Production for Mass Consumption," in *Technology in Western Civilization*, Vol. I, M. Kranzberg and C.W. Pursell, Editors, Oxford University Press, New York, 1967, pp. 678–692.
16. A. De Saint-Exupéry, *Wind, Sand and Stars*, Reynal and Hitchcock, New York, 1939.

CHAPTER 7 EXERCISES

7–1. Automation has raised fears in the minds of workers that they will be put out of work. This fear has been reenforced by articles and speeches which mention "workerless factories" or a "factory that runs itself." In fact, no plant as yet runs itself or is without workers. Nevertheless, automation does cause a displacement of workers from one task or skill to another new task or skill. The new task usually requires a more highly-skilled worker. Thus, unskilled workers may lose employment because of automation. Contact a manufacturing plant or business near your city and obtain information about its level of automation and the types of skills required by the workers in that plant. Review the data for the plant over the last decade, if possible.

7–2. The computer has added a new dimension to the possibilities for auto- mation. Locate an automated manufacturing plant in your region that uses a computer in the automation process. Request figures on productivity increases attributed to the use of computer automation.

7–3. A large number of individuals contributed to the development of rationalized factory operations. These operations used procedures which divided a manufacturing process into large number of small steps and utilized interchangeable parts. Richard Arkwright in Britain developed a rigidly-scheduled and highly disciplined factory process. In America, Oliver Evans built a mechanized grist mill in 1787 near Philadelphia. Eli Whitney used a process with interchangeable parts for producing muskets at the United States Arsenal at Harpers Ferry. Samuel P. Colt used the system in Connecticut to produce his famous pistol. Explore the work of Arkwright, Evans, Whitney and Colt which led to automotive assembly lines and other mass production lines of the late 1800s and early 1900s.

7–4. One estimate states that by 1980 the United States will have a $4,413 per capita GNP, with a 37½ hour work week, and a 48 week work year and it will provide retraining for 1% of the work force each year. With increased automation, society can choose to (1) reduce the work week; (2) add 1½ weeks vacation; (3) retrain 4% of the work force; or (4) lower the retirement age. Which of these would you choose, or in what combination?

7–5. The production of clothing remained almost entirely a hand operation until the introduction of the sewing machine in 1846. During the 1850s close to 100,000 sewing machines were sold in the United States.[15] The factory production of clothing grew rapidly between 1860 and 1900. In 1860 $88 million worth of ready-to-wear clothing was produced; this business grew to a $436 million industry by 1900. Examine the mass production of clothes and the necessary machinery which spurred this development.

7–6. The marketing and distribution of mass-produced goods is essential to the continuing growth of an industry. Examine. the effect of chain stores, mail-order houses such as Sears, Roebuck, and advertising on these functions. Machinery for mass production is not sufficient. The goods must be distributed inexpensively. What transportation systems permitted mass distribution of goods?

7–7. In 1793, Eli Whitney graduated from Yale University and took a job tutoring some boys in Georgia. His experience in Georgia led to his inventing the cotton gin that same year. Cotton grew profusely in Georgia but the fibers clung to the seed so that a slave working all day could clean only a single pound. The cotton gin multiplied the amount of cleaned cotton by a factor of fifty. Explore the social effect of this machine on the southern United States. In 1793 the cotton crop was 5 million pounds in the United States, but it grew to 35 million pounds by 1800. Did the cotton gin help to emancipate the slaves?

7–8. Man has written many songs about his relationship to his machines. One song, "John Henry," was written about 1870; it tells of a railroad man and his hammer (a machine) in competition with a new machine—the steam drill. The steam drill multiplied the force of its operator by means of the steam and outmoded a man and his hammer and spike. The song follows.

Find another song that tells of man and his machines. Is there a common theme to these songs?

When John Henry was a little baby,
Sittin' on his daddy's knee,
He grabbed himself a hammer and a piece of steel,
Said, "This hammer'll be the death of me, Lawd, Lawd,
This hammer'll be the death of me."

Now the captain said to John Henry,
"Gonna bring that steam drill round,
Gonna take that steam drill out on the job,
Gonna whop that steel on down, Lawd, Lawd,
Gonna whop that steel on down."

John Henry told his captain,
"A man ain't nuthin' but a man,
But before that steam drill beat me down
I'll die with my hammer in my hand, Lawd, Lawd,
I'll die with my hammer in my hand."

John Henry said to his shaker,
"Shaker, why don't you sing?
'Cause I'm throwin' twelve pounds from my hips on down,
Just listen to that cold steel ring, Lawd, Lawd,
Just listen to that cold steel ring."

The man that invented the steam drill,
He thought he was mighty fine,
But John Henry he made sixteen feet,
And the steam drill only made nine, Lawd, Lawd,
And the steam drill only made nine.
John Henry hammered on the mountain
Till his hammer was strikin' fire.
He drove so hard he broke his pore heart,
Then he laid down his hammer and died, Lawd, Lawd,
He laid down his hammer and died.

They took John Henry to the graveyard,
And they buried him in the sand,
And every locomotive comes roarin' by,
Says, "There lies a steel-driven' man, Lawd, Lawd,
There lies a steel-drivin' man."

8

THE TECHNOLOGICAL IMPERATIVE AND THE CONTROL OF TECHNOLOGY

TECHNOLOGY IS THE USE of the logical forces in the world for the benefit of society. Nevertheless, because of the rapid development of new technologies and the rapid social changes emanating from these technologies, many believe technology becomes its own force—and control of it, therefore, is often lost by society. It is said that we experience the *technological imperative*. The technological imperative is the force to implement a new technology because it is possible to do so. Man cannot, in the case of an imperative, decide not to implement the new technology, because it must and will be done. The technology itself is its own reason for existence. Jacques Ellul, in his famous book *La Tecnique*, states:[1] "It is possible, if the human being falters even momentarily in accommodating himself to the technological imperative, that he will be excluded from it completely even in his mechanical functions, much as he finds himself excluded from any participation in an automated factory."

COMPULSIONS

One theory suggested by Marshall McLuhan is that the effect of major technological changes upon society is to cause it to become numb. This temporary anesthesia takes the form of a refusal to admit the existence of the change and its accompanying effects. This numbness to change often leads to the further cliché "You cannot stop progress," used as justification for a new technology of limited benefit. Often technological change is equated with progress and progress confused with human purpose. When this confusion holds, then man does experience the technological imperative, for he has abdicated his role in his own society.

It is as if Americans, and Europeans also, have somehow lost control of the management of our human affairs—of the direction of our lives and destiny.[2] Archibald MacLeish says that what has happened is "a sort of technological *coup d'état*." Society was built with the technological means at hand. It was technology leading society, not the reverse; we were proud of technological achievements. However, these achievements were not altogether ours to direct, to control, but rather the process had somehow taken over, leaving the purpose to shift for itself. What part did men play in this self-fulfilling process? It is as if we lived in a world of technological compulsions.

Technology is a human tool, fashioned to be obedient to human wishes. However, science and technology can become forces of their own if lead by mindless men and women. Some may say that what it is possible for science to know, science must know; and what it is possible for technology to do, technology *will have done*.[2] If it is possible to split the atom, then the atom *will be split*. If it is possible to build an atomic bomb, then the atomic bomb will be built.

We must not ask where science and technology are taking us, but rather how we can manage science and technology so that they can help us get where we want to go. Technology is to serve man's purposes, not its own.

PHILOSOPHICAL VIEWPOINTS

Perhaps there are two opposed ways of looking at the dynamics of new technology. They may be termed the ontological and teleological viewpoints.[3] The ontological viewpoint is that invention and innovation are visible manifestations of a self-generating process, of the institution of technology having a life of its own. If we accept the first premiss, then technology is understood primarily in terms of a response to technological opportunities. Many hold this view; they seem to say that technology is an uncontrollable "sorcerer's apprentice."

The contrasting, teleological view, is that invention and innovation are social processes determined by social needs or economic needs of man. It is stated that if Edison had not invented the electric light, someone else would have, because its time had come. The electric light was a social and economic need. In other words, the market and society precede the technology and call for its development. Unfortunately, social needs are not always a reliable guide to new technological developments. In the opinion of Ayres, only when a social need is translated effectively into an economic need in the marketplace will it stimulate new technologies. Whenever a competitive market exists, the technology which develops will reflect the interaction between buyers and sellers. For instance, as buyers demand improved safety in autos, automobile manufacturers will provide safer cars. Of course, the government, through legislative or judicial means, can also establish the existence of a social need for a new technology. Control, in the teleological view, exists through the market or the government.

Technology responds to the pull of explicitly-defined objectives and missions and to the push of new discoveries and technological opportunities. One can cite examples of new technologies resulting from the pull of social needs as well as new technologies resulting from the push of technological opportunities. The three-point seat restraint belt is an example of response to a social need, while the xerography process is a response to the push of new technologies.

Social control of technology is possible through many mechanisms. However, some of them do not work effectively in the field of technology, which often seems to have too much momentum to take the time to debate its effects and its control effectively. Each member of the public is a secondary party to new technological developments and yet seems to have little direct influence on the choice of new technologies and the decision to proceed with new developments. Also, our traditional legal mechanisms are not suited for redressing wrongs from technology as they are for the case where two persons are involved. One may *blame* society or a large corporation, but how does one individual bring a suit against the society or the corporation?

REGULATION BY GOVERNMENT

A primary responsibility for the regulation of technology lies with government, by means of regulatory agencies such as the United States Federal Communications Commission or the United States Food and Drug Administration. Usually these agencies react to new technologies as they are presented to them; they do not anticipate or control change. An alternative is to encourage through the courts an increased amount of litigation between consumer advocate groups and the corporations. Again, this can occur only as the technology becomes available. Another possibility is to forecast new technological opportunities and to initiate early debate about the desirability of pursuing them. Industrial management should not only invest in sufficient research to ensure the availability of knowledge of the effects of a new technology, but should also share this knowledge with universities and the government.

Professional societies, such as the American Society of Mechanical Engineers, must be active in exploring the potential effects of new technologies and their control. This can be accomplished, in part, through the organizations like the American Society for Testing and Materials, which establishes standards of performance for materials and products. As far as possible our engineering standards should be placed on a performance basis rather than a design basis.[4]

We can state the performance level desired as a society and then let the new technologies be developed to meet this challenge. This approach focuses on social goals and is an incentive to examine many technologies to meet them.

THE ROLE OF EDUCATION

As Robert Hutchins observes, we cannot accept without a struggle the proposition that technology is autonomous and not subject to human guidance. Hutchins suggests the remedy is to redefine and restore liberal education.[12] The goal of education in a democracy, in his view, is to help people become human by helping them to use their minds. In addition, Hutchins suggests that the exercise of politics be revitalized along with a revival of political philosophy. The future leaders, while in college, must experience the great interaction between the disciplines. Thus, in his most controversial statement, Mr. Hutchins says, "independent engineering schools like M.I.T., Georgia Tech and California Tech should be stamped out."[12] Engineering students in universities would be exposed to the interaction of philosophy and politics and thus learn to seek the humane society.

POLITICAL CONTROL

The legal institutions of the western world function, in part, to assist in the control of technological advances. However, the judicial and legal institutions are not adapted to a modern technological society because they do not operate early enough in the development process. Rules of law, for example, usually come long after the potential for injury has been demonstrated. The characteristic of precedence is not inherent in our legal system, and the legal system does not forecast the future.[5] Nevertheless, the legal system has made an impact upon the control of innovations.

Also, the legislative branch has provided some important controls such as the Refuse Act section of the 1899 Rivers and Harbors Act, which is a powerful source of authority for combatting most forms of water pollution as they occur. The United States Congress established the National Environmental Policy Act of 1969 (83 Stat. 852) which requires all state and federal agencies to make environmental impact reports on any programs they propose which could have a significant impact on the environment prior to requesting funds other than planning funds.

Local communities are able to stem what is often considered a technological juggernaut. The residents of New York City were able to stop the development of a nuclear power plant station within the city limits. The city chose against the nuclear power station. A conventional fossil fuel plant, the Astoria plant, was sanctioned with a restricted output of electricity and the city and the region decided against a pumped storage system a few miles up the Hudson River in the unchanged Palisades.* The city has chosen not to pursue now power in these ways, and it has paid the cost in shortages of electric power each summer since. This is evidence that some communities can choose to control technology, if they want to pay the cost.

SELF-REGULATION OF TECHNOLOGY

Control by scientific and technology peer groups holds some potential for the control of new undesirable technologies. Often, however, members of a peer group experience a great amount of pressure to accept the new developments in their field. It is often said, "if we don't do it, somebody else will." Nevertheless, there are continuing important examples of practicing engineers who speak out against an imminent danger. This practice, popularly called "whistle blowing," is one which requires great courage on the part of an individual employed by a corporation when he finds he must challenge one of its developments.[6] In the private and Federal sector there is no protection, at present, beyond legal recourse for the employee who believes that his responsibility to society is greater than that to his immediate employer.

In the following article, Professor James Carroll examines the role of citizen participation in the public development, use and regulation of technology.† [7]

Participatory Technology

James D. Carroll

In this article I analyze the incipient emergence of participatory technology as a countervailing force to technological alienation in contemporary society. I interpret participatory technology as one limited aspect of a more general search for ways of making technology more responsive to the felt needs of the individual and of society. The term *participatory technology* refers to the inclusion of people in the social and technical processes of developing, implementing, and regulating a technology, directly and through agents under their control, when the people included assert that their interests will be substantially affected by the technology and when they advance a claim to a legitimate and substantial participatory role in its development or redevelopment and implementation. The basic notion underlying the concept is that participation in the public development, use, and regulation of technology is one way in which individuals and groups can increase their understanding of technological processes and develop opportunities to influence such processes in appropriate cases. Participatory technology is not an entirely new social phenomenon, but the evidence reviewed below suggests that its scope and impact may be increasing in contemporary society.

---

*Stopping the so-called "Storm King" project was a triumph for conservationists. The triumph may prove short-lived, however, in the face of what is called an "energy crisis."—Ed.

†From *Science*, Vol. 171, 19 February 1971, pp. 647 653. Copyright 1971 by the American Association for the Advancement of Science.

Participatory technology is one limited way of raising questions about the specific technological forms in terms of which social change is brought about. It is directed toward the development of processes and forums that are consistent with the expectations and values of the participatory individuals, who may resort to them in the absence of other means of making their views known. In participatory technology, however, as in other participatory processes, the opportunity to be heard is not synonomous with the right to be obeyed.

I here analyze two kinds of activities to illustrate some of the empirical referents of the concept of participatory technology.

Litigation

The first is the citizen lawsuit, directed toward the control and guidance of technology. As Sax indicates, "The citizen-initiated lawsuit is ... principally an effort to open the decision-making process to a wider constituency and to force decision-making into a more open and responsive forum. ... [The] courts are sought out as an instrumentality whereby complaining citizens can obtain access to a more appropriate forum for decision-making."

The courts, of course, rely heavily on adversary proceedings, various forms of which have been suggested as appropriate for handling scientific and technological issues involving the public interest. Not only can litigation restrict the use of technology, it can also lead to the modification and redevelopment of existing technology and stimulate the development of new technology to satisfy social values expressed in the form of legal norms, such as a right to privacy.

The legal response to cases involving technology has taken two forms. The first is an extension of those aspects of the legal doctrine of standing which determine who has a right to be heard in court on particular issues involving activities undertaken or regulated by public agencies. The second is a search by legal scholars, practicing lawyers, and judges for systems of conceptual correspondence in the terms of which scientific and technological developments and activities can be conceptualized and evaluated as changes in social values and norms that may warrant a legal response. The appropriate role of law in the regulation of genetic experimentation is an example.

An extension of the doctrine of standing has occurred in several recent cases involving technology, although the extension is not limited to such cases. In the words of the United States Supreme Court, "The question of standing is related only to whether the dispute sought to be adjudicated will be presented in an adversary context and in a form historically viewed as capable of judicial resolution." The basic question is "whether the interest sought to be protected by the complainant is arguably within the zone of interests to be protected or regulated by the statute or constitutional guarantee in question". The question of standing is a question not of whether a party should win or lose but of whether he should be heard.

The current extension of the doctrine is sometimes called the "private attorney general" concept. Under this concept a private citizen is allowed to present a case as an advocate of the public interest. A leading case is Scenic Hudson Preservation Conference v. Federal Power Commission, decided by the Second Circuit of the United States Court of Appeals on 29 December 1965. On 9 March 1965 the Federal Power Commission granted a license to Consolidated Edison Company to construct a pumped storage hydroelectric project on the west side of the Hudson River at Storm King Mountain in Cornwall, New York. A pumped storage plant generates electric energy for use during peak load periods by means of hydroelectric units driven by water from a headwater pool or reservoir. The Storm King Project, as proposed by Consolidated Edison, would have required the placement of overhead transmission lines on towers 100 to 150 feet (30 to 45 meters) high. The towers would have required a path some 125 feet wide through Westchester and Putnam counties from Cornwall to the Consolidated Edison's facilities in New York City—a distance of 25 miles (40 kilometers). The petitioners were conservation and other groups and municipalities who claimed that the project, as designed by Consolidated Edison and as approved by the Federal Power Commission, would destroy the character of the land and the beauty of the area.

The Federal Power Commission argued, among other things, that the petitioners did not have standing to obtain judicial review of the legality of the license because they "make no claim of any personal economic injury resulting from the Commission's action."

The Court of Appeals held that the petitioners were entitled to raise the issue of the legality of the license and the licensing procedure even though they might not have a personal economic interest in the question. The court reasoned that a citizen has an interest in actions that affect the nature of the environment, and that this interest is arguably within the zone of interests that are or should be protected by law. On the merits of the case, the court held that the Federal Power Commission was required to give full consideration to alternative plans for the generation of peak-load electricity, including a plan proposed by one of the petitioners for the use of gas turbines.

The Scenic Hudson case is significant because it set a precedent for the enlargement of the opportunity of citizens, acting as citizens and not as private parties, to secure judicial review of the actions of public agencies, and of actions of the interests these agencies often regulate, in cases involving technology as well as other matters. The decision supports the proposition that, in certain cases, citizens will be recognized in court as advocates of a public interest, on the grounds that, as members of the public, they have been or may be injured by the actions complained of. They need not claim that they have been or will be injured economically or otherwise as private persons.

The development of the "private attorney general" concept does not mean that substantive changes will automatically occur in the constitutional, statutory, and common law doctrines that regulate rights and duties pertaining to the development and use of science and technology. The work of analysts in the areas of law, science, and technology—analysts such as Patterson Frampton, Cowan, Miller, Cavers, Mayo and Jones, Korn, Green, Ferry, Wheeler, and others—indicates the difficulties of developing systems of conceptual correspondence between scientific and technological developments and legal concepts and doctrines. Scientific, technological, and legal systems often further different values and serve different purposes, and the reconciliation of conflicts in these values and purposes is only in part a juridical task. The "private attorney general" concept, however, does invite more active judicial scrutiny of such conflicts and may contribute to substantive changes in legal doctrine in the future in areas such as the computer and privacy; air and water supply and pollution; noise control; medical, genetic, and psychological experimentation; drug testing and use; nuclear energy and radiation; food purity and pesticides; and the control and handling of chemical and biological weapons.

While the legal form of citizen participation in the control and development of technology has severe limitations because it tends to be (i) reactive rather than anticipatory, (ii) controlled by restrictive rules of evidence, and (iii) subject to dilatory tactics, litigation has proven, over time, to be a significant element in the efforts of individuals and groups to influence the processes and institutions that affect them.

Ad Hoc Activity

A second form of participatory technology encompasses a variety of ad hoc activities of individuals and groups beyond the scope of structured processes of litigation and assessment. This form includes activist intellectualism of the sort undertaken by Carson, Nader, and Commoner; quasi-official action of the kind undertaken by Congressman R. D. McCarthy concerning chemical and biological warfare; political and informational activities of the sort undertaken by such groups as the Citizens' League Against the Sonic Boom, the Scientists' Institute for Public Information, the Sierra Club, Friends of the Earth, and Zero Population Growth; and sporadic activities of loose coalitions of individuals and groups energized by particular situations and issues.

Rather than attempt to survey such ad hoc activities, I here briefly describe and analyze an example of abortive participation that occurred in 1967 and 1968 in the initial efforts to develop a new town on the site of Fort Lincoln in Washington, D.C. In some ways the Fort Lincoln example is typical of problems that often arise in processes of citizen participation in urban development. In other ways the case is distinctive because the primary purpose of the Fort Lincoln project was to demonstrate on a national basis the potentials of technological and administrative innovation for urban development.

On 30 August 1967, President Johnson publicly requested several members of his administration and of the government of the District of Columbia to begin at once to develop a new community on the site of Fort Lincoln, which consists of 345 acres of nearly vacant land in the northeast section of Washington, D.C. The President explained the purpose of the project as the development of a community that would demonstrate the potentials of administrative and technological innovation in urban development. The Fort Lincoln project was conceptualized as the leading project in a national program to develop "new towns in town" on federally owned land in various cities throughout the country.

On 25 January 1968, Edward J. Logue, who had achieved national recognition as an urban development administrator in New Haven and Boston, was retained as principal development consultant for Fort Lincoln. In the following 10 months, Logue and his associates developed an ambitious and innovative plan that was based on, among other things, a thorough analysis of the potentials for technological innovation in the development of Fort Lincoln and on a proposal for an innovative educational system for the new community.

Fort Lincoln was a federal urban renewal project. Some form of citizen participation in urban renewal projects is required by law. Logue and the government officials involved in the Fort Lincoln project had had extensive experience with citizen participation in other urban development projects, including a model cities project in Washington, D.C. In developing the plans for Fort Lincoln, they made extensive efforts to fashion a participatory structure that would be acceptable to the citizens of the northeast section of Washington. For the most part they failed. Political activists in the area perceived the technical planning process as the locus of political opportunity and choice concerning such questions as the number of low-income families to be housed on the site. Although these activists disagreed over who could speak for the citizens, they agreed that the residents of the area should be granted funds to hire professionals to participate with and for them in the technical planning and development processes. At one point the Department of Housing and Urban Development offered to grant money for this purpose to the council that represented the citizens, but for various reasons the council rejected the offer.

The Nixon Administration suspended development of Fort Lincoln in September 1969, pending further study. One analyst has argued that the project was suspended because neither federal nor local officials believed that the development plan was either technologically or politically feasible. Other analysts have suggested that the project was suspended because members of the Nixon Administration regarded it as a personal undertaking of President Johnson's and as an example of the overly ambitious social engineering activities of "the Great Society."

The struggle over citizen participation diminished support for the project in the neighborhood and among its potential supporters in other areas of the city. No strong political constituency favored the project. The Nixon Administration could and did suspend it without antagonizing any strong or vocal interest group.

Fort Lincoln is one example of the extent to which technical planning and development processes can become the locus of political conflict when these processes are perceived as the de facto locus of political choice. It is also an example of some of the difficulties that can arise in the course of efforts to reconcile the dictates of administrative and technological reasoning with the dictates of the political thinking of participating individuals in particular situations.

Problems

Like many other participatory processes, participatory technology raises questions about the adequacy of the theory and practice of representative government.

According to traditional theories of American public life, citizens should express their demands for public action to their political and governmental representatives. Conflicting demands should be reconciled by persons elected or appointed to policy-making positions in which they are publicly accountable for their actions. Administrative and technical processes are not, in theory, the appropriate locus for the exercise of political influence and the reconciliation of political conflicts, because these processes are not usually structured as open political forums, and because most administrators and technical people are not directly accountable to electorates.

This theory of government is a prescriptive rather than a descriptive one. It does not correspond well with the realities of the exercise of political power in and through administrative and technical activities. Among other things, increases in population, the expansion of the public sector, and the increase in technological complexity have changed the number and, to some extent, the nature of demands and possibilities for governmental action in recent decades. While legislative bodies and individual elected officials continue to respond to some of these demands, many other demands are considered and resolved in administrative processes of limited visibility. The very act of translating most legislation into specific processes usually involves an exercise of political choice. Furthermore, agencies often invite demands upon themselves as a way of expanding the scope of their support and powers.

The politicalization of administration in this century, especially in response to the activities of interest groups, is a widely recognized phenomenon.

Participatory technology is an attempt to influence public agencies directly, and, through them, the quasipublic and private interests they often influence and regulate. Like other participatory processes, participatory technology in some of its forms circumvents traditional processes of expressing demands through elected representatives and of relying on representatives to take appropriate action.

As Professor Carroll points out, participatory technology includes the will of the people in the process of developing, implementing and regulating a new technology. This is accomplished, in part, through litigation, *ad hoc* activity and government.

THE SST CONTROVERSY

An interesting and important example of a new technology which moved with great force toward development in the United States was the Supersonic Transport (SST). The SST was not built because American citizens, acting through their congressional representatives, prevented it. We trace this decision, as an example of control of technology, in the following paragraphs.

A Federally-funded United States project was first advocated in the early 1960's as a result of a general conviction that the SST represented the next advance in commercial aviation. Some feared that the Soviet or Anglo-French domination of SST markets would give a damaging blow to American technological prestige and the competitive ability of one of the United State's strongest industries. The initial funding, under President Kennedy in 1963, was made with some knowledge of two major environmental problems of the SST, namely sonic boom and engine noise. The original proposal set forth a design objective of an SST with a sonic boom during acceleration of less than 2 psf (pounds per square foot) and during cruise of less than 1.5 psf. At 2.5 psf the sonic boom is like close thunder or an explosion. The engine noise was to be limited according to the specifications to that of a typical (1963) subsonic jet transport. As far as the economic analysis went at that time, it appeared difficult to present more than a marginal case for an economic advantage over the large subsonic jets.[8] In June 1963, President Kennedy announced the SST program with the statement, "The Congress and country should be prepared to invest the funds and effort necessary to maintain this nation's lead in long-range aircraft."

The Federal Aviation Administration accepted a design from Boeing Aircraft Corporation in 1969 for a swept-wing design. The design was for a large, more economical SST with a significant sonic boom. The Boeing SST design is shown in Figure 8–1 and the SST mock-up at the Seattle plant is shown in Figure 8–2. The design characteristics of the Boeing SST are given in Table 8–1. The cost of the research and development phase of the SST was $800 million. The cost of the two prototype airplanes was to be $400 million additional, with flight testing to be held in 1972. The production program was to follow successful flight testing, with the first delivery scheduled for 1978. The financing of the prototype plane required a $1,342 million investment by the federal government. The government was to recover all the money appropriated for the SST program through contract royalty payments. Boeing estimated that the government would recover its investment by the time of the sale of the 300th airplane and that SST sales to foreign airlines would total $12 billion by 1990. At the peak of the SST program, it was expected to employ 50,000 workers.

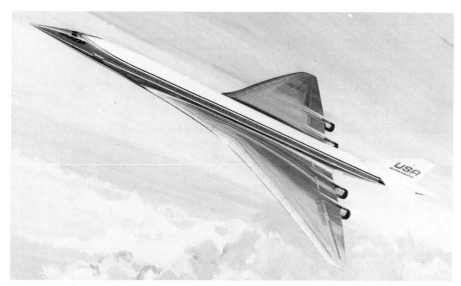

FIGURE 8-1   The Boeing SST, in an artist's conception. This aircraft would cruise over 12 miles above the earth's surface at nearly 2,000 miles per hour. Courtesy Boeing Aircraft Corporation.

FIGURE 8-2   A mockup of the Boeing SST. The sleek design of the craft belies the fact that nearly 300 passengers would be carried. Courtesy Boeing Aircraft Corporation.

TABLE 8-1

The Boeing SST design characteristics.

Maximum Design Weight: 750,000 pounds
Engines: Four General Electric GE 4/J, 67,000 pounds thrust
Length: 298'
Wingspan: 143' 5"
Payload: All Tourist Class . . . 298 seats
           Mixed Class . . . . . . 273 seats
Range: Paris to New York nonstop
Cruise Speed: 1,800 miles per hour
Trip Time: New York to London . . . 2 3/4 hours
Cruise Altitude: Over 60,000 feet

In 1964 the FAA had conducted a public acceptance test of sonic booms in which 300,000 citizens of Oklahoma City were subjected to booms of 1.3 psf eight times daily for five months. Seventy-three percent of those polled felt they could tolerate booms of this level, but thousands filed complaints and damage claims. Thus, when President Nixon recommended funding for the prototype plane on September 23, 1969, he announced the SST would not fly supersonically over the United States and would be used only as a transocean airplane by this country.

Nevertheless, the House of Representatives barely passed the SST appropriation by a vote of 176 to 162, and on December 3, 1970 the Senate voted down the appropriation by a vote of 52 to 41. Three months later both houses agreed to terminate the project. The SST was a national issue and the legislators responded to the pressures of the many groups in the United States opposing the SST. One member of the opposition, Richard L. Garwin, was quoted as saying, "the SST will produce as much noise as the simultaneous take-off of 50 jumbo jets." A well-financed campaign in opposition to the SST was organized by the Coalition Against the SST, which widely distributed statements and information. While without prototype tests all calculations were theoretical, the Coalition stated, "The disturbance at one mile from a subsonic jet is about the equivalent of the disturbance at 15 miles from the SST." Also, it was calcuated that the fuel used per passenger-mile was, for the SST, twice that of the Boeing 747, making the SST quite energy-inefficient.

Another issue arising from possible fleet use of the SST was related to the possibility of a significant reduction in the earth's ozone blanket. If this reduction occurred, would it increase the incidence of skin cancer? It now appears that SST fleet operation *would* result in the reduction of ozone in the atomsphere and this could potentially result in an increase in the incidence of skin cancer. However, the expected reduction in ozone is mainly caused by the introduction of Nitric Oxide from engine exhaust and the amounts of oxide could be markedly reduced by changes in engine design.[10]

Meanwhile, the Russian Tupolev (TU) 144 is scheduled to fly its first domestic route as well as Moscow-to-Tokoyo in 1974. The Concorde will be put into commercial service by Air France and British Airways in 1975. A prototype Concorde is shown in Figure 8–3 in a flight over Paris. The United States airlines have decided to postpone a decision to purchase the Concorde, which holds a price tag of $47.5 million.[10] Official measurements of the Concorde's noise level show that, in sound output, it is indistinguishable from most modern jet planes.[10] The sonic boom remains, and therefore the Concorde will be flown at subsonic speeds over land. The design characteristics of the Concorde and TU–144 are given in Table 8–2.

On Sunday, June 3, 1973 the TU–144 was shown in flight at the Paris Air Shown. Tragically, the TU–144 completed a demonstration of its flight capabilities before 100,000 spectators only to have the airplane fall apart and crash following a touch-and-go landing manuever. The plane approached the strip with the wheels down and then zoomed off in a steep climb. At about 1,800 feet the plane began to vibrate and plunged toward the earth. A wing ripped off and the

FIGURE 8–3   Concorde 001—the first prototype—is shown in flight over Paris. This aircraft is capable of traveling between San Francisco and Tokyo in 6½ hours. The Concorde has a flying range of about 4,000 miles. It is 203 feet long and has a wing span of 84 feet. Courtesy of British Aircraft Corporation.

TABLE 8-2

Design characteristics of the Concorde and the TU-144.

	Russian TU-144	Anglo–French Concorde
Maximum Weight	395,000 pounds	340,000 pounds
Engine	NK144: 44,000 lbs. thrust	Rolls Royce: 38,000 lbs. thrust
Payload	140 seats	120 seats
Cruise Speed	1550 miles per hour	1,400 miles per hour
Trip Time New York to London	3 1/4 hours	3 1/3 hours
Cost of Project (estimated)	$2,500 million	$2,000 million

plane exploded, killing six crew members and 13 onlooking Frenchmen.[11]

American proponents of a newly-revised SST program have advocated a second-generation SST made of Titanium and capable of carrying 350 passengers at a cruising speed of 1,750 miles per hour.[10] This new design, developed to eliminate the environmental effects of the first SST Boeing design, would cost the United States government over $2 billion. This machine would have to overcome the problems of (1) sonic booms; (2) take-off noise; (3) the effect on the ozone layer; and (4) the economics of such an expensive airplane. Nevertheless, the issue of the SST as a new technology will not die easily. The Concorde and the TU–144 may be competitors for the international air travel and airplane markets, a market many in the United States would not be willing to give up easily. It is too early to know whether the cancellation of the United States SST was wisdom or folly. Nevertheless, the technological imperative of the SST illustrates the cliché, "If we don't do it, someone else will." Perhaps in this case the people showed that they could control technology. Whether the control was wise only time will tell.

CHAPTER 8 REFERENCES

1. J. Ellul, *The Technological Society*, Random House, New York, 1964.
2. A. MacLeish, "The Great American Frustration," *Saturday Review*, July 13, 1968, pp. 13–16.
3. R.V. Ayres, "The Forces That Change Technology," *The Futurist*, Oct. 1969, pp. 132–33.
4. L.M. Branscomb, "Taming Technology," *Science*, March 12, 1971, pp. 972–977.

5. M.S. Baram, "Social Control of Science and Technology," *Science*, May 7, 1971, pp. 535–539.
6. R. Nader, P. Petkas and K. Blackwell, *Whistle-Blowing*, Grossman Publishers, New York, 1972.
7. J.D. Carroll, "Participatory Technology," *Science*, Feb. 19, 1971, pp. 648–654.
8. J. Primack and F. Von Hippel, "Scientists, Politics and the SST: A Critical Review," *Bulletin of the Atomic Scientists*, April, 1972, pp. 24–30.
9. F.S. Johnson, "SST's, Ozone, and Skin Cancer," *Astronautics and Aeronautics*, July 1973, pp. 16–21.
10. J.S. Buts, Jr., "The Inevitable SST," *World*, Aug. 21, 1973, pp. 14–20.
11. "The Fall of the TU–144," *Newsweek*, June 18, 1973, pp. 86–87.
12. R. Hutchins, "Stamp Out Engineering Schools," *Engineer*, April, 1968, pp. 17–19.
13. T.B. Sheridan, "Citizen Feedback: New Technology for Social Choice," *Technology Review*, Oct., 1972, pp. 41–50.
14. R.D. English and D.I. Bolef, "Project Sanguine," *The New Republic*, Oct. 20, 1973, pp. 13–15.
15. K. Shifferd, "The Fight Against Project Sanguine," in *Patient Earth*, J. Harte and R.H. Socolow, editors, Holt, Rinehart and Winston, Inc., New York, 1971.
16. J. Mattill, "Productivity: The Key to Success," *Technology Review*, Nov., 1973, pp. 66–67.

CHAPTER 8 EXERCISES

8–1. Effective democratic governance requires extensive communication from the citizen to his representatives. Citizens can go to the polls once or twice a year and vote for a slate of candidates and bond issues. Also, citizens may write letters to their elected representatives. Technology may be used in the future to enhance the ways citizens may respond to choices such as whether or not to build the SST or the Alaskan pipeline. The computer and electronic communications media may be used to establish a system for citizen feedback.[13] Discuss various forms that citizen feedback systems might take and the advantages and disadvantages of each form.

8–2. The United States Navy has proposed for several years to build a very large antenna for radio transmission to its submarines while they are under water at sea. The antenna would cover 1,300 square miles in the shape of an ellipse of 35 to 40 miles. The grid antenna would be buried about six feet below the surface. The radio transmits at a very low frequency, about 60 hertz (cycles per second). Because of this low frequency, the system can transmit to submerged nuclear submarines in the case of an attack.

The antenna system, called Project Sanguine, is operated at low power and is buried to avoid the effect of a nuclear attack. Much concern has been shown over the possible ecological damage and electrical interference which might be caused by the large antenna currents. The system has been proposed to be built in Wisconsin and Texas.[14, 15] Discuss the ecological and economic effects of such a large system. Would you like to have such a system built near your city? Is this system an imperative in light of the United States dependence on nuclear submarines with ICBM missiles acting as nuclear deterrents to enemy attack?

8–3. One criticism of the Anglo-French Concorde SST is that it will be economically unproductive.[16] Productivity is lower for the Concorde since it has a lower utilization, due in part to associated scheduling difficulties and little if any increase in seated passengers over other aircraft. Review aircraft productivity and consider why all airlines do not use the 747 B or DC 8, both of which have high productivity.

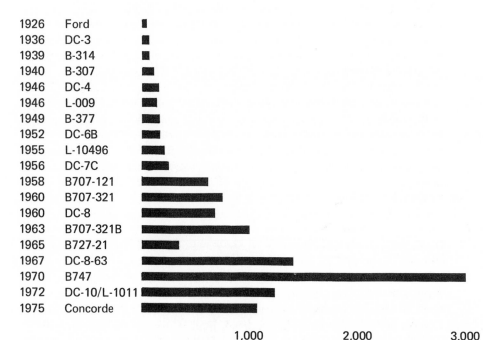

1926	Ford
1936	DC-3
1939	B-314
1940	B-307
1946	DC-4
1946	L-009
1949	B-377
1952	DC-6B
1955	L-10496
1956	DC-7C
1958	B707-121
1960	B707-321
1960	DC-8
1963	B707-321B
1965	B727-21
1967	DC-8-63
1970	B747
1972	DC-10/L-1011
1975	Concorde

1,000   2,000   3,000

Passenger-seat-miles or equivalent cargo-miles per year

8–4. In his book *The Doomsday Syndrome*, John Maddox, editor of *Nature* magazine, criticizes the literature that states that technology directly causes undesirable consequences and is uncontrollable. He states:*

Of course, there are serious social problems to be tackled. To the extent that these have been created by changes that have come about in recent decades, they can be partially laid at the door of technology. The fact that ocean beaches are now so much more crowded than half a century ago is a consequence of the development of motor cars and of the way in which large numbers of people have been enabled to own them privately. But would it ever have been sensible to ask that internal combustion engines should not have been invented for the sake of avoiding overcrowding at the beaches? Is it not preferable to enjoy the other benefits of the invention but to regulate the crowding of beaches by other means? And in any case, where such developments are in question, is it not entirely misleading to suggest that the automobile industry has grown to its present size for reasons connected with the character of technology and not because a need for the products became apparent? On issues like this, the doomsday literature is dishonest.

Do you believe that technology is controllable? Did the availability of autos fill a real need? Has it created other problems such as crowding at beaches? What approach would you recommend for the control of new technologies?

8-5. The quality of life on an assembly line has not changed much since the system was introduced in the Chicago slaughterhouse a century ago. More steps may be automated, but the basic principle remains the same: a worker must perform one or more operations in a prescribed number of seconds or minutes in a recurring sequence. As education and economic security have increased these conditions have become intolerable to many workers. This has resulted in wildcat strikes, walkouts, rapid turnover of manpower, and absenteeism arising in part from boredom and monotony. There has been a recent movement to provide job enrichment. The goal is to provide workers with increased responsibility. Study the life of an assembly-line worker and suggest ways by which the workers can be relieved from the tedium of the machine.

8-6. A national data center for information storage has been proposed for the United States. Large data centers become feasible because of the availability of large, high-speed digital computers. Is it inevitable that such a center be developed as a result of the availability of the technology? What measures of protection would you suggest?

8-7. What are the potential technological imperatives that the United States may face in the next decade? Consider the effects of the small "minicomputer" and the laser as potential technologies with significant imperative effects.

8-8. Do you believe that the United States should reconsider the development of an SST with federal funds for support? Do you believe it is inevitable that half of the overseas flights will be by SST by the year 1990?

9

TECHNOLOGY AND SAFETY

WITHIN THE PAST DECADE the need for safe, reliable devices and machines has been reaffirmed. Ralph Nader exposed the problems of an unsafe vechicle, the Corvair, in his book *Unsafe at Any Speed*. Of course, increased safety is a benefit of a technological improvement, but it has an associated cost. Improved technological performance requires the investment of societal resources in order to achieve the improved performance.

SAFETY AND NUCLEAR POWER GENERATION

A current example of public discussion of the acceptable public risk attributed to technology is that concerning nuclear power plants. The first risk to man comes from radiation brought about by a concentration of pollutants. Second, there is the relationship among the design, operation, size and number of potential pollutant sources and the resulting exposures to the public. The setting of a standard of pollutant, or risk, allowable then determines the level of safety acceptable. Thus the determination of an allowable level of risk is the key item.

For example, a nuclear reactor power plant is a source of radioactivity if the radiation is not properly contained within the reactor. Reactors are designed with containment a primary goal.

It is estimated that a continuous exposure of the United States populace to 100 millirem (units of radiation exposure) per year would result in approximately 3,600 additional cancer deaths per year, or an increase of 1.2%.[1] Current United States standards say in effect that under no circumstances should the exposure to a large portion of the public from man-made, nonmedical radiation sources exceed about 167 millirems per year. The level of exposure to radiation from medical and dental x-rays may be higher than this amount for many people, but it is taken by our society, at present, as an acceptable risk. Reactors in which radiation is properly contained do not emit significant radiation.

A present concern with nuclear power plants is that accidents or sabotage may result in uncontrolled exposure of radiation to a large population. The operational level of radiation from a power plant is set by standard as 500 millirem maximum for a brief period or 500/3 millirem/year average to a large sample of the population. In general, it can be shown that nuclear power plants operate well within these standards.[1] The frequency and severity of nuclear plant accidents is a subject under continuing study.[2] So far there has been no experience with nuclear accidents—which is certainly good; however, this leaves us without actuarial data. Estimates at present are that the frequency of an accident leading to a case of more than 1,000 deaths is 10^5 reactor years, or once in 100,000 reactor years. Thus, if there were 1,000 reactor power plants in the United States, we might expect an accident exceeding that level of 1,000 deaths once every 100 years. Is that an acceptable level of risk? (What risk do we incur with airplane travel or automobile travel?)

To the present time there have been only 170 reactor years of experience, so the figures given above are simply estimates. Many safety features of the nuclear reactor have not been fully put to the test of an actual accident, so we are unsure of the actual risks. Many are concerned with the potential performance of the emergency core-cooling system, which is an integral part of the safety system.* The safety of nuclear reactor power plants will continue to be an issue over the next decade as we construct many new plants throughout the United States and Europe.

What are we willing to pay for increased safety? In recent years, Dr. Chauncey Starr has led in the development of quantitative measures of benefit in new technology relative to the cost of accidental deaths arising from the technology. Starr's article on the issue of safety, and the attendant costs, follows.**

Social Benefit versus Technological Risk

Chauncey Starr

The evaluation of technical approaches to solving societal problems customarily

* If a core-cooling system fails, might the resulting high temperatures not result in what some engineers grimly call the China syndrome?
** From *Science*, Vol. 165, pp. 1232-1238, Sept. 19, 1969. Copyright 1969 by the American Association for the Advancement of Science.

involves consideration of the relationship between potential technical performance and the required investment of societal resources. Although such performance-versus-cost relationships are clearly useful for choosing between alternative solutions, they do not by themselves determine how much technology a society can justifiably purchase. This latter determination requires, additionally, knowledge of the relationship between social benefit and justified social cost. The two relationships may then be used jointly to determine the optimum investment of societal resources in a technological approach to a social need.

Technological analyses for disclosing the relationship between expected performance and monetary costs are a traditional part of all engineering planning and design. The inclusion in such studies of *all* societal costs (indirect as well as direct) is less customary, and obviously makes the analysis more difficult and less definitive. Analyses of social value as a function of technical performance are not only uncommon but are rarely quantitative. Yet we know that implicit in every nonarbitrary national decision on the use of technology is a trade-off of societal benefits and societal costs.

In this article I offer an approach for establishing a quantitative measure of benefit relative to cost for an important element in our spectrum of social values—specifically, for accidental deaths arising from technological developments in public use. The analysis is based on two assumptions. The first is that historical national accident records are adequate for revealing consistent patterns of fatalities in the public use of technology. (That this may not always be so is evidenced by the paucity of data relating to the effects of environmental pollution.) The second assumption is that such historically revealed social preferences and costs are sufficiently enduring to permit their use for predictive purposes.

In the absence of economic or sociological theory which might give better results, this empirical approach provides some interesting insights into accepted social values relative to personal risk. Because this methodology is based on historical data, it does not serve to distinguish what is "best" for society from what is "traditionally acceptable."

Maximum Benefit at Minimum Cost

The broad societal benefits of advances in technology exceed the associated costs sufficiently to make technological growth inexorable. Shef's socioeconomic study has indicated that technological growth has been generally exponential in this century, doubling every 20 years in nations having advanced technology. Such technological growth has apparently stimulated a parallel growth in socioeconomic benefits and a slower associated growth in social costs.

The conventional socioeconomic benefits—health, education, income—are presumably indicative of an improvement in the "quality of life." The cost of this socioeconomic progress shows up in all the negative indicators of our society—urban and environmental problems, technological unemployment, poor physical and mental health, and so on. If we understood quantitatively the causal relationships between specific technological developments and societal values, both positive and negative, we might deliberately guide and regulate technological developments so as to achieve maximum social benefit at minimum social cost. Unfortunately, we have not as yet developed such a predictive system analysis. As a result, our society historically has arrived at acceptable balances of technological benefit and social cost empirically—by trial, error, and subsequent corrective steps.

In advanced societies today, this historical empirical approach creates an increasingly critical situation, for two basic reasons. The first is the well-known difficulty in changing a technical subsystem of our society once it has been woven into the economic, political, and cultural structures. For example, many of our environmental-pollution problems have known engineering solutions, but the problems of economic readjustment, political jurisdiction, and social behavior loom very large. It will take many decades to put into effect the technical solutions we know today. To give a specific illustration, the pollution of our water resources could be completely avoided by means of engineering systems now available, but public interest in making the economic and political adjustments needed for applying these techniques is very limited. It has been facetiously suggested that, as a means of motivating the public, every community and industry should be required to place its water intake downstream from its outfall.

In order to minimize these difficulties, it would be desirable to try out new developments in the smallest social groups that would permit adequate assessment. This is a common practice in market-testing a new product or in field-testing a new drug. In both these cases, however, the experiment is completely under the control of a single company or agency, and the test information can be fed back to the controlling group in a time that is short relative to the anticipated commercial lifetime of the product. This makes it possible to achieve essentially optimum use of the product in an acceptably short time. Unfortunately, this is rarely the case with new technologies. Engineering developments involving new technology are likely to appear in many places simultaneously and to become deeply integrated into the systems of our society before their impact is evident or measurable.

This brings us to the second reason for the increasing severity of the problem of obtaining maximum benefits at minimum costs. It has often been stated that the time required from the conception of a technical idea to its first application in society has been drastically shortened by modern engineering organization and management. In fact, the history of technology does not support this conclusion. The bulk of the evidence indicates that the time from conception to first application (or demonstration) has been roughly unchanged by modern management, and depends chiefly on the complexity of the development.

However, what *has* been reduced substantially in the past century is the time from first use to widespread integration into our social system. The techniques for *societal diffusion* of a new technology and its subsequent exploitation are now highly developed. Our ability to organize resources of money, men, and materials to focus on new technological programs has reduced the diffusion-exploitation time by roughly an order of magnitude in the past century.

Thus, we now face a general situation in which widespread use of a new technological development may occur before its social impact can be properly assessed, and before any empirical adjustment of the benefit-versus-cost relation is obviously indicated.

It has been clear for some time that predictive technological assessments are a pressing societal need. However, even if such assessments become available, obtaining maximum social benefit at minimum cost also requires the establishment of a relative value system for the basic parameters in our objective of improved "quality of life." The empirical approach implicitly involved an intuitive societal balancing of such values. A predictive analytical approach will require an explicit scale of relative social values.

For example, if technological assessment of a new development predicts an increased per capita annual income of x percent but also predicts an associated accident probability of y fatalities annually per million population, then how are these to be compared in their effect on the "quality of life"? Because the penalties or risks to the public arising from a new development can be reduced by applying constraints, there will usually be a functional relationship (or trade-off) between utility and risk, the x and y of our example.

There are many historical illustrations of such trade-off relationships that were empirically determined. For example, automobile and airplane safety have been continuously weighed by society against economic costs and operating performance. In these and other cases, the real trade-off process is actually one of dynamic adjustment, with the behavior of many portions of our social systems out of phase, due to the many separate "time constants" involved. Readily available historical data on accidents and health, for a variety of public activities, provide an enticing stepping-stone to quantitative evaluation of this particular type of social cost. The social benefits arising from some of these activities can be roughly determined. On the assumption that in such historical situations a socially acceptable and essentially optimum tradeoff of values has been achieved, we could say that any generalizations developed might then be used for predictive purposes. This approach could give a rough answer to the seemingly simple question "How safe is safe enough?"

The pertinence of this question to all of us, and particularly to governmental regulatory agencies, is obvious. Hopefully, a functional answer might provide a basis for establishing performance "design objectives" for the safety of the public.

Voluntary and Involuntary Activities

Societal activities fall into two general categories—those in which the individual participates on a "voluntary" basis and those in which the participation is "involuntary," imposed by the society in which the individual lives. The process of empirical optimization of benefits and costs is fundamentally similar in the two cases—namely, a reversible exploration of available options—but the time required for empirical adjustments (the time constants of the system) and the criteria for optimization are quite different in the two situations.

In the case of "voluntary" activities, the individual uses his own value system to evaluate his experiences. Although his eventual trade-off may not be consciously or analytically determined, or based upon objective knowledge, it nevertheless is likely to represent, for that individual, a crude optimization appropriate to his value system. For example, an urban dweller may move to the suburbs because of a lower crime rate and better schools, at the cost of more time spent traveling on highways and a higher probability of accidents. If, subsequently, the traffic density increases, he may decide that the penalties are too great and move back to the city. Such an individual optimization process can be comparatively rapid (because the feedback of experience to the individual is rapid), so the statistical pattern for a large social group may be an important "realtime" indicator of societal trade-offs and values.

"Involuntary" activities differ in that the criteria and options are determined not by the individuals affected but by a controlling body. Such control may be in the hands of a government agency, a political entity, a leadership group, an assembly of authorities or "opinion-makers," or a combination of such bodies. Because of the complexity of large societies, only the control group is likely to be fully aware of all the criteria and options involved in their decision process. Further, the time required for feedback of the experience that results from the controlling decisions is likely to be very long. The feedback of cumulative individual experiences into societal communication channels (usually political or economic) is a slow process, as is the process of altering the planning of the control group. We have many examples of such "involuntary" activities, war being perhaps the most extreme case of the operational separation of the decision-making group from those most affected. Thus, the real-time pattern of societal trade-offs on "involuntary" activities must be considered in terms of the particular dynamics of approach to an acceptable balance of social values and costs. The historical trends in such activities may therefore be more significant indicators of social acceptability than the existent tradeoffs are.

In examining the historical benefit-risk relationships for "involuntary" activities, it is important to recognize the perturbing role of public psychological acceptance of risk arising from the influence of authorities or dogma. Because in this situation the decision-making is separated from the affected individual, society has generally clothed many of its controlling groups in an almost impenetrable mantle of authority and of imputed wisdom. The public generally assumes that the decision-making process is based on a rational analysis of social benefit and social risk. While it often is, we have all seen after-the-fact examples of irrationality. It is important to omit such "witchdoctor" situations in selecting examples of optimized "involuntary" activities, because in fact these situations typify only the initial stages of exploration of options.

Quantitative Correlations

With this description of the problem, and the associated caveats, we are in a position to discuss the quantitative correlations. For the sake of simplicity in this initial study, I have taken as a measure of the physical risk to the individual the fatalities (deaths) associated with each activity. Although it might be useful to include all injuries (which are 100 to 1000 times as numerous as deaths), the difficulty in obtaining data and the unequal significance of varying disabilities would introduce inconvenient complexity for this study. So the risk measure used here is the statistical probability of fatalities per hour of exposure of the individual to the activity considered.

The hour-of-exposure unit was chosen because it was deemed more closely related to the individual's intuitive process in choosing an activity than a year of exposure would be, and gave substantially similar results. Another possible alternative, the risk per activity, involved a comparison of too many dissimilar units of measure; thus, in comparing the risk for various modes of transportation, one could use risk per hour, per mile, or per trip. As this study was directed toward exploring a methodology for determining social acceptance of risk, rather than the safest mode of transportation for a particular trip, the simplest common unit—that of risk per exposure hour—was chosen.

The social benefit derived from each activity was converted into a dollar equivalent, as a measure of integrated value to the individual. This is perhaps the most uncertain aspect of the correlations because it reduced the "quality-of-life" benefits of an activity to an overly simplistic measure. Nevertheless, the correlations seemed useful, and no better measure was available. In the case of the "voluntary" activities, the amount of money spent on the activity by the average involved individual was assumed proportional to its benefit to him. In the case of the "involuntary" activities, the contribution of the activity to the individual's annual income (or the equivalent) was assumed proportional to its benefit. This assumption of roughly constant relationship between benefits and monies, for each class of activities, is clearly an approximation. However, because we are dealing in orders of magnitude, the distortions likely to be introduced by this approximation are relatively small.

In the case of transportation modes, the benefits were equated with the sum of the monetary cost to the passenger and the value of the time saved by that particular mode relative to a slower, competitive mode. Thus, airplanes were compared with automobiles, and automobiles were compared with public transportation or walking. Benefits of public transportation were equated with their cost. In all cases, the benefits were assessed on an annual dollar basis because this seemed to be most relevant to the individual's intuitive process. For example, most luxury sports require an investment and upkeep only partially dependent upon usage. The associated risks, of course, exist only during the hours of exposure.

Probably the use of electricity provides the best example of the analysis of an "involuntary" activity. In this case the fatalities include those arising from electrocution, electrically caused fires, the operation of power plants, and the mining of the required fossil fuel. The benefits were estimated from a United Nations study of the relationship between energy consumption and national income; the energy fraction associated with electric power was used. The contributions of the home use of electric power to our "quality of life"—more subtle than the contributions of electricity in industry—are omitted. The availability of refrigeration has certainly improved our national health and the quality of dining. The electric light has certainly provided great flexibility in patterns of living, and television is a positive element. Perhaps, however, the gross-income measure used in the study is sufficient for present purposes.

Information on acceptance of "voluntary" risk by individuals as a function of income benefits is not easily available, although we know that such a relationship must exist. Of particular interest, therefore, is the special case of miners exposed to high occupational risks. In Fig. 9–1, the accident rate and the severity rate of mining injuries are plotted against the hourly wage. The acceptance of individual risk is an exponential function of the wage, and can be roughly approximated by a third-power relationship in this range. If this relationship has validity, it may mean that several "quality of life" parameters (perhaps health, living essentials, and recreation) are each partly influenced by any increase in available personal resources, and that thus the increased acceptance of risk is exponentially motivated. The extent to which this relationship is "voluntary" for the miners is not obvious, but the subject is interesting nevertheless.

Risk Comparisons

The results for the societal activities studied, both "voluntary" and "involuntary," are assembled in Fig. 9–2. Also shown in Fig. 9–2 is the third-power relationship between risk and benefit characteristic of Fig. 9–1. For comparison, the average risk of death from accident and from disease is shown. Because the average number of fatalities from accidents is only about one-tenth the number from disease, their inclusion is not significant.

Several major features of the benefitrisk relations are apparent, the most obvious being the difference by several orders of magnitude in society's willingness to accept "voluntary" and "involuntary" risk. As one would expect, we are loath to let others do unto us what we happily do to ourselves.

The rate of death from disease appears to play, psychologically, a yardstick role in determining the acceptability of risk on a voluntary basis. The risk of death in most sporting activities is surprisingly close to the risk of death from disease—almost as though, in sports, the individual's subconscious computer adjusted his courage and made him take risks associated with a fatality level equaling but not exceeding the statistical mortality due to involuntary exposure to disease. Perhaps this defines the demarcation between boldness and foolhardiness.

In Fig. 9–2 the statistic for the Vietnam war is shown because it raises an interesting point. It is only slightly above the average for risk of death from disease. Assuming that some long-range societal benefit was anticipated from this war, we find that the related risk, as seen by society as a whole, is not substantially different from the average nonmilitary risk from disease. However, for individuals in the military service age group (age 20 to 30), the risk of death in Vietnam is about ten times the normal mortality rate

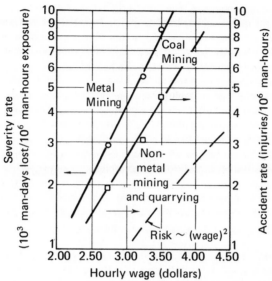

FIGURE 9-1   Mining accident rates plotted relative to incentive.

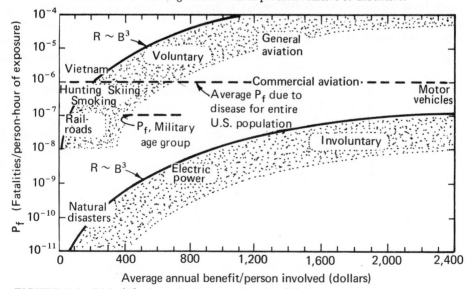

FIGURE 9-2   Risk (R) plotted relative to benefit (B) for various kinds of voluntary and involuntary exposure.

(death from accidents or disease). Hence the population as a whole and those directly exposed see this matter from different perspectives. The disease risk pertinent to the average age of the involved group probably would provide the basis for a more meaningful comparison than the risk pertinent to the national average age does. Use of the figure for the single group would complicate these simple comparisons, but that figure might be more significant as a yardstick.

The risks associated with general aviation, commercial aviation, and travel by motor vehicle deserve special comment. The latter originated as a "voluntary" sport, but in the past half century the motor vehicle has become an essential utility. General aviation is still a highly voluntary activity. Commercial aviation is partly voluntary and partly essential and, additionally, is subject to government administration as a transportation utility.

Travel by motor vehicle has now reached a benefit-risk balance, as shown in Fig. 9–3. It is interesting to note that the present risk level is only slightly below the basic level of risk from disease. In view of the high percentage of the population involved, this probably represents a true societal judgment on the acceptability of risk in relation to benefit. It also appears from Fig. 9–3 that future reductions in the risk level will be slow in coming, even if the historical trend of improvement can be maintained.

Commercial aviation has barely approached a risk level comparable to that set by disease. The trend is similar to that for motor vehicles, as shown in Fig. 9–4. However, the percentage of the population participating is now only 1/20 that for motor vehicles. Increased public participation in commercial aviation will undoubtedly increase the pressure to reduce the risk, because, for the general population, the benefits are much less than those associated with motor vehicles. Commercial aviation has not yet reached the point of optimum benefit-risk trade-off.

For general aviation the trends are similar, as shown in Fig. 9–5. Here the risk levels are so high (20 times the risk from disease) that this activity must properly be considered to be in the category of adventuresome sport. However, the rate of risk is decreasing so rapidly that eventually the risk for general aviation may be little higher than that for commercial aviation. Since the percentage of the population involved is very small, it appears that the present average risk levels are acceptable to only a limited group.

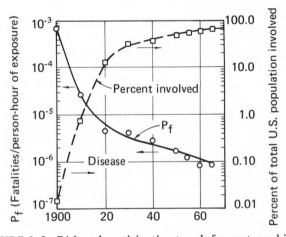

FIGURE 9-3   Risk and participation trends for motor vehicles.

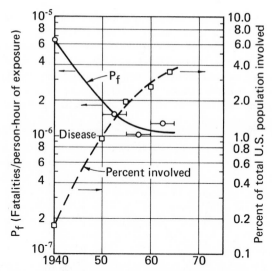

FIGURE 9-4   Risk and participation trends for certified air carriers.

The similarity of the trends in Figs. 9–3 through 9–5 may be the basis for another hypothesis, as follows: the acceptable risk is inversely related to the number of people participating in an activity.

The product of the risk and the percentage of the population involved in each of the activities of Figs 9–3 through 9–5 is plotted in Fig. 9–6. This graph represents the historical trend of total fatalities per hour of exposure of the population involved. The leveling off of motor-vehicle risk at about 100 fatalities per hour of exposure of the participating population may be significant. Because most of the U.S. population is involved, this rate of fatalities may have sufficient public visibility to set a level of social acceptability. It is interesting, and disconcerting, to note that the trend of fatalities in aviation, both commercial and general, is uniformly upward.

Public Awareness

Finally, I attempted to relate these risk data to a crude measure of public awareness of the associated social benefits (see Fig. 9–7). The "benefit awareness" was arbitrarily defined as the product of the relative level of advertising, the square of the percentage of population involved in the activity, and the relative usefulness (or importance) of the activity to the individual. Perhaps these assumptions are too crude, but Fig. 9–7 does support the reasonable position that advertising the benefits of an activity increases public acceptance of a greater level of risk. This, of course could subtly produce a fictitious benefit-risk ratio—as may be the case for smoking.

Atomic Power Plant Safety

I recognize the uncertainty inherent in the quantitative approach discussed here, but the trends and magnitudes may nevertheless be of sufficient validity to warrant their use in determining national "design objectives" for technological activities. How would this be done?

Let us consider as an example the introduction of nuclear power plants as a principal source of electric power. This is an especially good example because the technology has been primarily nurtured, guided, and regulated by the government, with industry undertaking the engineering development and the diffusion into public use. The government specifically maintains responsibility for public safety. Further, the engineering of nuclear plants permits continuous reduction of the probability of accidents, at a substantial increase in cost. Thus, the trade-off of utility and potential risk can be made quantitative.

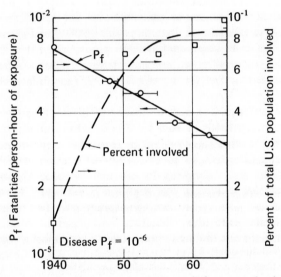

FIGURE 9-5   Risk and participation trends for general aviation.

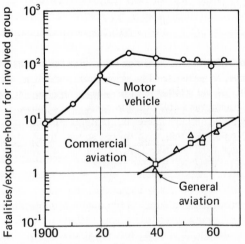

FIGURE 9-6   Group risk plotted relative to year.

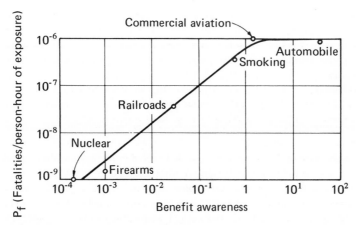

FIGURE 9-7   Accepted risk plotted relative to benefit awareness.

Moreover, in the case of the nuclear power plant the historical empirical approach to achieving an optimum benefit-risk trade-off is not pragmatically feasible. All such plants are now so safe that it may be 30 years or longer before meaningful risk experience will be accumulated. By that time, many plants of varied design will be in existence, and the empirical accident data may not be applicable to those being built. So a very real need exists now to establish "design objectives" on a predictive-performance basis.

Let us first arbitrarily assume that nuclear power plants should be as safe as coal-burning plants, so as not to increase public risk. Figure 9-2 indicates that the total risk to society from electric power is about 2×10^{-9} fatality per person per hour of exposure. Fossil fuel plants contribute about ⅕ of this risk, or about 4 deaths per million population per year. In a modern society, a million people may require a million kilowatts of power, and this is about the size of most new power stations. So, we now have a target risk limit of 4 deaths per year per million-kilowatt power station.

Technical studies of the consequences of hypothetical extreme (and unlikely) nuclear power plant catastrophes, which would disperse radioactivity into populated areas, have indicated that about 10 lethal cancers per million population might result. On this basis, we calculate that such a power plant might statistically have one such accident every 3 years and still meet the risk limit set. However, such a catastrophe would completely destroy a major portion of the nuclear section of the plant and either require complete dismantling or years of costly reconstruction. Because power companies expect plants to last about 30 years, the economic consequences of a catastrophe every few years would be completely unacceptable. In fact, the operating companies would not accept one such failure, on a statistical basis, during the normal lifetime of the plant.

It is likely that, in order to meet the economic performance requirements of the power companies, a catastrophe rate of less than 1 in about 100 plant-years would be needed. This would be a public risk of 10 deaths per 100 plant-years, or 0.1 death per year per million population. So the economic investment criteria of the nuclear plant user—the power company—would probably set a risk level 1/200 the present socially accepted risk associated with electric power, or 1/40 the present risk associated with coal-burning plants.

An obvious design question is this: Can a nuclear power plant be engineered with a predicted performance of less than 1 catastrophic failure in 100 plant-years of operation? I believe the answer is yes, but that is a subject for a different occasion. The principal point is that the issue of public safety can be focused on a tangible, quantitative, engineering design objective.

This example reveals a public safety consideration which may apply to many other activities: The economic requirement for the protection of major capital investments may often be a more demanding safety constraint than social acceptability.

Conclusion

The application of this approach to other areas of public responsibility is self-evident. It provides a useful methodology for answering the question "How safe is safe enough?" Further, although this study is only exploratory, it reveals several interesting points. (i) The indications are that the public is willing to accept "voluntary" risks roughly 1000 times greater than "involuntary" risks. (ii) The statistical risk of death from disease appears to be a psychological yardstick for establishing the level of acceptability of other risks. (iii) The acceptability of risk appears to be crudely proportional to the third power of the benefits (real or imagined). (iv) The social acceptance of risk is directly influenced by public awareness of the benefits of an activity, as determined by advertising, usefulness, and the number of people participating. (v) In a sample application of these criteria to atomic power plant safety, it appears that an engineering design objective determined by economic criteria would result in a design-target risk level very much lower than the present socially accepted risk for electric power plants.

Perhaps of greatest interest is the fact that this methodology for revealing existing social preferences and values may be a means of providing the insight on social benefit relative to cost that is so necessary for judicious national decisions on new technological developments.

CHAPTER 9 REFERENCES

1. T.J. Connolly, "Nuclear Energy and the Environment," *Proceedings of the American Society for Engineering Education Annual Conference*, June 28, 1973, Iowa State University.
2. C. Starr, M.A. Greenfield, D.F. Hausknecht, "A Comparison of Public Health Risks: Nuclear and Oil-Fired Power Plants," *Nuclear News*, Vol. 15, No. 10, Oct. 1972, pp. 37–45.
3. C. Starr, "Social Benefits *versus* Technological Risk," *Science*, Vol. 165, Sept. 19, 1969, pp. 1232–1238.
4. "How Much is Health Worth?" *Business Week*, Nov. 3, 1973, pp. 36.
5. J. Dunster, "Costs and Benefits of Nuclear Power," *New Scientist*, Oct. 18, 1973, pp. 192–194.
6. "How Safe is This Super Underground?" *New Scientist*, Oct. 11, 1973, pp. 96–99.
7. V.J. Yannacone, Jr., *Energy Crisis: Danger and Opportunity*, West Publishing Co., St. Paul, Minn., 1974, Ch.5.

CHAPTER 9 EXERCISES

9-1. In Figure 9-1, Dr. Starr indicates that the risk that an individual is willing to assume may be proportional to the third power of the wages he can expect to receive. In other words, a worker will take up hazardous work if he can expect to receive high wages. Examine this concept for mine workers, especially in light of recent laws to establish safety conditions for miners in the United States.

9-2. As shown in Figure 9-3, the risk for an individual using a motor vehicle has declined over the past decades. Can we expect this risk to continue to decline? Notice that only recently has the risk become less than that due to disease. Compare the risk involved in aviation with that of automobile transportation. (The "energy crisis" which began to be noticed by the public in 1973 may have a profound effect on your response.)

9-3. Examine Figure 9-7 which portrays the risk of several activities *versus* a figure representing a measure of the benefit related to the activity. Review Starr's definition of *benefit*. Define another measure of benefit and obtain your own plot of risk *versus* benefit.

9-4. Dr. Bertram Dinman, medical director of Aluminum Company of America, has posed a troubling environmental issue: Can we protect the chronically ill from the effects of air pollution regardless of the cost of the protection?[4] Under the 1970 Clean Air Act, the Environmental Protection Agency sets standards to protect public health. Does this imply safeguarding the general public only, or those with respiratory diseases such as chronic bronchitis and emphysema also? At every level of Sulfur Dioxide in the air someone is adversely affected, since there is no threshold in the case of respiratory diseases. The cost of abating air pollution grows exponentially as zero pollution is approached. Dinman suggests isolating those ill from air pollution by moving them to less affected areas. Discuss the risks and costs of averting air pollution for the chronically ill, and suggest a level of safety you would be willing to propose for air-pollution standards.

9-5. The most controversial auto safety devices are the passive restraint devices. These do not require the rider to take any action (such as buckling a seat belt). One such passive restraint is the air bag. This is a porous cloth bag folded in front of the auto passenger. It inflates upon collision. Examine the arguments for and against the required installation of air bags in new automobiles. What are the costs and the benefits?

9-6. The drinking of alcoholic beverages is involved in approximately half of the auto fatalities. There are several proposals for devices to keep drinking drivers off the road. One device attempts to lock the ignition when the air exhaled by the driver indicates alcohol. Another uses a dashboard device to display a code number and a keyboard. A number is flashed and the driver must enter the number accurately or the ignition will not start. Explore devices to reduce the number of drunk drivers and examine the cost-benefits, if possible.

9–7. The nuclear power plant is planned to be the major supplier of electricity in the United States and Europe by the year 2000. There is currently a debate over the level of radiation that society can tolerate. Many believe that any increase in radiation will increase the corresponding undesirable effects. The question is: what degree of harm can we live with?

One view is that society cannot afford to reduce radiation levels below the point of diminishing returns. In other words, improvement in safety standards should not be implemented when the measurable cost of extra safety begins to exceed the estimated improvements gained from these safety measures.[5] Do you agree with this approach? Is this the approach you take with your home or automobile?

What is your view of the use of x-rays for dental and physical examinations?

9–8. The proposed tunnel linking Britain and France would be 49 kilometers long and would cost over $1 billion. It is planned that every 2 ½ minutes an 800 meter-long train would go through the tunnel at 145 km/hour (90 miles/hour). The trains would go from London to Paris in 3 hours and 40 minutes. The trains would be very frequent and safety measures would have to be strict. Fire, derailment or breakdown pose serious threats. Communication and fire detection systems are central to safety. A derailment detection system may be required.[6] Consider safety measures you would propose for the tunnel. Would you permit dangerous chemical cargoes to pass through the tunnel?

9–9. During the last 25 years the aviation industry has improved its safety record. Aviation safety is supervised, in part, by the Federal Aviation Administration. What is the equivalent agency for highway safety? Has the record of highway safety been as good as for commercial aviation?

9–10. The risk to the public health caused by thermal or nuclear power plants is a very important factor in the decision whether to build many more plants of either kind in the United States. It is interesting to compare the potential effects of oil fires on public health as contrasted with the potential effects of a nuclear accident.[7] During the past few years several oil fires have occurred in the Unites States at large power plants using petroleum fuels. If a large power plant using petroleum experienced an oil fire it would release Sulfur Dioxide over a wide region. Explore the risks inherent in the use of many new large oil-fueled power plants and compare them with the risks of a nuclear plant using similar conservative assumptions and the medical estimates of public-health effects of an accident.

10

TECHNOLOGY AND THE TRAGEDY
OF THE COMMONS

ONE OF THE GREAT PROBLEMS of the twentieth century arises from the relationship between man and his environment. As Lynn White, Jr. discusses in his article on this subject (see Chapter 13), man looks upon his environment as something which is subject to him. In the view of many, the earth is to be manipulated for the benefit of mankind. This is commonly seen as the viewpoint of the engineers, who are said to believe that the universe is an object to be completely controlled by human desire. It is an exaggeration to state that engineers, or for that matter most of our citizens, believe that unfettered control and use of the environment and its materials are proper goals of mankind. Nevertheless, such a statement is frequently used as the basis for a definition of engineering. We must recall that engineering is the use of materials and forces of nature for the benefit of mankind. It is the latter part of the definition that is often forgotten or not carried through to its full consequences. The benefit of mankind is achieved only when nature and the environment are left whole. Of what benefit is it to man to produce massive amounts of steel at the expense of the pollution of a lake, such as Lake Erie, or a river?

SELF-AGGRANDIZEMENT

Yet some believe that if everyone goes ahead and does what he feels is to his own best interests, then everything will somehow work out in the end. They think that effects created by every self-aggrandizing individual will never intersect and intermesh in such a way as to lead to dysfunctional and negative consequences.[1] So, many believe, the total results will work out for the good if everybody acts in his own interest. It is this view, coupled with a lack of thought on what is the ultimate benefit of man, that has led to harmful applications of technologies.

But what can be done to show a man that his individual gain may be at the ultimate cost of others? The gain in his driving an auto built without pollution controls (and thus more powerful) will result in some deterioration of the environment. Furthermore, if many take the individualistic view, then air pollution is bound to rise significantly. And yet why should one individual find it to his interest *not* to use a resource—in this case the air? For the resource is infinite, in his view, even though all the drivers in Los Angeles contribute enough air pollution to show that fresh air in the Los Angeles basin is indeed not infinite, but actually very limited indeed.

CONCEPT OF THE COMMONS

The *commons* consists of all those attributes of the earth which humans use jointly.[5] There is no exclusive ownership right, but rather some form of common-usage rights. One example of common usage can be seen in the case of the oceans. Another form of commons is the atmosphere. These common sources of goods have been seen as free to human use in the past.

In the following articles Professor Hardin and Professor Crowe discuss what is called the *tragedy of the commons*, which occurs when each man is locked into a system that compels him to increase his well-being, often at the expense of his fellow citizens of the commons. For example, worn-out tires discarded in the oil fields of Texas in 1939 are shown in Figure 10–1. In this case the commons may be our common atmosphere, land, energy resources—whatever. The steelmaker or auto driver finds that his share of the cost of the wastes he discharges into the commons is less than the cost of purifying his wastes before releasing them. Since this is true for all on the commons, how do we avoid polluting our commons? Hardin and Crowe here discuss legislative, technological and administrative paths to the solution of the problems of the tragedy of the commons.

The Tragedy of the Commons*

Garrett Hardin

At the end of a thoughtful article on the future of nuclear war, Wiesner and York concluded that: "Both sides in the arms race are ... confronted by the dilemma of steadily increasing military power and steadily decreasing national security. *It is our considered professional judgment that this dilemma has no technical solution.* If the great powers

*From *Science*, Vol. 162, pp. 1243–1248, 13 Dec. 1968. Copyright 1968 by the American Association for the Advancement of Science.

FIGURE 10-1   Worn-out tires deposited in oil fields in Texas, 1939. Courtesy of the
Library of Congress.

continue to look for solutions in the area of science and technology only, the result will be
to worsen the situation."

I would like to focus your attention not on the subject of the article (national security
in a nuclear world) but on the kind of conclusion they reached, namely that there is no
technical solution to the problem. An implicit and almost universal assumption of
discussions published in professional and semipopular scientific journals is that the
problem under discussion has a technical solution. A technical solution may be defined as
one that requires a change only in the techniques of the natural sciences, demanding little
or nothing in the way of change in human values or ideas of morality.

In our day (though not in earlier times) technical solutions are always welcome.
Because of previous failures in prophecy, it takes courage to assert that a desired technical
solution is not possible. Wiesner and York exhibited this courage; publishing in a science
journal, they insisted that the solution to the problem was not to be found in the natural
sciences. They cautiously qualified their statement with the phrase, "It is our considered
professional judgment. ... " Whether they were right or not is not the concern of the
present article. Rather, the concern here is with the important concept of a class of human
problems which can be called "no technical solution problems," and, more specifically,
with the identification and discussion of one of these.

It is easy to show that the class is not a null class. Recall the game of tick-tack-toe.

Consider the problem, "How can I win the game of tick-tack-toe?" It is well known that I cannot, if I assume (in keeping with the conventions of game theory) that my opponent understands the game perfectly. Put another way, there is no "technical solution" to the problem. I can win only by giving a radical meaning to the word "win." I can hit my opponent over the head; or I can drug him; or I can falsify the records. Every way in which I "win" involves, in some sense, an abandonment of the game, as we intuitively understand it. (I can also, of course, openly abandon the game—refuse to play it. This is what most adults do.)

The class of "No technical solution problems" has members. My thesis is that the "population problem," as conventionally conceived, is a member of this class. How it is conventionally conceived needs some comment. It is fair to say that most people who anguish over the population problem are trying to find a way to avoid the evils of overpopulation without relinquishing any of the privileges they now enjoy. They think that farming the seas or developing new strains of wheat will solve the problem —technologically. I try to show here that the solution they seek cannot be found. The population problem cannot be solved in a technical way, any more than can the problem of winning the game of tick-tack-toe.

What Shall We Maximize?

Population, as Malthus said, naturally tends t grow "geometrically," or, as we would now say, exponentially. In a finite world this means that the per capita share of the world's goods must steadily decrease. Is ours a finite world?

A fair defense can be put forward for the view that the world is infinite; or that we do not know that it is not. But, in terms of the practical problems that we must face in the next few generations with the foreseeable technology, it is clear that we will greatly increase human misery if we do not, during the immediate future, assume that the world available to the terrestrial human population is finite. "Space" is no escape.

A finite world can support only a finite population; therefore, population growth must eventually equal zero. (The case of perpetual wide fluctuations above and below zero is a trivial variant that need not be discussed.) When this condition is met, what will be the situation of mankind? Specifically, can Bentham's goal of "the greatest good for the greatest number" be realized?

No—for two reasons, each sufficient by itself. The first is a theoretical one. It is not mathematically possible to maximize for two (or more) variables at the same time. This was clearly stated by von Neumann and Morgenstern, but the principle is implicit in the theory of partial differential equations, dating back at least to D'Alembert (1717–1783).

The second reason springs directly from biological facts. To live, any organism must have a source of energy (for example, food). This energy is utilized for two purposes: mere maintenance and work. For man, maintenance of life requires about 1600 kilocalories a day ("maintenance calories"). Anything that he does over and above merely staying alive will be defined as work, and is supported by "work calories" which he takes in. Work calories are used not only for what we call work in common speech; they are also required for all forms of enjoyment, from swimming and automobile racing to playing music and writing poetry. If our goal is to maximize population it is obvious what we must do: We must make the work calories per person approach as close to zero as possible. No gourmet meals, no vacations, no sports, no music, no literature, no art. ... I think that everyone will grant, without argument or proof, that maximizing population does not maximize goods. Bentham's goal is impossible.

In reaching this conclusion I have made the usual assumption that it is the acquisition of energy that is the problem. The appearance of atomic energy has led some to question this assumption. However, given an infinite source of energy, population growth still produces an inescapable problem. The problem of the acquisition of energy is replaced by the problem of its dissipation, as J. H. Fremlin has so wittily shown. The arithmetic signs in the analysis are, as it were, reversed; but Bentham's goal is still unobtainable.

The optimum population is, then, less than the maximum. The difficulty of defining the optimum is enormous; so far as I know, no one has seriously tackled this problem. Reaching an acceptable and stable solution will surely require more than one generation of hard analytical work—and much persuasion.

We want the maximum good per person; but what is good? To one person it is wilderness, to another it is ski lodges for thousands. To one it is estuaries to nourish ducks for hunters to shoot; to another it is factory land. Comparing one good with another is, we usually say, impossible because goods are incommensurable. Incommensurables cannot be compared.

Theoretically this may be true; but in real life incommensurables *are* commensurable. Only a criterion of judgment and a system of weighting are needed. In nature the criterion is survival. Is it better for a species to be small and hideable, or large and powerful? Natural selection commensurates the incommensurables. The compromise achieved depends on a natural weighting of the values of the variables.

Man must imitate this process. There is no doubt that in fact he already does, but unconsciously. It is when the hidden decisions are made explicit that the arguments begin. The problem for the years ahead is to work out an acceptable theory of weighting. Synergistic effects, nonlinear variation, and difficulties in discounting the future make the intellectual problem difficult, but not (in principle) insoluble.

Has any cultural group solved this practical problem at the present time, even on an intuitive level? One simple fact proves that none has: there is no prosperous population in the world today that has, and has had for some time, a growth rate of zero. Any people that has intuitively identified its optimum point will soon reach it, after which its growth rate becomes and remains zero.

Of course, a positive growth rate might be taken as evidence that a population is below its optimum. However, by any reasonable standards, the most rapidly growing populations on earth today are (in general) the most miserable. This association (which need not be invariable) casts doubt on the optimistic assumption that the positive growth rate of a population is evidence that it has yet to reach its optimum.

We can make little progress in working toward optimum poulation size until we explicitly exorcize the spirit of Adam Smith in the field of practical demography. In economic affairs, *The Wealth of Nations* (1776) popularized the "invisible hand," the idea that an individual who "intends only his own gain," is, as it were, "led by an invisible hand to promote ... the public interest". Adam Smith did not assert that this was invariably true, and perhaps neither did any of his followers. But he contributed to a dominant tendency of thought that has ever since interfered with positive action based on rational analysis, namely, the tendency to assume that decisions reached individually will, in fact, be the best decisions for an entire society. If this assumption is correct it justifies the continuance of our present policy of laissez-faire in reproduction. If it is correct we can assume that men will control their individual fecundity so as to produce the optimum population. If the assumption is not correct, we need to reexamine our individual freedoms to see which ones are defensible.

Tragedy of Freedom in a Commons

The rebuttal to the invisible hand in population control is to be found in a scenario first sketched in a little-known pamphlet in 1833 by a mathematical amateur named William Forster Lloyd (1794–1852). We may well call it "the tragedy of the commons," using the word "tragedy" as the philosopher Whitehead used it.: "The essence of dramatic tragedy is not unhappiness. It resides in the solemnity of the remorseless working of things." He then goes on to say, "This inevitableness of destiny can only be illustrated in terms of human life by incidents which in fact involve unhappiness. For it is only by them that the futility of escape can be made evident in the drama."

The tragedy of the commons develops in this way. Picture a pasture open to all. It is to be expected that each herdsman will try to keep as many cattle as possible on the commons. Such an arrangement may work reasonably satisfactorily for centuries because tribal wars, poaching, and disease keep the numbers of both man and beast well below the carrying capacity of the land. Finally, however, comes the day of reckoning, that is, the day when the long-desired goal of social stability becomes a reality. At this point, the inherent logic of the commons remorselessly generates tragedy.

As a rational being, each herdsman seeks to maximize his gain. Explicitly or implicitly, more or less consciously, he asks, "What is the utility *to me* of adding one more animal to my herd?" This utility has one negative and one positive component.

1) The positive component is a function of the increment of one animal. Since the herdsman receives all the proceeds from the sale of the additional animal, the positive utility is nearly $+1$.

2) The negative component is a function of the additional overgrazing created by one more animal. Since, however, the effects of overgrazing are shared by all the herdsmen, the negative utility for any particular decision-making herdsman is only a fraction of -1.

Adding together the component partial utilities, the rational herdsman concludes that the only sensible course for him to pursue is to add another animal to his herd. And another; and another. ... But this is the conclusion reached by each and every rational herdsman sharing a commons. Therein is the tragedy. Each man is locked into a system that compels him to increase his herd without limit—in a world that is limited. Ruin is the destination toward which all men rush, each pursuing his own best interest in a society that believes in the freedom of the commons. Freedom in a commons brings ruin to all.

Some would say that this is a platitude. Would that it were! In a sense, it was learned thousands of years ago, but natural selection favors the forces of psychological denial. The individual benefits as an individual from his ability to deny the truth even though society as a whole, of which he is a part, suffers. Education can counteract the natural tendency to do the wrong thing, but the inexorable succession of generations requires that the basis for this knowledge be constantly refreshed.

A simple incident that occurred a few years ago in Leominster, Massachusetts, shows how perishable the knowledge is. During the Christmas shopping season the parking meters downtown were covered with plastic bags that bore tags reading: "Do not open until after Christmas. Free parking courtesy of the mayor and city council." In other words, facing the prospect of an increased demand for already scarce space, the city fathers reinstituted the system of the commons. (Cynically, we suspect that they gained more votes than they lost by this retrogressive act.)

In an approximate way, the logic of the commons has been understood for a long time, perhaps since the discovery of agriculture or the invention of private property in real estate. But it is understood mostly only in special cases which are not sufficiently generalized. Even at this late date, cattlemen leasing national land on the western ranges demonstrate no more than an ambivalent understanding, in constantly pressuring federal authorities to increase the head count to the point where overgrazing produces erosion and weed-dominance. Likewise, the oceans of the world continue to suffer from the survival of the philosophy of the commons. Maritime nations still respond automatically to the shibboleth of the "freedom of the seas." Professing to believe in the "inexhaustible resources of the oceans," they bring species after species of fish and whales closer to extinction.

The National Parks present another instance of the working out of the tragedy of the commons. At present, they are open to all, without limit. The parks themselves are limited in extent—there is only one Yosemite Valley—whereas population seems to grow without limit. The values that visitors seek in the parks are steadily eroded. Plainly, we must soon cease to treat the parks as commons or they will be of no value to anyone.

What shall we do? We have several options. We might sell them off as private property. We might keep them as public property, but allocate the right to enter them. The allocation might be on the basis of wealth, by the use of an auction system. It might be on the basis of merit, as defined by some agreed-upon standards. It might be by lottery. Or it might be on a first-come, first-served basis, administered to long queues. These, I think, are all the reasonable possibilities. They are all objectionable. But we must choose—or acquiesce in the destruction of the commons that we call our National Parks.

Pollution

In a reverse way, the tragedy of the commons reappears in problems of pollution. Here it is not a question of taking something out of the commons, but of putting something in—sewage, or chemical, radioactive, and heat wastes into water; noxious and dangerous fumes into the air; and distracting and unpleasant advertising signs into the line of sight. The calculations of utility are much the same as before. The rational man finds that his share of the cost of the wastes he discharges into the commons is less than the cost of purifying his wastes before releasing them. Since this is true for everyone, we are locked into a system of "fouling our own nest," so long as we behave only as independent, rational, free-enterprisers.

The tragedy of the commons as a food basket is averted by private property, or something formally like it. But the air and waters surrounding us cannot readily be fenced, and so the tragedy of the commons as a cesspool must be prevented by different means, by coercive laws or taxing devices that make it cheaper for the polluter to treat his pollutants than to discharge them untreated. We have not progressed as far with the solution of this problem as we have with the first. Indeed, our particular concept of private property, which deters us from exhausting the positive resources of the earth, favors pollution. The owner of a factory on the bank of a stream—whose property extends to the middle of the stream—often has difficulty seeing why it is not his natural right to muddy the waters flowing past his door. The law, always behind the times, requires elaborate stitching and fitting to adapt it to this newly perceived aspect of the commons.

The pollution problem is a consequence of population. It did not much matter how a lonely American frontiersman disposed of his waste. "Flowing water purifies itself every 10 miles," my grandfather used to say, and the myth was near enough to the truth when he was a boy, for there were not too many people. But as population became denser, the natural chemical and biological recycling processes became overloaded, calling for a redefinition of property rights.

How To Legislate Temperance?

Analysis of the pollution problem as a function of population density uncovers a not generally recognized principle of morality, namely: *The morality of an act is a function of the state of the system at the time it is performed.* Using the commons as a cesspool does not harm the general public under frontier conditions, because there is no public; the same behavior in a metropolis is unbearable. A hundred and fifty years ago a plainsman could kill an American bison, cut out only the tongue for his dinner, and discard the rest of the animal. He was not in any important sense being wasteful. Today, with only a few thousand bison left, we would be appalled at such behavior.

In passing, it is worth noting that the morality of an act cannot be determined from a photograph. One does not know whether a man killing an elephant or setting fire to the grassland is harming others until one knows the total system in which his act appears. "One picture is worth a thousand words," said an ancient Chinese; but it may take 10,000 words to validate it. It is as tempting to ecologists as it is to reformers in general to try to persuade others by way of the photographic shortcut. But the essense of an argument cannot be photographed: it must be presented rationally—in words.

That morality is system-sensitive escaped the attention of most codifiers of ethics in the past. "Thou shalt not ... " is the form of traditional ethical directives which make no allowance for particular circumstances. The laws of our society follow the pattern of ancient ethics, and therefore are poorly suited to governing a complex, crowded, changeable world. Our epicyclic solution is to augment statutory law with administrative law. Since it is practically impossible to spell out all the conditions under which it is safe to burn trash in the back yard or to run an automobile without smog-control, by law we delegate the details to bureaus. The result is administrative law, which is rightly feared for an ancient reason—*Quis custodiet ipsos custodes?*—"Who shall watch the watchers themselves?" John Adams said that we must have "a government of laws and not men." Bureau administrators, trying to evaluate the morality of acts in the total system, are singularly liable to corruption, producing a government by men, not laws.

Prohibition is easy to legislate (though not necessarily to enforce); but how do we legislate temperance? Experience indicates that it can be accomplished best through the mediation of administrative law. We limit possibilities unnecessarily if we suppose that the sentiment of *Quis custodiet* denies us the use of administrative law. We should rather retain the phrase as a perpetual reminder of fearful dangers we cannot avoid. The great challenge facing us now is to invent the corrective feedbacks that are needed to keep custodians honest. We must find ways to legitimate the needed authority of both the custodians and the corrective feedbacks.

· · ·

Pathogenic Effects of Conscience

The long-term disadvantage of an appeal to conscience should be enough to condemn it; but has serious short-term disadvantages as well. If we ask a man who is exploiting a commons to desist "in the name of conscience," what are we saying to him? What does he hear?—not only at the moment but also in the wee small hours of the night when, half asleep, he remembers not merely the words we used but also the nonverbal communication cues we gave him unawares? Sooner or later, consciously or sub-consciously, he senses that he has received two communications, and that they are contradictory: (i) (intended communication) "If you don't do as we ask, we will openly

condemn you for not acting like a responsible citizen"; (ii) (the unintended communication) "If you *do* behave as we ask, we will secretly condemn you for a simpleton who can
be shamed into standing aside while the rest of us exploit the commons."

Everyman then is caught in what Bateson has called a "double bind." Bateson and
his co-workers have made a plausible case for viewing the double bind as an important
causative factor in the genesis of schizophrenia. The double bind may not always be so
damaging, but it always endangers the mental health of anyone to whom it is applied. "A
bad conscience," said Nietzsche, "is a kind of illness."

To conjure up a conscience in others is tempting to anyone who wishes to extend his
control beyond the legal limits. Leaders at the highest level succumb to this temptation.
Has any President during the past generation failed to call on labor unions to moderate
voluntarily their demands for higher wages, or to steel companies to honor voluntary
guidelines on prices? I can recall none. The rhetoric used on such occasions is designed to
produce feelings of guilt in noncooperators.

For centuries it was assumed without proof that guilt was a valuable, perhaps even an
indispensable, ingredient of the civilized life. Now, in this post-Freudian world, we doubt
it.

Paul Goodman speaks from the modern point of view when he says: "No good has
ever come from feeling guilty, neither intelligence, policy, nor compassion. The guilty do
not pay attention to the object but only to themselves, and not even to their own interests,
which might make sense, but to their anxieties".

One does not have to be a professional psychiatrist to see the consequences of
anxiety. We in the Western world are just emerging from a dreadful two-centuries-long
Dark Ages of Eros that was sustained partly by prohibition laws, but perhaps more
effectively by the anxiety-generating mechanisms of education. Alex Comfort has told the
story well in *The Anxiety Makers*; it is not a pretty one.

Since proof is difficult, we may even concede that the results of anxiety may
sometimes, from certain points of view, be desirable. The larger question we should ask is
whether, as a matter of policy, we should ever encourage the use of a technique the
tendency (if not the intention) of which is psychologically pathogenic. We hear much talk
these days of responsible parenthood; the coupled words are incorporated into the titles of
some organizations devoted to birth control. Some people have proposed massive
propaganda campaigns to instill responsibility into the nation's (or the world's) breeders.
But what is the meaning of the word responsibility in this context? Is it not merely a
synonym for the word conscience? When we use the word responsibility in the absence of
substantial sanctions are we not trying to browbeat a free man in a commons into acting
against his own interest? Responsibility is a verbal counterfeit for a substantial *quid pro
quo*. It is an attempt to get something for nothing.

If the word responsibility is to be used at all, I suggest that it be in the sense Charles
Frankel uses it. "Responsibility," says this philosopher, "Is the product of definite social
arrangements." Notice that Frankel calls for social arrangements—not propaganda.

Mutual Coercion Mutually Agreed upon

The social arrangements that produce responsibility are arrangements that create
coercion, of some sort. Consider bank-robbing. The man who takes money from a bank
acts as if the bank were a commons. How do we prevent such action? Certainly not by
trying to control his behavior solely by a verbal appeal to his sense of responsibility.
Rather than rely on propaganda we follow Frankel's lead and insist that a bank is not a

commons; we seek the definite social arrangements that will keep it from becoming a commons. That we thereby infringe on the freedom of would-be robbers we neither deny nor regret.

The morality of bank-robbing is particularly easy to understand because we accept complete prohibition of this activity. We are willing to saay "Thou shalt not rob banks," without providing for exceptions. But temperance also can be created by coercion. Taxing is a good coercive device. To keep downtown shoppers temperate in their use of parking space we introduce parking meters for short periods, and traffic fines for longer ones. We need not actually forbid a citizen to park as long as he wants to; we need merely make it increasingly expensive for him to do so. Not prohibition, but carefully biased options are what we offer him. A Madison Avenue man might call this persuasion; I prefer the greater candor of the word coercion.

Coercion is a dirty word to most liberals now, but it need not forever be so. As with the four-letter words, it dirtiness can be cleansed away by exposure to the light, by saying it over and over without apology or embarrassment. To many, the word coercion implies arbitrary decisions of distant and irresponsible bureaucrats; but this is not a necessary part of its meaning. The only kind of coercion I recommend is mutual coercion, mutually agreed upon by the majority of the people affected.

To say that we mutually agree to coercion is not to say that we are required to enjoy it, or even to pretend we enjoy it. Who enjoys taxes? We all grumble about them. But we accept compulsory taxes because we recognize that voluntary taxes would favor the conscienceless. We institute and (grumblingly) support taxes and other coercive devices to escape the horror of the commons.

An alternative to the commons need not be perfectly just to be preferable. With real estate and other material goods, the alternative we have chosen is the institution of private property coupled with legal inheritance. Is this system perfectly just? As a genetically trained biologist I deny that it is. It seems to me that, if there are to be differences in individual inheritance, legal possession should be perfectly correlated with biological inheritance—that those who are biologically more fit to be the custodians of property and power should legally inherit more. But genetic recombination continually makes a mockery of the doctrine of "like father, like son" implicit in our laws of legal inheritance. An idiot can inherit millions, and a trust fund can keep his estate intact. We must admit that our legal system of private property plus inheritance is unjust—but we put up with it because we are not convinced, at the moment, that anyone has invented a better system. The alternative of the commons is too horrifying to contemplate. Injustice is preferable to total ruin.

It is one of the peculiarities of the warfare between reform and the status quo that it is thoughtlessly governed by a double standard. Whenever a reform measure is proposed it is often defeated when its opponents triumphantly discover a flaw in it. As Kingsley Davis has pointed out, worshippers of the status quo sometimes imply that no reform is possible without unanimous agreement, an implication contrary to historical fact. As nearly as I can make out, automatic rejection of proposed reforms is based on one of two unconscious assumptions: (i) that the status quo is perfect; or (ii) that the choice we face is between reform and no action; if the proposed reform is imperfect, we presumably should take no action at all, while we wait for a perfect proposal.

But we can never do nothing. That which we have done for thousands of years is also action. It also produces evils. Once we are aware that the status quo is action, we can then compare its discoverable advantages and disadvantages with the predicted advantages and disadvantages of the proposed reform, discounting as best we can for our lack of experience. On the basis of such a comparison, we can make a rational decision which will not involve the unworkable assumption that only perfect systems are tolerable.

Recognition of Necessity

Perhaps the simplest summary of this analysis of man's population problems is this: the commons, if justifiable at all, is justifiable only under conditions of low-population density. As the human population has increased, the commons has had to be abandoned in one aspect after another.

First we abandoned the commons in food gathering, enclosing farm land and restricting pastures and hunting and fishing areas. These restrictions are still not complete throughout the world.

Somewhat later we saw that the commons as a place for waste disposal would also have to be abandoned. Restrictions on the disposal of domestic sewage are widely accepted in the Western world; we are still struggling to close the commons to pollution by automobiles, factories, insecticide sprayers, fertilizing operations, and atomic energy installations.

In a still more embryonic state is our recognition of the evils of the commons in matters of pleasure. There is almost no restriction on the propagation of sound waves in the public medium. The shopping public is assaulted with mindless music, without its consent. Our government is paying out billions of dollars to create supersonic transport which will disturb 50,000 people for every one person who is whisked from coast to coast 3 hours faster. Advertisers muddy the airwaves of radio and television and pollute the view of travelers. We are a long way from outlawing the commons in matters of pleasure. Is this because our Puritan inheritance makes us view pleasure as something of a sin, and pain (that is, the pollution of advertising) as the sign of virtue?

Every new enclosure of the commons involves the infringement of somebody's personal liberty. Infringements made in the distant past are accepted because no contemporary complains of a loss. It is the newly proposed infringements that we vigorously oppose; cries of "rights" and "freedom" fill the air. But what does "freedom" mean? When men mutually agreed to pass laws against robbing, mankind became more free, not less so. Individuals locked into the logic of the commons are free only to bring on universal ruin; once they see the necessity of mutual coercion, they become free to pursue other goals. I believe it was Hegel who said, "Freedom is the recognition of necessity."

The most important aspect of necessity that we must now recognize, is the necessity of abandoning the commons in breeding. No technical solution can rescue us from the misery of overpopulation. Freedom to breed will bring ruin to all. At the moment, to avoid hard decisions many of us are tempted to propagandize for conscience and responsible parenthood. The temptation must be resisted, because an appeal to independently acting consciences selects for the disappearance of all conscience in the long run, and an increase in anxiety in the short.

The only way we can preserve and nurture other and more precious freedoms is by relinquishing the freedom to breed, and that very soon. "Freedom is the recognition of necessity"—and it is the role of education to reveal to all the necessity of abandoning the freedom to breed. Only so, can we put an end to this aspect of the tragedy of the commons.

The Tragedy of the Commons Revisited*

Beryl L. Crowe

There has developed in the contemporary natural sciences a recognition that there is a subset of problems, such as population, atomic war, and environmental corruption, for which there are no technical solutions. There is also an increasing recognition among contemporary social scientists that there is a subset of problems, such as population, atomic war, environmental corruption, and the recovery of a livable urban environment, for which there are no current political solutions. The thesis of this article is that the common area shared by these two subsets contains most of the critical problems that threaten the very existence of contemporary man.

The importance of this area has not been raised previously because of the very structure of modern society. This society, with its emphasis on differentiation and specialization, has led to the development of two insular scientific communities—the natural and the social—between which there is very little communication and a great deal of envy, suspicion, disdain, and competition for scarce resources. Indeed, these two communities more closely resemble tribes living in close geographic proximity on university campuses than they resemble the "scientific culture" that C.P. Snow placed in contrast to and opposition to the "humanistic culture".

Perhaps the major problems of modern society have, in large part, been allowed to develop and intensify through this structure of insularity and specialization because it serves both psychological and professional functions for both scientific communities. Under such conditions, the natural sciences can recognize that some problems are not technically soluble and relegate them to the nether land of politics, while the social sciences recognize that some problems have no current political solutions and then postpone a search for solutions while they wait for new technologies with which to attack the problem. Both sciences can thus avoid responsibility and protect their respective myths of competence and relevance, while they avoid having to face the awesome and awful possibility that each has independently isolated the same subset of problems and given them different names. Thus, both never have to face the consequences of their respective findings. Meanwhile, due to the specialization and insularity of modern society, man's most critical problems lie in limbo, while the specialists in problem-solving go on to less critical problems for which they can find technical or political solutions.

In this circumstance, one psychologically brave, but professionally foolhardy soul, Garrett Hardin, has dared to cross the tribal boundaries in his article "The tragedy of the commons". In it, he gives vivid proof of the insularity of the two scientific tribes in at least two respects: first, his "rediscovery" of the tragedy was in part wasted effort, for the knowledge of this tragedy is so common in the social sciences that it has generated some fairly sophisticated mathematical models, second, the recognition of the existence of a subset of problems for which science neither offers nor aspires to offer technical solutions is not likely, under the contemporary conditions of insularity, to gain wide currency in the social sciences. Like Hardin, I will attempt to avoid the psychological and professional benefits of this insularity by tracing some of the political and social implications of his proposed solution to the tragedy of the commons.

The commons is a fundamental social institution that has a history going back

*From *Science*, Vol. 166, pp. 1103–1107, 28 Nov. 1969. Copyright 1969 by the American Association for Advancement of Science.

through our own colonial experience to a body of English common law which antidates the Roman conquest. That law recognized that in societies there are some environmental objects which have never been, and should never be, exclusively appropriated to any individual or group of individuals. In England the classic example of the commons is the pasturage set aside for public use, and the "tragedy of the commons" to which Hardin refers was a tragedy of overgrazing and lack of care and fertilization which resulted in erosion and underproduction so destructive that there developed in the late 19th century an enclosure movement. Hardin applies this social institution to other environmental objects such as water, atmosphere, and living space.

The cause of this tragedy is exposed by a very simple mathematical model, utilizing the concept of utility drawn from economics. Allowing the utilities to range between a positive value of 1 and a negative value of 1, we may ask, as did the individual English herdsman, what is the utility to me of adding one more animal to my herd that grazes on the commons? His answer is that the positive utility is near 1 and the negative utility is only a fraction of minus 1. Adding together the component partial utilities, the herdsman concludes that it is rational for him to add another animal to his herd; then another, and so on. The tragedy to which Hardin refers develops because the same rational conclusion is reached by each and every herdsman sharing the commons.

Assumptions Necessary To Avoid the Tragedy

In passing the technically insoluble problems over to the political and social realm for solution, Hardin has made three critical assumptions: (i) that there exists, or can be developed, a "criterion of judgment and a system of weighting ... " that will "render the incommensurables ... commensurable ... " in real life; (ii) that, possessing this criterion of judgment, "coercion can be mutually agreed upon," and that the application of coercion to effect a solution to problems will be effective in modern society; and (iii) that the administrative system, supported by the criterion of judgment and access to coercion, can and will protect the commons from further desecration.

If all three of these assumptions were correct, the tragedy which Hardin has recognized would dissolve into a rather facile melodrama of setting up administrative agencies. I believe these three assumptions are so questionable in contemporary society that a tragedy remains in the full sense in which Hardin used the term. Under contemporary conditions, the subset of technically insoluble problems is also politically insoluble, and thus we witness a full-blown tragedy wherein "the essence of dramatic tragedy is not unhappiness. It resides in the remorseless working of things."

The remorseless working of things in modern society is the erosion of three social myths which form the basis for Hardin's assumptions, and this erosion is proceeding at such a swift rate that perhaps the myths can neither revitalize nor reformulate in time to prevent the "population bomb" from going off, or before an accelerating "pollution immersion," or perhaps even an "atomic fallout."

• • •

Erosion of the Myth of the Monopoly of Coercive Force

In the past, those who no longer subscribed to the values of the dominant culture were held in check by the myth that the state possessed a monopoly on coercive force. This myth has undergone continual erosion since the end of World War II owing to the success of the strategy of guerrilla warfare, as first revealed to the French in Indochina, and later conclusively demonstrated in Algeria. Suffering as we do from what Senator

Fulbright has called "the arrogance of power," we have been extremely slow to learn the lesson in Vietnam, although we now realize that war is political and cannot be won by military means. It is apparent that the myth of the monopoly of coercive force as it was first qualified in the civil rights conflict in the South, then in our urban ghettos, next on the streets of Chicago, and now on our college campuses has lost its hold over the minds of Americans. The technology of guerrilla warfare has made it evident that, while the state can win battles, it cannot win wars of values. Coercive force which is centered in the modern state cannot be sustained in the face of the active resistance of some 10 percent of its population unless the state is willing to embark on a deliberate policy of genocide directed against the value dissident groups. The factor that sustained the myth of coercive force in the past was the acceptance of a common value system. Whether the latter exists is questionable in the modern nation-state. But, even if most members of the nation-state remain united around a common value system which makes incommensurables for the majority commensurable, that majority is incapable of enforcing its decisions upon the minority in the face of the diminished coercive power of the governing body of the nation-state.

. . .

How Can Science Contribute to the Saving of the Commons?

It would seem that, despite the nearly remorseless working of things, science has some interim contributions to make to the alleviation of those problems of the commons which Hardin has pointed out.

These contributions can come at two levels:

1) Science can concentrate more of its attention on the development of technological responses which at once alleviate those problems and reward those people who no longer desecreate the commons. This approach would seem more likely to be successful than the " ... fundamental extension in morality ... " by administrative law; the engagement of interest seems to be a more reliable and consistent motivator of advantage-seeking groups than does administrative wrist-slapping or constituency pressure from the general public.

2) Science can perhaps, by using the widely proposed environmental monitoring systems, use them in such a way as to sustain a high level of "symbolic disassurance" among the holders of generalized interests in the commons—thus sustaining their political interest to a point where they would provide a constituency for the administrator other than those bent on denuding the commons. This latter approach would seem to be a first step toward the " ... invention of the corrective feedbacks that are needed to keep custodians honest." This would require a major change in the behavior of science, however, for it could no longer rest content with development of the technology of monitoring and with turning the technology over to some new agency. Past administrative experience suggests that the use of technology to sustain a high level of "dis-assurance" among the general population would also require science to take up the role and the responsibility for maintaining, controlling, and disseminating the information.

Neither of these contributions to maintaining a habitable environment will be made by science unless there is a significant break in the insularity of the two scientific tribes. For, if science must, in its own insularity, embark on the independent discovery of "the tragedy of the commons," along with the parameters that produce the tragedy, it may be too slow a process to save us from the total destruction of the planet. Just as important, however, science will, by pursuing such a course, divert its attention from the production of technical tools, information, and solutions which will contribute to the political and social solutions for the problems of the commons.

Because I remain very suspicious of the success of either demands or pleas for fundamental extensions in morality, I would suggest that such a conscious turning by both the social and the natural sciences is, at this time, in their immediate self-interest. As Michael Polanyi has pointed out, " ... encircled today between the crude utilitarianism of the philistine and the ideological utilitarianism of the modern revolutionary movement, the love of pure science may falter and die". The sciences, both social and natural, can function only in a very special intellectual environment that is neither universal or unchanging, and that environment is in jeopardy. The questions of humanistic relevance raised by the students at M.I.T., Stanford Research Institute, Berkeley, and wherever the headlines may carry us tomorrow, pose serious threats to the maintenance of that intellectual environment. However ill-founded *some* of the questions raised by the new generation may be, it behooves us to be ready with at least some collective, tentative answers—if only to maintain an environment in which both sciences will be allowed and fostered. This will not be accomplished so long as the social sciences continue to defer the most critical problems that face mankind to future technical advances, while the natural sciences continue to defer those same problems which are about to overwhelm all mankind to false expectations in the political realm.

CHAPTER 10 REFERENCES

1. V. Ferkiss, "Toward the Creation of Technological Man," *The Futurist*, Feb. 1970, pp. 11–12.
2. G. Hardin, "The Tragedy of the Commons," *Science*, Dec. 13, 1968, pp. 1243–1248.
3. B.L. Crowe, "The Tragedy of the Commons Revisited," *Science*, Nov. 28, 1969, pp. 1103–1107.
4. W. Marx, "Sewage: The Surprising Resource," *The Nation*, May 7, 1973, pp. 588–591.
5. S. Edmunds and J. Letey, *Environmental Administration*, McGraw-Hill Book Co., New York, 1973, Chapter 4.

CHAPTER 10 EXERCISES

10–1. Perhaps you have visited Yosemite or Yellowstone National Park during the vacation season. As Dr. Hardin mentions, the parks are finite, while those who visit them increase in number each year. What system of admission would you institute for controlling access to these and other overcrowded parks? Would you use a reservation system? What basis would you use in a reservation system for selecting who is admitted?

10–2. Dr. Hardin states that the morality of an act is a function of the state of the system at the time it is performed. Can you find contemporary examples of acts with little effect such as that of the plainsmen in 1800 killing a bison to obtain only the tongue for food?

10–3. Dr. Hardin suggests mutual coercion as a method of maintaining a socially desirable commons. What mutual coercion, mutually agreed upon, can you cite currently in use to maintain a desirable state of the commons?

10–4. Dr. Crowe states, "Under contemporary conditions, the subset of technically insoluble problems is also politically insoluble." The tragedy, in this view, resides in the remorseless working of things. Reflect on these views and record your criticisms, if any, of these viewpoints.

10–5. Dr. Crowe states that the technology of guerilla warfare has demonstrated that, while the state can win battles, it cannot win wars of values. This contributes to an erosion of the myth of the monopoly of coercive force as a solution to the tragedy of the commons. Review this viewpoint and develop your own view with respect to the viability of mutual coercion.

10–6. Dr. Crowe suggests that science rather than politics may assist most in the amelioration of the tragedy of the commons. Give several examples of the use of science for this purpose.

10–7. Many workers commute to work in their automobiles on crowded, congested roads. The relation of each driver to the use of the road may be analyzed in terms of the commons. It is of benefit to the driver to use the highway into the city, although when too many of his fellows join him, the usefulness of the road may be destroyed. Discuss the use of autos for commuting and suggest alternatives which could ameliorate this problem.

10–8. The use of septic tanks for the sewage of weekend cabins around Lake Tahoe, California grew over the period 1940–1960. As the shores of the lake became lined with second homes, the lake became a common sewer. Eventually, it became necessary to pass a law in California and Nevada to ban the use of septic tanks and require the development of municipal and regional waste disposal systems with connecting sewer lines to each home. Discuss the use of Lake Tahoe as a "commons" and suggest alternatives to a legal solution of the problem.

10–9. Pollution is often the most conspicuous product of a sewage treatment plant. However, reclamation of the water from sewage can be beneficial.[4] Water can be reclaimed several times over before reaching the ocean. Sewage treatment can provide useful water. The Golden Gate Park in San Francisco was originally irrigated with treated sewage. Treated sewage can be used to produce lakes, greenbelts, orchards and fresh water. Consider the value of reclaimed water to the commons. The use of reclaimed water from sewage may provide new added benefits to the common good. Review the reclamation of water from sewage and discuss its utility.

10–10. As a result of the energy shortage of the 1970s, mining companies plan to strip-mine the 128 million acres of arid and semi-arid land in the northwestern United States underlain by coal. A large part of this coal lies on public lands. In addition, the conversion of the coal into power and gas requires large amounts of water. Furthermore, to reclaim the stripped land requires water. Also, stripping operations may have repercussions in the streams and aquifers serving the ranches.

In the states of Montana and Wyoming 42 power plants are planned; they would produce 50,000 megawatts by the end of the century. At peak activity these plants would consume 855,000 acre-feet of water each year. With an added gasification plant, this amount could rise to over 1.5 million acre-feet. What are the implications of this large use of water in these states? To whom does the water of Montana and Wyoming belong? Who should decide on the allocation of the available water in these semi-arid regions?

11

TECHNOLOGY, SOCIAL VALUES AND SOCIAL RESPONSIBILITY

DECISION-MAKING IN A TECHNOLOGICAL world must include judgments of what is of social value. Decision-making is a process of relating judgments of fact to judgments of value for the sake of arriving at judgments about action. All decisions encompassing new technological devices and systems incorporate judgments about their value. But which values are adequate for this day? The challenge of change has struck many of the old values and is said to have made them obsolete. The new facts and new actions dynamically influence the organization of modern life and what it values. The judgment between the value-preferences of flexibility and freedom on the one hand and responsibility on the other hand come into play in the purchase of a second car or the decision to construct a new electric power plant, for example. What a man does depends both upon what he knows or believes and what he wants or values.[1]

Many factors, such as interests, aversions, attractions, pleasures, duties and moral obligations influence human behavior. The process of making values, or valuing, appears to be central to the interaction of man and technology.[15]

Values are those conceptions of desirable states of affairs that are utilized in selective conduct as criteria for preferences or choice, or as justification for proposed or actual behavior.[20] The values of a society are its established ideals for life.

In modern society there has been a tendency to believe that science is the universal ethic and not just a limited method for finding the truth in some inquiries. This approach, often called *scientism*, is an application of what has been called the scientific method to the study and manipulation of all events and the adoption of this as the keystone of a world view. This approval illustrates the difficulty of viewing any phenomena in terms of human or social needs, wants or values. The category of *value*, unlike the category of empirical fact, has no recognized place in the scientistic world view. The value-free aspect of scientism has been widely advocated. Thus we have the rise of the expert, the technocrat, the person who possess the facts.

MAKING ROOM FOR HUMAN VALUES

Yet there is room in our decision-making process for the introduction of human values and social responsibility. The pragmatism of the marketplace or the rationale of facts are not alone sufficient to move man to action. He must continue to ask not only, "Can we do it?" but also, "Should we do it?" Science and technology are value-laden in their origin and in their application, and it is up to those who are involved in the process to dedicate themselves to the highest human values.

In the modern world, the range of value models is very large. One of the most impressive capabilities of modern technology is the capacity to exhibit remote occurences and transmit information to a person by means of a communication system such as the television network. Is this ability "good" for everybody? The answer may be no.

An example of the clash of two-value systems is that of the Amish sect and the State of Wisconsin.[1] The Amish provide their own schooling for their children through the eighth grade and wish to maintain a society free from certain technological devices such as automobiles and television. The State of Wisconsin requires all children to attend school until age sixteen and has endeavored through the courts to send the Amish children to public high schools. The State argues that it has a responsibility to liberate these children by exposing them to a wide range of values and technologies. But the Amish know that this exposure could result in socialization of their children to a culture other than the one they value.[1] Similar restrictions on the use of technology, such as television, are used by the government of South Africa.

BALANCING VALUES

Each time an action is taken, a balance of values must be achieved which must contrast the short-term effects with the long-term effects. For example, the short-term benefits of the development of a new technology, such as the SST, may be greatly outweighed by the long-term costs of such a development. An ambitious technological project is now being executed in Brazil. A 3,000 mile highway is being cut through the Amazon jungle in an attempt to open up the heart of the continent for the development of natural resources. In this case, the short-term sacrifices in the diversion of capital and labor may pay long-term rewards from the increased availability of raw materials and land. The long-term effects may also have a significant effect on the cultural life of the people who live in the Amazon basin and the Amazon rain forest.

As we noted in Chapter 10, individual prudence may inexorably produce collective disaster. Nevertheless, it is difficult to impose a collective value. It is as if your neighboring farmer urges you not to add another cow to the commons because he can attest that a big herd is nothing but problems. Similarly, the United States and Europe can tell African nations, for example, that environmental problems preclude continuing expansion of the number of automobiles and steel mills, but these nations have their own interests and values. Also, attempts to tell nations without a nuclear weapon capability that they should not develop a capability is met with scorn. It is difficult to get an individual or a nation to accept the interests and values of competitors, so that when they make decisions, they will take "our" values into consideration. If we exercised a free choice and valued an expanded nuclear arsenal, why should they now adopt another value? Again, individual choice can lead, if unchecked by collective wisdom, to collective disaster.

TECHNOLOGY AS BALANCING AGENT

Technology serves human values and purposes in the view of many authors. For example, in Harvey Cox's *The Secular City*, Cox states that man's technological capabilities enable him to build large functional social structures (secular cities) in ways that enhance personal individuality and dignity.[2] Cox believes that, historically and logically, religious and personal values determine the development of technology. However, even he has reservations. In Cox's latest book, *The Feast of Fools*, he states that technology has "damaged the inner experience of Western man." He continues, "Technology need not be the enemy of the spirit ... but it should be a means to man's fulfillment, not the symbol or goal of that fulfillment itself."[3]

That technology in some way determines values is also a view that is strongly held.[4] The deterministic view is perhaps best evidenced by the words over the portals of the 1933 Chicago World's Fair: "Science explores; Technology executes; Man conforms."

TECHNIQUE

Marshall McLuhan, in his statement "the medium is the message," maintains that the tool itself becomes its end or purpose. As McLuhan states: "The medium, or process, of our time—electric technology—is reshaping and restructuring patterns of social interdependence and every aspect of our social life."[5] Jacques Ellul, in his book *The Technological Society*, also supports the view that technology itself sets its own purposes and determines society.[6] The single controlling value of society, in Ellul's view, may become, if it is not already, the technique, the efficiency of technology. By "technique" Ellul means not only a procedure that produces complicated and efficient machines, but also a style of conduct that governs all our personal and social lives. Technique is primarily a method of solving problems which asks only: What is the best way to solve this particular problem? According to Ellul, technique has no concern for values such as truth and justice, nor does it consider the long-term effects of the method chosen. It concentrates on the immediate solution of immediate problems. Each technical solution leads to another problem which leads to another technical solution and so on until it all becomes uncontrollable. Ellul believes that our only hope for rescuing our own personhood is to embrace a concept of man which is non-scientific and not captured by technology.

THE ESSENTIAL NEUTRALITY

And yet technology like science can be seen as value-free or neutral. The proponents of this view hold that technology itself is neutral and that it is only the way technology is used that determines whether it is good or bad. John von Neumann, the brilliant computer engineer, has written: "Technology—like science—is neutral all through, providing only means of control applicable to any purpose, indifferent to all."[15, 16]

Many authors and critics of technology and its applications are not satisfied with the view that technology is neutral. Rather, they agree with Herbert Marcuse, who has observed, "The traditional notion of the "neutrality" of technology can no longer be maintained. Technology as such cannot be isolated from the use to which it is put; the technological society is a system of domination which operates already in the concept and construction of technique."[17]

PERMANENCE

The question of the dominance of our society by machines is raised. Are the values of the machine, productivity, and efficiency paramount? In her book *The Human Condition*, Hanna Arendt writes: "The question is not so much whether we are masters or slaves of our machines, but whether machines will still serve the world and its things, or if, on the contrary, they and the automatic motion of their processes have begun to rule and even destroy world and things."[18]

Yet many authors confronting the age of technology and complexity develop an optimistic view. Pierre Teilhard de Chardin, French philosopher and scientist, describes this new, technological "growth on our horizon, of a true 'ultra human'

... which will have the effect, not of dehumanizing us through mechanization, but of super humanizing us by the intensification of our powers of understanding and love."[19]

POWER

Many authors see technology as aiding the centralizing of power into the hands of the powerful, thus rendering obsolete the value of individual freedom and choice. One author[7], believes that *laissez innover*, that is, the freedom to innovate, is now the premier ideology of the American society. In his view technology not only conquers nature, but also man. With unchecked freedom to pursue new technologies, man is overcome. The power must be removed from the elite ideologues and returned to the democracy.

A new man is emerging in a new culture, according to Victor Ferkiss, Charles Reich and Theodore Roszak.[8, 9, 10, 11] This new existential revolution of personal consciousness is on the individual level; it places its value on individual man. Ferkiss, in *Technological Man*, offers the future hope of the new man in a new partnership with nature. The new man understands the interconnectedness of things as well as his individual potential. The culture to be countered, according to Roszak, is that of technocracy.[10] The myth of objective consciousness, Roszak argues, sustains technocracy and distinguishes it from the counterculture. Objective consciousness "is alienated life promoted to its most honorific status as the scientific method." Reductionism, the desire to reduce all things to terms that objective consciousness might master, is the ultimate value of the technological society, in Roszak's opinion.[10] Man's increasing concern for selfawareness and meditation gives evidence of the new culture, in the author's view, and hopes for a new society in which people do not turn to science and technology for answers.

SOURCES OF VALUE PRIORITIES

While personal freedom for the individual makes a man truly human, almost every man has a set of multiple values. Differences in value systems in different societies result partly from different priorities which these societies assign to this set of multiple values.[4] A communist society gives priority to social equality at the expense, if necessary, of individual liberty. Liberal democratic societies assign the opposite priority. It is therefore consistent with both the American and Russian value systems that in Russia there is one automobile for every 45 citizens, while in the United States there is one for every two citizens. When setting a series of priorities, they will normally not be all equally achievable. Choice in the multiple values is not easy. As Alfred North Whitehead said, "Seek simplicity, but distrust it."

Professor Leo Marx* has written extensively about technology and literature.[24] In the following excerpt from an article, Marx discusses several American authors and their visions of man and his relation to the technological society and nature.[12]

In the best known American fables—I am thinking, for example, of Thoreau's *Walden*, Melville's *Moby Dick* and Mark Twain's *Huckleberry Finn*—the symbolic landscape is inseparable from the action or narrative structure, which may be divided into three movements: the retreat, the exploration of nature, and the return.

First, then, the retreat. The action begins with the heronarrator's withdrawal from a relatively complex, organized community from which he is alienated. Here life seems to be dominated by an oppressively mechanistic system of value, a preoccupation with the routine means of existence and an obliviousness of its meaning or purpose. Here, Thoreau says, men have become the tools of their tools. Unable to relate his inward experience to his environment, the narrator retreats in the direction of nature.

In the second, or central, movement he explores the possibilities of a simpler, more harmonious, way of life. At some point, invariably, there is an idyllic interlude when the beauty of the visible world inspires him with a sense of relatedness to the invisible order of the universe. During this episode, which can only be described as a moment of religious exaltation, he enjoys an unusual feeling of peace and harmony, free of anxiety, guilt, and conflict. But the possibilities of a life beyond the borders of ordinary society prove to be limited, and two characteristic kinds of episode help to define those limits.

In one, which may be called the interrupted idyll, the peace and harmony of the retreat into the middle landscape is shattered by the sudden, often violent intrusion of a machine, or of a force or person closely associated, in the figurative design, with the new industrial power. (Recall the scene in which the shriek of the locomotive destroys Thoreau's revery at Walden Pond; or when Ahab's violent declaration of purpose, which he associates with mechanized power, follows Ishmael's pantheistic masthead dream; or the decisive moment when the steamboat smashes into the raft in *Huckleberry Finn*.) The second characteristic limiting episode occurs when the narrator's retreat carries him close to or into untouched, untrammeled nature, and though his exposure to the wilderness often proves to be a spiritual tonic, evoking an exhilarating sense of psychic freedom, it also arouses his fear. For he soon comes to recognize that an unchecked recoil from civilization may destroy him—either in the sense of extinguishing his uniquely human traits or in the quite literal sense of killing him. He discovers, in short, that there are two hostile forces which impinge, from opposite sides of the symbolic landscape, upon the gardenlike scene of his retreat: one is the expanding power of civilization, and the other is the menacing anarchy of wild nature.

These insights lead, however indirectly, to the third and final phase of the action: the return. Having discovered the limited possibilities of withdrawal, above all its transience, the narrator now returns, or seems to be on the point of returning, to society. But the significance of this movement, which is also the ending of the work, is clouded by ambiguity. Has the hero been redeemed? Is he prepared to take up, once again, the common life? What is he able to bring back, as it were, from his exploration of the natural

*Excerpted from "Pastoral Ideals and City Troubles" by Leo Marx. Reprinted from *The Fitness of Man's Environment, Smithsonian Annual II*, 1968, by permission of the author and The Smithsonian Institution.

environment? Though he apparently acknowledges that society is inescapable, he usually remains a forlorn and lonely figure. Our most admired American fables seldom, if ever, depict a satisfying, wholehearted return, and in the closing sentences of one of them—*Huckleberry Finn*—the protagonist already has begun a new retreat, as if to suggest an unending cycle of withdrawal and return.

Perhaps this portrayal of man's retreat from his mechanized society, his exploration of nature, and his return to remake a new society is the story of mankind in the twentieth century. Some will find retreat from a confined world necessary and they will experience a self-realization in the new sub-culture. And perhaps they will return to the technological society with refreshed values and the ability to lead in the hard task of decision-making in a society where man shapes technology to fit his concept of what is desirable, good and needed.

THE MACHINE IN ART

The concept that the machine can beneficially share the landscape is perhaps best exemplified by the American paintings of Inness and Sheeler and the poems of Emerson and Whitman. Instead of causing disharmony, the machine is seen as a unifying device in the landscape. A beautiful example of a painting of the machine incorporated within the pastoral ideal is shown in Figure 11–1. The beauty of the industrial plant illustrates the American ideal of technological progress of the late ninteenth century America.

WHO IS RESPONSIBLE?

Social responsibility goes beyond the individual, however, to the corporate entities of our society. Recently corporations have reexamined their role beyond the simple goal of profit for the owners. For an increasing number of companies it is not enough to make a profit; they also wish to help clear the air and water, to provide jobs for minorities, and in general to help enhance the quality of life for everyone. Some examples of corporations exercising their social responsibility are: executives of one firm counseling minority-owned businesses; establishment of affirmative action programs for minority hiring; and training school dropouts for entry-level jobs.[13] One corporation president has said, "The fulfillment of valid, rational human needs in a viable economic way is becoming as much a concern as profit." While much of this is pleasing rhetoric, this view may well become a fully-accepted role for business and many stockholders many demand such action.

With the added explicit goal of social responsibility for the corporations, they are finding that they are being held accountable for what they are trying to do. While profit and loss statements account for the financial record, the success of a corporation in the social realm is less easy to account for. Social accounting

FIGURE 11-1 *American Landscape* by Charles Sheeler, 1930. Collection, The Museum of Modern Art. Gift of Abby Aldrich Rockefeller.

and the corporate social audit is a new concept within recent years.[25] It is proposed that a firm include its social audit with its finacial audit in its annual report to the stockholders. An audit would include an inventory of a firm's activities with social impact and an evaluation of these programs by an outside expert. Finally, in its annual report the firm would assess how these programs mesh with the objectives of the firm itself and society. Several firms, operating in a manner similar to certified public accountants, have assisted as independent social auditors for large corporations. While one cannot often put a dollar figure on social performance, it can be reported and evaluated. Social accounting is a healthy move toward open accountability in social responsibility.

Engineers, as individuals, hold a great responsibility in the development and implementation of new technologies. Engineers are often guided by a personal code of ethics and some who hold a license as a professional engineer are guided by a code of ethics issued by the licensing state. However, this code of ethics typically is concerned only with business relations among practitioners.

Professor Charles Susskind has proposed an engineer's Hippocratic Oath analogous to the principles attributed to the school of Hippocrates of Cos (c. 460 to 320 B.C.), the Greek physician.[14] Like medicine, engineering is concerned with improving the human condition. Also, the practitioners are to be professional in education and practice. Perhaps an oath of ethics should be developed, agreed upon and administered to new engineering practitioners. One version of an engineer's Hippocratic Oath follows:[14]

AN ENGINEER'S HIPPOCRATIC OATH*

I solemnly pledge myself to consecrate my life to the service of humanity. I will give to my teachers the respect and gratitude which is their due; I will be loyal to the profession of engineering and just and generous to its members; I will lead my life and practice my profession in uprightness and honor; whatever project I shall undertake, it shall be for the good of mankind to the utmost of my power; I will keep for aloof from wrong, from corruption, and from tempting others to vicious practice; I will exercise my profession solely for the benefit of humanity and perform no act for a criminal purpose, even if solicited, far less suggest it; I will speak out against evil and unjust practice wheresoever I encounter it; I will not permit considerations of religion, nationality, race, party politics, or social standing to intervene between my duty and my work; even under threat, I will not use my professional knowledge contrary to the laws of humanity; I will endeavor to avoid waste and the consumption of nonrenewable resources. I make these promises solemnly, freely, and upon my honor.

In a recent case study of the BART transit district and a problem of ethical behavior by engineers and the firm, a rather stark picture of responsible actions by three engineers is recorded.[21] This case study is particularly important in light of the fact that the BART system has not yet proved able to operate at its specified speeds or levels of safety and that use of the tube under San Francisco Bay was delayed by several years. The case report adequately illustrates the need for institutional protections for engineers who exercise their public responsibility.*

The BART Case:
Ethics and the Employed Engineer

Stephen H. Unger

Introduction

There has been an upsurge of discussion recently about the status of engineering as a profession, the obligations of the engineer toward the public, and the relationship of the engineer to his employer. Some very important facets of these questions are illuminated by the fate of three engineers employed by BART (Bay Area Rapid Transit).

A few words about the structure of BART will be useful as a background. ... BART is a fast (80 MPH top speed), modern rail transit system, with 34 stations and 75 miles of track, serving the counties of San Francisco, Alameda, and Contra Costa. Ownership and control is vested in the Bay Area Rapid Transit District (BARTD), created by public

*From *Understanding Technology* by Charles Susskind. Copyright 1973, The Johns Hopkins University Press. With permission.

* From IEEE Committee on Social Implications of Technology *Newsletter*, Sept. 1973, pp. 6–8. With permission of the author and publisher.

statute in 1957 and govemed by a 12 person Board of Directors, 4 from each county. It is financed by public funds.

Construction began about 1963 and the overall-cost is now estimated at about 1.5×10^9 dollars[1]. Partial revenue service commenced, between Oakland and Fremont, on September 11, 1972, almost 3 years behind schedule.

A consortium of 3 engineering firms, referred to as Parsons, Brinkerhoff-Tudor-Bechtel (PBTB), was retained by BART to direct and engineer the construction of the system. They in turn contracted out various phases of the operation to other firms. In particular, Westinghouse Electric Corporation, on the basis of competitive bidding, was awarded (in 1967) a 26 million dollar contract to design, install and operationally qualify the Automated Train Control (ATC) System.

Bart itself has an engineering staff whose functions include system maintenance and operation, surveillance and status checking of construction, approval of design changes and general investigation of problem situations.

The following account is based on a collection of over 40 documents including letters, memos, newspaper articles and reports, ranging in length from a few paragraphs to over 100 pages. These were acquired principally through correspondence. Because it was not feasible to interview the participants (even by phone), certain details have not been clarified. However, these are not important enough to affect the overall picture that emerges. The same is true for a few pieces of information that were given to the writer in confidence; these only serve to reinforce the impressions created by other information.

The Events

Holger Hjortsvang, a systems engineer in the BART Maintenance Section since 1966, and a Senior Member of IEEE (Institute of Electrical and Electronic Engineers), was involved with the ATC system. He became, over a period of years, increasingly concerned with the way the development of this system was progressing. He felt that BART had not internal structure adequate to monitor this phase of the project, relying instead on PBTB, who were also not set up to oversee this task. In part as a result of his having been sent to work for 10 months with the Westinghouse Computer Systems group responsible for ATC, Hjortsvang had grave doubts about the success of this phase of the project. He expressed these concerns to his superiors both orally and in a series of five written memorandums dating back as for as April 1969. In one of these reports criticizing the ATC system, he predicted a mean time between failures (each stopping a train) of 3 1/2 hours when the system was in full operation. There was no significant response from his management.

Max Blankenzee, a programmer analyst working with Hjortsvang since 1971, had a similar experience. His memos to his superiors criticizing various aspects of the ATC development drew only vague verbal responses and warnings not to become a "troublemaker."

Meanwhile, in BART's Construction Section, Robert Bruder, an electrical engineer monitoring various phases of the project since 1969, was growing increasingly disturbed about the "unprofessional" manner in which the installation and testing of control and communications equipment was being supervised by both BART and PBTB, as well as the obviously unrealistic opening dates being released to the public. His management was also not responsive to his expressed concerns.

Toward the end of 1971 the 3 engineers decided that in the public interest they must take steps to have their concerns dealt with seriously. Accordingly they made contact with Mr. Daniel Helix, a member of the BART Board of Directors, told him about the problems they were encountering, and gave him some written material. Mr. Helix expressed interest and was persuaded that action was needed. He conferred with two other board members and gave copies of a report on the subject to the entire board and the top management of BART.

The next step (and the elapsed time here is not clear) was the release to the press by Mr. Helix of the news of the controversy. This was followed by a public meeting on February 24 (or February 25) of the BART board at which presentations were made by Edward Burfine, a consulting engineer engaged by either Helix or the 3 engineers (possibly both—another unclear point) to present the criticisms of the handling of the ATC development, and by representatives of PBTB and Westinghouse in defense of their approach. The board voted 10 to 2 (one source said 8 to 2) in support of BART's management, in effect rejecting the criticisms.

Apparently the identities of the 3 who initiated the controversy had not been made public, and BART's management now proceeded to identify them. On the 2nd or 3rd days of March, Hjortsvang, Blankenzee and Bruder were given the options of resigning or being fired. Upon refusing to resign they were summarily dismissed with no written reasons being given.

On February 23 (just prior to the public meeting of the board) Bruder, a member of the California Society of Professional Engineers, telephoned CSPE President, William F. Jones, outlined the situation as it then stood, and asked for support. Mr. Jones immediately contacted the Diablo Chapter of CSPE (to which Bruder belonged) and, along with the leaders of that Chapter, initiated a thorough study of the situation. Subsequent to the discharge of the 3 engineers, Jones (on March 13) attempted to reach Mr. B. R. Stokes, BART's General Manager. (All accounts attribute the firings to Stoke's initiative.) Jones was never able to reach Stokes. He did speak to Chief Engineer David Hammond, who expressed surprise that CSPE should be interested in the situation. BART's top management declined to meet with CSPE.

Requests by the fired engineers for hearings on their case, or even for written statements of the charges justifying their dismissals, met with no response, and in fact BART has refused to issue any explanation to anyone. (Of 3 letters of inquiry I wrote to various BART managers—including Stokes—who were involved in the case, only one reply has been received. This was a refusal by Blankenzee's superviser to provide any explanation, on the grounds of pending legal action.)

A full investigation of the firings, the conduct of the 3 engineers and of the substance of their concerns about the BART project was then undertaken by CSPE. President Jones stated that be and other CSPE members (Gilbert A. Verdugo, State Director Diablo Chapter CSPE, and Roy W. Anderson, Chairman of CSPE's Transportation Safety Committee, also played major roles) involved in the case were "convinced that the three engineers acted in the best interest of the public welfare in disclosing to the BART Board of Directors problems regarding train control, systems management and contractual procedures." He also stated that "a large volume of most distressing information on the employment practices of BART, and on its apparent disregard for public safety, has been gathered."

On June 19, 1972 a report of CSPE's findings authored by Roy Anderson and entitled: "The BART inquiry" was submitted to the Califormia State Senate. At about the same time, the Diablo Chapter of CSPE circulated a public petition calling for a wide ranging investigation of BART by the State Legislature (a number of specific charges were made, but the case of the fired engineers, and employment practices in general were not mentioned). CSPE also took some tentative steps toward a court action on behalf of the fired three, but never did follow through on this.

The State Legislature did investigate, producing what is known as the "Post Report". It acknowledges the CSPE report as its starting point. Several instances of mismanagement of the project are pointed out although no mention is made of employment practices or of the 3 men whose initial warnings led directly to the Legislature's investigation. The Post Report, a further study by a special panel of distinguished engineers, and several other independent studies all confirmed, in general outline, the concerns expressed by Bruder, Hjortsvang and Blankenzee. A great deal of information pointing to poor engineering design was uncovered.

A more dramatic confirmation occurred on October 2, 1972 when a BART train over-ran the station at Fremont as a result of an ATC failure and several passengers were injured. This occurred just 3 weeks after the initiation of partial revenue service.

At this writing, the BART ATC is still under a cloud, with the trains being controlled ultimately in the traditional manner. The 3 engineers are now suing BART for damages totalling $885,000. They charge breach of contract and deprivation of constitutional rights. Blankenzee also charged that BART officials intervened on several occasions to discourage prospective employers from hiring him on the grounds that he was a "troublemaker".

Comments and Conclusions

The code of ethics of the NSPE states that the engineer "will regard his duty to the public welfare as paramount", and that "he will notify the proper authority of any observed conditions which endanger public safety and health." The Employment Guidelines approved by many engineering societies, including IEEE, and published in the May 73 issue of *Spectrum* are also highly relevant.* The facts related above indicate that Hjortsvang, Blankenzee and Bruder acted in a manner fully consistent with the letter and spirit of this code and guidelines, a conclusion also attested to by the CSPE. There is no indication that they did anything in this situation that could reasonably be called improper. When they felt it necessary to depart from normal administrative channels, they addressed themselves to the BART Board of Directors, an action difficult to interpret as irresponsible. (An interesting sidelight on the cautious approach of at least one of the 3 was provided by reporter Justin Roberts of the Contra Costa Times. He stated that he has met Robert Bruder some months prior to the firings, and having heard, from other sources, of trouble in BART, "attempted to pump him." "He politely but firmly rebuffed my efforts." Only after the matter became public knowledge, did Bruder speak to the press.)

Dr. Willard H. Wattenburg, a consultant who looked into the matter, referred to Holger Hjortsvang as "one very honest engineer" who was "ruthlessly sacrificed."

Nevertheless, having performed an obvious public service in the highest tradition of engineering, the considerable personal sacrifices of Blankenzee, Bruder, and Hjortsvang have been largely ignored in the reports that subsequently validated their claims. Only the CSPE showed any concern for them, and this group was apparently unable to take effective action on their behalf.

Unfortunately, the BART case is not a unique example of employed engineers being

*Objective #3: The responsibility of the professional employee to safeguard the public interest must be recognized and shared by the professional employee and employer alike.

forced to choose between compromising their ethics or seriously jeopardizing their careers. It is imperative that the engineering profession develop institutional means for eliminating such dilemmas.

Terms of employment #2: The professional employee should have due regard for the safety, life, and health of the public and fellow employees in all work for which he/she is responsible. Where the technical adequacy of a process or product is involved he/she should protect the public and his/her employer by withholding approval of plans that do not meet accepted professional standards and by presenting clearly the consequences to be expected if his/her professional judgment is not followed.

SUMMARY

In this chapter we have considered the importance of social values and social responsibility in a technological world. A balance of values must be achieved, each time an action is taken, which contrasts the short-term effects and the long-term effects. With the development of several new sub-cultures in the United States, new systems of values are emerging; no longer can the United States hold that there exists one unified system of values. Thus, it becomes more difficult to decide whether to proceed with large public projects. Whose value system shall be used in this process?

The social responsibility of the firm and the engineer is a cornerstone to the future of our society. New mechanisms must be developed that guarantee the engineer who is an employee of a firm the right and obligation to speak out on a project, if he can demonstrate existing or potential undesirable side effects for society. An Engineer's Hippocratic Oath may be one of the answers to the problem of social responsibility for the engineer.

CHAPTER 11 REFERENCES

1. K.E Schiebe, "Five Views on Values and Technology," *IEEE Transactions on Systems, Man and Cybernetics*, Vol. SMC-2, No. 5, Nov. 1972, pp. 566-571.
2. H. Cox, *The Secular City*, Macmillan, New York, 1966.
3. H. Cox, *The Feast of Fools*, Harper and Row, New York, 1971.
4. D.W. Shriver, Jr., "Man and His Machines: Four Angles of Vision," *Technology and Culture*, Vol. 13, No. 4, Oct. 1972, pp. 531-555.
5. M. McLuhan, *Understanding Media: The Extensions of Man*, McGraw-Hill Book Co., New York, 1964.
6. J. Ellul, *The Technological Society*, Alfred Knopf, Inc., New York, 1965.
7. J. McDermott, "Technology: The Opiate of the Intellectual," *New York Review of Books*, July 31, 1969, pp. 25–35.

8. V.C. Ferkiss, *Technological Man: The Myth and the Reality*, Braziller, Inc., New York, 1969.
9. T. Roszak, *The Making of a Counter-culture*, Doubleday and Co. New York, 1969.
10. T. Roszak, *Where the Wasteland Ends*, Doubleday and Co., New York, 1972.
11. C.A. Reich, *The Greening of America*, Random House, Inc., New York 1970.
12. L. Marx, "The Fitness of Man's Environment," *Smithsonian Annual II*, 1968.
13. C.N. Stabler, "For Many Corporations Social Responsibility is Now a Major Concern," *Wall Street Journal*, Oct. 26, 1971, p. 1.
14. C. Susskind, *Understanding Technology*, Johns Hopkins University Press, Baltimore, Maryland, 1973, p. 118.
15. J.D. Horgan, "Technology and Human Values: The Circle of Action," *Mechanical Engineering*, August 1973, pp. 19–22.
16. J. von Neumann, "Can We Survive Technology?" *Fortune*, June 1955, pp. 106-108.
17. H. Marcuse, *One-Dimensional Man*, Beacon Press, Boston, 1964.
18. H. Arendt, *The Human Condition*, University of Chicago Press, Chicago, 1958.
19. P. Teilhard de Chardin, *The Future of Man*, Harper and Row, New York, 1964.
20. R.M. Williams, "Individual and Group Values," *Annals of the American Academy of Political and Social Science*, Vol. 371, 1967, pp. 20–37.
21. S.H. Unger, "The BART Case; Ethics and the Employed Engineer," IEEE Committee on Social Implications of Technology *Newsletter*, Sept. 1973, pp. 6–8.
22. A. Toffler, *Future Shock*, Random House, Inc., New York, 1970.
23. S. de Beauvoir, *Les Belles Images*, Gallimard, Paris, 1966.
24. L. Marx, *The Machine in the Garden*, Oxford University Press, New York, 1964.
25. "The First Attempts at a Corporate Social Audit," *Business Week*, Sept. 23, 1972.

CHAPTER 11 EXERCISES

11-1. Review the proposed Engineer's Hippocratic Oath and revise it or expand it to fit your own objectives for such an oath.
11-2. What institutional safeguards would you propose in order to guarantee an engineer's obligation to speak out on a project his firm may be pursuing? What is his obligation to work within his firm so far as possible prior to making any public statements?
11-3. The mechanisms by which technological change effects value change may

be either direct or indirect. New technologies create new opportunities and new options and can lead to a new hierarchy of values. The dominant values of individualism, achievement, success, progress, efficiency, practicality and rationaltiy have been suited to the development and application of technology. Examine the effects of modern technology on these values. Has the influence of technology been direct or indirect?

11-4. Communications systems and the computer have had significant effects on our society. Examine the values of privacy and individuality in light of these developments.

11-5. Perhaps one of the most poetic uses of the progressive idiom in our language is Walt Whitman's "Passage to India" written in 1868. It begins:

> Singing my days,
> Singing the great achievements of the present,
> Singing the strong light works of engineers,
> Our modern wonders, (the antique ponderous Seven outvied,)
> In the Old World the east the Suez Canal,
> The New by its mighty railroad spann'd,
> The seas inlaid with eloquent gentle wires;
> Yet first to sound, and ever sound, the cry with thee
> O soul,
> The Past! the Past! the Past!

In this poem, Whitman speaks of the new railroads and lands as the new wonders of the world. Find other poems that speak of the progress of technology and the machine.

11-6. Examine the works of European and American painters of the period 1850–1950. Can you find a painting that illustrates the painter's feelings toward technology in modern society? Try to view the painting "The Lackawanna Valley" by George Inness, which is displayed in the National Gallery of Art, Washington, D.C.

11-7. In his book *Future Shock*, Alvin Toffler says, "We have reached a dialectical turning point in the technological development of our society." He goes on to say, "Technology, far from restricting our individuality, will multiply our choices—and our freedom—exponentially."[22] Individuality in western civilization has been a paramount value. Review the impact of technology on individual freedom and state whether you find man has been limited or made free by technology.

11-8. The mass production of impermanent items such as paper plates and modular trailer housing has led some to call modern society "the throw-away society." Many also believe we are a people with no sense of history, since we do not consider things to be permanent. Whole neighborhoods are removed for urban renewal or new highways. Is there a sense of impermanence in your city? Is the value of history and knowledge of our historical roots lost to man because of technology's influence on values?

11-9. Organizations are put together like factories to run efficiently. Is man over-organized, and has he lost the meaning of his individuality? Explore the effect of large organizations on the value of the individual in industrialized nations.

11-10. A critic of technology's influence on modern life, Simone de Beauvoir, states:[23]

"In every country of the world, socialist or capitalist, man is crushed by technology, alienated from his work, enslaved and brutalized. This has happened because man has multiplied his needs rather than containing them. Instead of seeking an abundance which does not exist, and may never exist, he should content himself with a vital minimum, as is still the case in certain very poor communities in Sardinia and Greece, for example, where technology has not penetrated and money has not corrupted. There people know an austere happiness because certain values have been preserved: truly human values of dignity, fraternity, and generosity which give life a special flavor. So long as we continue to create new needs, we multiply frustrations. When did this downfall begin? The day that we began to prefer science to wisdom and utility to beauty. With the Renaissance came rationalism, capitalism, and all the isms of science. So be it, But now that we have arrived at this point, what can we do? Try to revive wisdom and a taste for beauty in ourselves and around ourselves. Only a moral revolution, not a social, political, or technological one, can bring man back to his lost truth."

Consider de Beauvoir's view. What do you think are the views of the poor in Sardinia and Greece toward the benefits of technology? Are the values of dignity and fraternity necessarily submerged by technology?

11-11. The Code of Ethics of the American Society of Civil Engineers states that it is unprofessional for an engineer "to exert undue influence or to offer, solicit or accept compensation for the purpose of affecting negotiations for an engineering engagement." Any complaint alleging misconduct or unethical practice on a part of a member of the Society is investigated by the Committee on professional Conduct. Some civil engineers were accused of bribery in Maryland in 1973 in a case associated with the resignation of Vice-President Agnew. Discuss the usefulness of a code of ethics in maintaining a profession with high standards of integrity and honesty.

The National Society of Professional Engineers has suggested legislation to limit political contributions by architects and engineers to $100 for the campaign of a candidate for public office during a two-year period before an election. Discuss this measure in terms of its usefulness toward insuring a better relation between engineer and office holder.

12

THE SOCIAL CONSEQUENCES OF TECHNOLOGY

THE MECHANICAL ENGINEERS OF THE MIDDLE AGEsinvented an ingenious device for training knights in the difficult art of jousting. It was called the *quintain*. It consisted of a wooden figure, usually representing a Saracen, fixed to a post. One arm held a shield and the other arm held a bag of sand. The arms were formed by a single beam pivoted on the top of the post, as shown in Figure 12–1.

The knight was supposed to charge the quintain at full gallop and hit the head with the point of his lance. If he caught it dead center he scored a point. If his lance wobbled to one side, so that if he hit the shield, the quintain would rotate with great speed and hit him in the back of the neck with the sandbag as he went past.

FEEDBACK AS CONSEQUENCES

The world of technology also contains quintains; in this case, painful and rapid feedback as a result of our errors in attempting to introduce beneficial technologies into the world. The social consequences of a new technology are often unforseen, or we miss the mark introducing an innovation, and the resulting side effects are rapidly deleterious.

FIGURE 12-1   The quintain, a practice device for knights with rapid feedback if an
error was made.

Perhaps in the past thirty years Americans have become overconfident in their technology. Overconfidence leads to faulty judgments or too-rapid implementation of new technologies. Perhaps we assume that our science and technology will always be excellent. [1] However, we are bound to miss and err in the implementation or design phase. The consequences of such errors in judgment, design or the setting of objectives can be significant.

As McLuhan and Nevitt note in the following excerpt, effects merge rapidly with causes:[2]*

CAUSALITY IN THE ELECTRIC WORLD

Marshall McLuhan and Barrington Nevitt
The situation is complicated and its difficulties are enhanced by the impossibility of saying everything at once.

In today's ECO-world of electric information that flows unceasingly upon us from every side, we all encounter the predicament of *Alice in Wonderland*. Now effects merge with causes instantly through speedup, while "software" etherealizes "hardware" by design. All rigid distinctions between thinker and doer, observer and observed, object and subject are being eroded by the "rim-spin" of electric media. Old ground rules and human perceptions are being transformed by this new resonant surround where nothing is stable but change itself. But like water to a fish, the environment we live in remains hidden. Only children and artists see "the emperor's new clothes."

Continuity in Discontinuity

Today, metamorphosis by *chiasmus*—the reversal-of-process caused by increasing its speed, scope, or size—is visible everywhere for anyone to see. The chiasmus of speedup is

*Reprinted from *Technology and Culture*, Vol. 14, No. 1 (January 1973) by permission of the Society for the History of Technology.

slowdown. Perhaps first noted by ancient Chinese sages in *I Ching* or *The Book of Changes*, the history of chiastic patterns is traced through classical Greek and Hebrew literature by Nils W. Lund in *Chiasmus in the New Testament*. Computer programmers have also learned that "information overload leads to pattern recognition" as breakdown becomes breakthrough: from "bits" to "bytes" to "whole" again.

Speeding up the components of any visually ordered structure or continuous space pattern will lead to breaking its connections and destroying its boundaries. They explode into the resonant gaps or interfaces that characterize the discontinuous structure of acoustic space. The visual perspective becomes an acoustic wraparound. Repetition of any visual pattern or modular form creates a mosaic with nonvisual effects, as the single photographic point of view becomes a multiple, iconic re-presentation. History becomes "mythic" through time compression and juxtaposition of events as past, present, and future merge in electric *nowness*.

Novelty Causes Antiquity

Every innovation, whether of "hardware" product or "software" information, is an extension of man. Novelty becomes cliché through use. And constant use creates a new hidden environment while simultaneously pushing the old invisible ground into prominence, as a new figure, clearly visible for the first time. Every innovation scraps its immediate predecessor and retrieves still older figures; it causes floods of antiques or nostalgic art forms and stimulates the search for "museum pieces." In such cliché-archetype patterns, the new continually recreates the old as novelty regenerates antiquity. Ancient cults and old jalopies are revived for "inner" satisfactions as we explore "outer" spaces. The motor car retrieved the countryside, scrapped the inner core of the city, and created suburban megalopolis. Invention is the mother of necessities, old and new.

Thus what the technologist, in McLuhan's words, "normally sees is either a replay of past scientific experience or an up-and-coming threat in his 'rear-view mirror.' Coming events cast their shadows before them? Don't look back; they may be gaining on you!"

The consequences are consciously made accessible either through invention or new causes. Computers, for example, in McLuhan's view, are used as agents to sustain precomputer effects: effects precede causes.[2]

FACING DELETERIOUS CONSEQUENCES

Man may be blind or dazed to the effects of his own technologies. The individual man shuts himself away from the public consequences of his own technologies. Henry Ford, amazed by the consequences of the auto on rural America, felt that he should try to save a least a model of it, so he recreated a miniscule figure of that old world in Greenfield Village in Dearborn.[3]

The deleterious consequences of modern technology are the topic of Lewis Mumford's recent book *The Myth of the Machine*.[4] Mumford recommends that in order to avoid such effects of technology in the future, we "overthrow the megamachine" in order to overcome "wholesale miscarriages of techniques." These miscarriages, in Mumford's opinion, include mass production, automation, space programs and computers. He finds the consequences of these technologies to outweigh any benefits. Few people would state this case as strongly.

Perhaps the student of the consequences of new technologies develops his own view of the qualities of these effects and then ultimate meaning. With unforeseen consequences and technological error, one might say with Jean-Paul Sartre, "The only meaning you will find in an accidental universe is the one you impress upon it." To the extent he finds the consequences accidental, the person experiencing them impresses the meaning upon them. Thus, one man's benefit is perceived by another as an unforeseen and unfortunate consequence.

WHAT WORKS?

How do we know if an innovation or technology has worked? Practically all technologies work in that they distribute some benefits to the people they aim to help. At the same time, no technology works out exactly as its proponents hoped; there are always unanticipated effects. Perhaps one must strike a balance of benefits and unforeseen consequences. Also, one must confront the disturbing paradox that insofar as technology is responsible for undesirable consequences and our present predicament, it may also be the means necessary for escaping the misfortunes of the undesired effects. Thus we might have a cause-and-effect process with an unending sequence of steps. The issue is, perhaps, whether the sequence is leading to a quality of human life that can be said to be *improved*.

Machines can surely be called a mixed blessing to mankind. Most inhabitants of industrialized countries who have employment can earn all the food, clothing, housing and medical care necessary for a healthy life with a 40 hour work week, usually in safe and comfortable surroundings. Machines have also given men freedom to travel, to communicate, to entertain and to live in confortable surroundings. Nevertheless, the by-products of these very technologies are pollution of the air and water; accidents in travel; the replacement of personal attention in shops and places of business by mechanized systems and the loneliness of high-rise dwellings; litter; ugliness of mechanical housing and the destruction of wild life.[5] As a result of the communications revolution, electronic eavesdropping has become so prevalent that nobody can be assured of his privacy. As Gene Marine states in his book *America the Raped*: "The point is that the engineers—all of those who take the engineering approach, build the bridge and get the people and the cars from one side of the river to the other and to hell with the side effects—are shaping the nation unchecked, molding the land and murdering thousands of its inhabitants, raping America while the rest of us look the other way. *Theirs is a rape from which America can never, never recover*."[6]

And yet if, as some maintain, we have sown the seeds of our own destruction by developing technologies with disastrous side-effects, is it not also true that it is ignorance of the causes of these effects that leads to their appearance? It may be more essential that an understanding of these causes be developed; this is a natural role for science. New technology, through more efficient utilization of natural resources and conservation of such resources, may alter the consequences of our current spend-thrift use.[7] Again, will it not be new technologies based on new science that will enable us to overcome environmental pollution and the mismanagement of waste?

THE SLOTH'S WAY

Of course, another available policy that can be used to avoid the social consequences of a technology is a policy of inaction. The best way to insure against the risk that DDT may be harmful to people is not to use DDT. The best way to avoid the side effects of power plants is not to build them. While this policy could produce social stagnation, it is being advocated by some. What are the potential consequences of fertilizing human eggs in test tubes, so-called genetic engineering? Are the consequences therapeutic for man or subversive? Is a policy of inaction best in this case? A decision for inaction is possible, as we noted in the case of the SST.

We concurrently face the portents of disaster and the potentiality of fulfillment. The fact that ocean beaches are now so much more crowded than half a century ago is a consequence of the wide availability of the auto. But would it be sensible not to have developed the motor car in order to avoid this side effect?[8]

THE ASWAN EXAMPLE; LIMITS TO SUCCESS

The consequences of new technologies are difficult to balance with the benefits. The Aswan Dam in Egypt is a case in point. The dam caused a large lake to develop behind the dam. It inundated many of Egypt's masterpieces of antiquity, such as the temple dedicated to the Goddess Isis, shown in Figure 12–2. Yet the dam provides water for irrigation and electricity on a significant scale. A further consequence, foreseen at this time, is the eventual silting of the dam and the loss of the benefit of the dam. Is a half-century of a fertile plain and available electricity a sufficient benefit to outweigh the consequences?

The approach of using technology to overcome the consequences of preceeding technologies may have its ultimate limit. Chemists observe Le Chatelier's Principle: if some change is made in one of the changeable characteristics of a chemical system whose several parts are more or less in some kind of relationship, the side effects of the change will tend to cancel out the change intended.[8] It is at this point that portions of our technology of our technology may be operating.

FIGURE 12-2   This temple, dedicated to the goddess Isis, is one of several ancient
Egyptian monuments that lie permanently half-submerged by the Nile.
Courtesy of Authenticated News International.

AFFLUENCE, POVERTY, EXPECTATIONS

Perhaps the factor of affluence leads to continuing rising expectations, and
thus the side effects of technology appear to be continually worse as our quality
of life improves. If one examines the writings of the 1930s, one finds that the
affluence of the United States today exceeds that envisioned forty years ago.[9]
Nevertheless, not many of us feel well-off. Perhaps this is the continuing
revolution of rising expectations. Technological innovations, coupled with rising
affluence, makes many things much cheaper. But it also makes other things very
expensive! Technologlogy has provided inexpensive refrigerators, freezers, telephones, air travel—all contributions to our high standard of living. But once one
has experienced these benefits and learned to take them for granted, then other
effects become more noticeable. We notice the parts of life unalterably changed
by technology. As Irving Kristol observes: "What this comes down to is the fact
that human values are inevitably shaped by human memories, whereas the
"values" offered by technological innovation are shaped by emerging technological possibilities. There is no fault or blame here—humanity cannot obtain its
values, nor can technology achieve its ends, in any other way. But there is enough
incompatibility in this partnership to make for a persistent, gnawing frustration."[9]

This frustration is particularly gnawing for the poor, who are only now gaining some of the sought-after benefits of technology. Inactivity with respect to technology could sentence the poor to a life of continuing poverty. Perhaps we need more transportation, but few automobiles; more housing, but less urban sprawl. Again we return to the possibility of using technology wisely to assure benefits to the many.

The full development of automated* factories for mass production and robots which can be programmed to do a wide variety of jobs can be seen to have very significant consequences to workers and employment. Yet these devices may provide items of necessity for the poor of the world. Professor Thring estimates that we can replace all the people now doing routine work in factories by one tenth as many maintenance and repair men servicing automated factories.[5] This may not be possible before the end of this century, however. The possibility of significant increased output per man-hour worked could result in widely-available goods at a low price. If this can occur with greatly increased production of goods, then employment could remain the same with a greatly increased productivity. If the market does not expand for one industry, then workers would be displaced, one would hope to another industry. The fear remains that increasing unemployment will be a consequence of increasingly automated factories.

Alternatives are, of course, decreased work weeks and shared jobs. The transition to a further automated society could greatly benefit the poor who have a job and income and thus can increasingly share in the wealth of goods. Automation may materially disadvantage the person who loses his job because of it, and thus is unable to avail himself of the benefits.

CYBERNETIC EFFECTS ON SOCIETY

Cybernetics is the theory of control and communication which can be applied to machines and animals. The signers of a statement entitled "The Triple Revolution," circulated in 1964, envisioned a very significant effect of the cybernetic revolution. By the word *cybernation* the signers referred to a force on society brought about by the combination of the computer and the automated self-regulating machine. The authors stated, "Cybernation is already reorganizing the economic and social system to meet its own needs." Automation was thus seen as a potential force in society sufficient to dislocate a significant percentage of workers from their jobs.

As yet no fully automated industry has developed. Nevertheless, the question of whether cybernation has had a significant effect on employment is not yet clear. Over the past decade unemployment has ranged from 3.3% to 5.5% of the work force, and has been influenced by factors such as military spending and the state of the economy as much as by cybernation.

---

*Automation is defined as the control of a process or mechanism by automatic rather than human means, especially by electric devices.

LONG-TERM DISAPPOINTMENTS

Attempts to influence society beneficially have often resulted in unforeseen negative consequences; Professor Jay Forrester of M.I.T. observes that this can be traced to the counter-intuitive character of social systems.[11] He notes that there is likely to be a fundamental conflict between the short-term and long-term consequences. A technology or program that produces improvement may at a later time result in a degraded situation. One example he cites is a road overcrowded with traffic. To improve the road, we decide to widen the road and increase the number of lanes. While this action will increase the traffic flow in the short run, it results in the long run in more drivers being attracted to the bigger road and more business and housing moving out along the road. The result may be said to be counter-intuitive: that is, it doesn't work the way people might have expected at first.

The new developments in information storage and transfer, often lumped together as the "information revolution," are based on computer and electronic communications technology. Advances in the storage, retrieval, processing and distribution of information may be the source of significant changes in the last third of the twentieth century. The use of computer data banks and their influence on privacy may have inportant consequences for mankind. The merging of computer and communications technology may result in personal computers (the size of a briefcase), artificial intelligence, three-dimensional television and library automation. A technology that will allow masses of citizens to have discussions with each other and to reach group decisions without leaving their homes may be realized in the near future. A model for this kind of electronic town meeting has been presented recently.[12] An electronic town meeting would provide a system for citizens' participation and could have an immense effect on our political system. The use of cable TV and national communications satellites could provide large as well as small communities with the potential for citizen participation similar to that of the traditional small New England town meeting.

But technological improvements in computers and communications have an equal potential for eroding individual rights. Dr. Jerome Weisner, president of M.I.T., stated recently, "I used to suspect that it would be much easier to guard against a malicious oppressor than to avoid being slowly but most surely dominated by an information Frankenstein of our own creation. Watergate has demonstrated I was clearly not worried enough about improper uses of technology. ... The great danger is that ... 1984 could come to pass ... because each step in the development of an 'information tyranny' has appeared to be constructive and useful ... and has occurred without specific overt decisions, without high-level encouragement or support and totally independent of malicious intent."[13]

Weisner then goes on to say that a "threat comes from the surveillance which modern technology subjects us to .... Surveillance systems ... are so

common that many people just assume that their telephones are monitored . . . .
The effect (of government spying activity) has been to intimidate many
individuals and make them draw back from perfectly legal political and social
activities."[13] To eradicate this threat, Weisner believes it is necessary to provide
new safeguards through the legislative and legal systems.

A CREATIVE, HUMANE WAY OF LIFE

Conviviality, a civilized and humane style of life, is advocated by Ivan Illich
as a contrast to life under technology.[14] By "conviviality" Illich means
"autonomous and creative intercourse among persons, and the intercourse of
persons with their environment; and this in contrast with the conditioned
response of persons to the demands made upon them by others, and by a
man-made environment. I consider conviviality to be individual freedom realized
in personal interdependence and, as such, an intrinsic ethical value."

This way of life is the opposite of the constant thrust toward industrial
productivity and its destructive side effects. The undesirable side effects of
technology, according to Illich, are specialization of functions, institutionaliza-
tion of values, and centralization and bureaucratization of power.[15] A convivial
society serves politically-related individuals rather than managers.

Tools foster conviviality to the extent to which they can be used easily by
anybody; a telephone is an example. In Illich's view, people must learn to live
within bounds, to abstain from unlimited progeny, consumption and use. This is
possible, but only if the world is restructured to emphasize conviviality and a set
of limits is established governing the growth of the service sector as well as the
production of industrial goods.

LIMITING ILL EFFECTS

The deleterious consequences of technological development are to be
avoided so that the costs do not outweigh the original benefits. But how are we to
achieve the goal of limiting the side effects of new technologies? There appear to
be three primary approaches that may be adopted in an attempt to limit the
undesired social consequences of technology. The first approach is that of using
technology itself to overcome the undesired consequence. Then, as we have seen,
this approach leads to a continuing series of causes and effects, one would hope
decreasing in impact as the sequence is followed. Therefore, when a new
technology results in an undesired side effect, a new innovation is developed to
overcome the side effect to be ameliorated. And so the sequence of technology,
side effect and technology goes on. This approach can work in many instances,
but it can be costly of resources and unfortunate in its tendency to result in an
unsettled society. This approach is often used with the introduction of a new

model of automobile. An unforeseen side effect is discovered in use and the auto is recalled so that a new additional device can be added to the auto to overcome the side effect.

A second possibility for the avoidance or limitation of the social consequences of new technologies is to limit the growth of new technologies and in some cases actually to prohibit the introduction of new technologies. This approach, which avoids unforeseen consequences by attempting to avoid change, has been applied in some cases and is widely advocated. The decision not to construct a prototype SST is an example of a case of avoiding potential side effects of a new technology by not pursuing that technology. The limiting of new technologies to a prescribed growth or operating level is commonly advocated. One example is the case of the construction of new nuclear power plants for the generation of electricity. The growth of the number of new plants has been severely curtailed in recent years. In addition, there has been a strong movement to limit the operating level of the nuclear power plants now in service.

Another possible approach to the avoidance of undesirable social consequences of technology is to assess the consequences of many alternative schemes prior to implementation and then to choose the most desirable alternative. This new approach, called *technology assessment*, is discussed in Chapter 18.

CHAPTER 12 REFERENCES

1. P.H. Abelson, "Overconfidence in American Technology," *Science*, March 21, 1969, p. 18.
2. M. McLuhan and B. Nevitt, "Causality in the Electric World," *Technology and Culture*, Jan. 1973, pp. 1–18.
3. F.D. Williamson, "Through a Rearview Mirror-Darkly," *Technology and Culture*, Jan. 1973, pp. 22–29.
4. L. Mumford, *The Myth of the Machine: The Pentagon of Power*, Harcourt, Brace and Jovanovich, New York, 1970.
5. M. Thring, "Society and the Engineer," *Electronics and Power*, April 5, 1973, pp. 127–128.
6. G. Marine, *America the Raped: The Engineering Mentality and the Devastation of a Continent*, Simon and Schuster, New York, 1969.
7. H. Brooks, "Can Science Survive in the Modern Age?" *Science*, Oct. 1, 1971, pp. 21–29.
8. J. Maddox, *The Doomsday Syndrome*, McGraw-Hill Book Co., New York, 1972.
9. I. Kristol, "The Frustrations of Affluence," *Wall Street Journal*, July 20, 1973, p. 14.
10. "The Triple Revolution," *Liberation*, April, 1964, pp. 1–7.

11. J. Forrester, "Counterintuitive Behavior of Social Systems," *Technology Review*, Jan. 1971, pp. 38–43.

12. A. Etzioni, "Minerva: An Electronic Town Hall," *Policy Sciences*, Vol. 3, 1972, pp. 457–474.

13. P. Hirsch, "1973 International Communications Association Conference; Communications Technology Continues to Develop Faster than the Industry's Conscience," *Datamation*, August, 1973, pp. 68–69.

14. I. Illich, *Tools for Conviviality*, Harper and Row, Inc. New York, 1973.

15. I. Illich, "Convivial Tools," *Saturday Review of Education*, May 1973, pp. 63–67.

16. R.A. Baker, *Second-Order Consequences*, The M.I.T. Press, Cambridge, Mass., 1969.

17. M.V. Jones, "How Cable Television May Change Our Lives," *The Futurist*, Oct. 1973, pp. 196–201.

CHAPTER 12 EXERCISES

12–1. The more industrialized nations are moving rapidly toward a time when cybernetion will result in decreased employment (or working hours) and increased affluence. If we can no longer find our primary sense of identity as productive workers, how shall man relate to work and leisure?

12–2. The welfare and way of life of the people of the United States was altered in the last half of the nineteenth century by the widespread introduction of the railroad. Discuss the consequences of the introduction of the railroad throughout the United States.

12–3. The industrial revolution was not so much a technological revolution as a social revolution. The effects of the industrialization of Europe and the United States during the period 1780–1880 were enormous. Discuss the economic benefits and costs, the population shifts and the changes in the social matrix.

12–4. CATV, which is the distribution of broadcast television signals as well as locally originated programming and other services by means of a coaxial cable, permits two-way interactive communication between the viewer and the originating company location. It is moving into more and more homes. Unlike broadcast television, CATV makes possible a vast array of two-way information services such as computer-aided instruction and electronically-delivered newspapers edited for the individual reader.[17] CATV may reach a market penetration of over 50% by 1980 if the present growth rate continues. Consider the possible consequences of the availability of electronic mail and an electronic newspaper delivered via CATV.

12–5. So-called labor-saving devices in the home have provided free time, and technologies for limiting conception have made possible smaller families.

Women thus have wider opportunities for participation in all sectors of society. Although women have been freed to hold jobs outside the home, social mores are slow to change. Explore the consequences to society of the new household devices and widely–available contraceptive devices, particularly as they effect the role of women in today's society.

12–6. Electronic mail may be the way we shall overcome the large delays in delivery of our mail to distant points. Now it can take up to six days to have a letter delivered to a nearby city. The use of the telephone system to deliver mail within one day would be a great improvement. Letters and mail sent over the telephone lines may be as inexpensive as $.50 for a 30 second message. The transportation of hard copy priority documents will probably be the first use of electronic mail. Regular United States Postal Service mail will probably maintain the majority of the mail delivery for several decades. It is expected that over 80 billion pieces of first class mail will be handled each year by the United States Postal Service by 1980. Analyze the social consequences of the introduction of an electronic mail system using the existing telephone network at an inexpensive rate.

12–7. The automobile may be viewed as a status symbol, an object of fantasy, an important source of employment and an indispensible element in the national income accounts. The automobile has had a profound effect on the quality of life in developed nations. One tenth of the non-governmental labor force in the United States is directly dependent upon the production, sales and service of automobiles for its employment. It costs most families over $1,500 each year to own, maintain and operate an automobile. Many cities devote about 25% of the urban area to roads, parking lots, and other auto-serving functions. Examine the full social consequences in the United States of our dependence on the automobile for personal transportation. What would be the consequences if a publically-controlled auto rental agency held a monopoly over the availability of autos in a large city such as New York?

13

TECHNOLOGY AND ECOLOGY

THE NEWLY-REDISCOVERED RELATIONSHIP OF MAN as a partner to others in the adventure of all living things underlies much of the science of ecology, which is the branch of science that deals with the relation of living things to their environment and to each other. Ecology as a formal scientific discipline was introduced in America in the beginning of the twentieth century, when serious ecological concern began to be asserted itself by conservationists such as John Muir. The great drought of the southwestern dust bowl of the 1930s as well as disastrous floods elsewhere helped to shape the awareness of the public to the need for conservation. As Adlai Stevenson said in his last speech, "We travel together, passengers on a little space ship, dependent on its vulnerable supplies of air and soil ... preserved from annihilation only by the care, the work, and I will say the love, we give to our fragile craft." While rushing forward with our new technological accomplishments we may have forgotten the earth and the fact that we should regard it properly as the source of our life.

POLLUTION AS ALTERATION OF NATURAL STATES

Environmental pollution is the unfavorable alteration of our surroundings, largely as by-products of man's actions, through changes in energy patterns, radiation levels, chemical and physical constitution and changes in numbers of organisms. Pollutants are byproducts of an industrialized society; they evidence themselves as gases, particulates, pesticides, sewage, solid wastes, excessive heating of rivers and lakes, and many other results of industrial activity.

The most controversial question raised—and left unanswered—is definition of what constitutes an *unfavorable* alternation. As Robert Ingersoll said on behalf of many of his generation in 1880, "I want the sky to be filled with the smoke of American industry, and upon that cloud of smoke will rest forever the bow of perpetual promise. That is what I am for."[1] In many parts of the world today, "smoke means jobs." Perhaps that could be said of Pittsburgh, Pennsylvania, a portion of which is shown in Figure 13–1.

In the United States the consumption per person of natural resources and the production of solid wastes is quite high; see Table 13–1. The 210 million Americans produce 231 million metric tons of solid waste each year. Furthermore, as shown in Table 13–2, this consumption of resources is growing each year. Electric power consumption, with a growth rate of over 7%, doubled in one decade.

The environmental costs of producing electric power serve to illustrate the side effects on the environment of our growing use of electricity. These environmental costs of producing one million watts (one megawatt) of electricity in a steam power plant using a fossil fuel such as coal are shown in Figure 13–2.

FIGURE 13-1   Steel and coke works in Pittsburgh. Courtesy Joan Sydlow, *Business Week*.

TABLE 13-1

Annual per capita consumption for the United States.

Item	Consumption per capita annually
Oil	3 cubic meters (800 gallons)
Natural gas	3,000 cubic meters
Coal	2 metric tons
Energy in all forms	100,000 kilowatt-hours
Steel	600 kilograms
Newsprint	40 kilograms
Copper	7 kilograms
Packaging materials	250 kilograms
Solid wastes production	1.1 metric tons

TABLE 13-2

Average annual rate of growth in the United States during the period 1950–1970.

Item	Annual Rate of Growth (Average)
Population	1.6
Energy consumption	3.2
Industrial production	4.0
Motor fuel consumption	4.5
Electric power consumption	7.6
Bauxite consumption in aluminum production	7.7

A coal-burning plant produces side effects in mining, the air, the water, and the land. Some environmental effects of the use of electric power use can be illustrated by a drawing of New York City in the 1880s, shown in Figure 13–3. The problem is obviously not a new one.

To the ecologist there is a system of systems in the world, all interrelated to each other. Each is wholly embedded in the structure of the natural process. Many persons see the ecosystem as extrinsic to man. They see the ecosystem, or nature, as something to be used and hence managed.

In the following article, Professor Lynn White, Jr. traces the historical roots of our uses of nature. White, among others, notes that western man has believed that the physical world was given by God to man to manage.[6] This view is perhaps best expressed in Genesis 1, Verse 28 and Psalm 8, Verse 4 where man is named to dominate: He is to rule the earth, which has been subjected to him. We are not in nature, but above nature. Perhaps, White believes, the patron saint of ecology should be Saint Francis of Assisi, who strove to establish a democracy of all God's creatures.[2]*

* From *Science*, Vol. 155, March 10, 1967, pp. 1203–1207. Copyright 1967 by the American Association for the Advancement of Science.

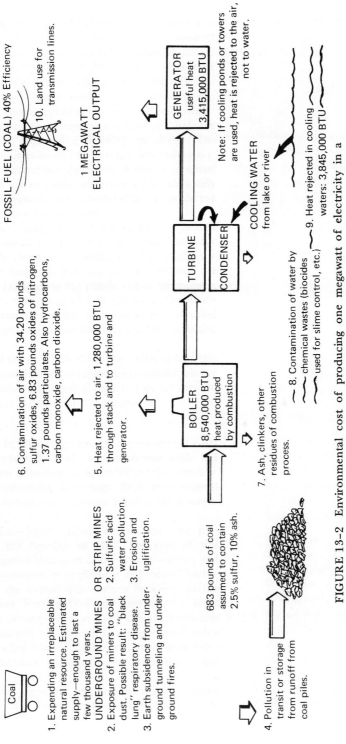

FIGURE 13-2 Environmental cost of producing one megawatt of electricity in a steam electric power station using coal (40% efficiency).

FIGURE 13-3 A street scene in the early 1880s in New York City. Broadway is shown as viewed from Maiden Lane. Courtesy of American Telephone and Telegraph Company.

The Historical Roots of Our Ecologic Crisis

Lynn White, Jr.

A conversation with Aldous Huxley not infrequently put one at the receiving end of an unforgettable monologue. About a year before his lamented death he was discoursing on a favorite topic: Man's unnatural treatment of nature and its sad results. To illustrate his point he told how, during the previous summer, he had returned to a little valley in England where he had spent many happy months as a child. Once it had been composed of delightful grassy glades; now it was becoming overgrown with unsightly brush because the rabbits that formerly kept such growth under control had largely succumbed to a disease, myxomatosis, that was deliberately introduced by the local farmers to reduce the rabbits' destruction of crops. Being something of a Philistine, I could be silent no longer, even in the interests of great rhetoric. I interrupted to point out that the rabbit itself had been brought as a domestic animal to England in 1176, presumably to improve the protein diet of the peasantry.

All forms of life modify their contexts. The most spectacular and benign instance is doubtless the coral polyp. By serving its own ends, it has created a vast undersea world favorable to thousands of other kinds of animals and plants. Ever since man became a numerous species he has affected his environment notably. The hypothesis that his fire-drive method of hunting created the world's great grasslands and helped to exterminate the monster mammals of the Pleistocene from much of the globe is plausible, if not proved. For 6 millennia at least, the banks of the lower Nile have been a human artifact rather than the swampy African jungle which nature, apart from man, would have made it. The Aswan Dam, flooding 5000 square miles, is only the latest stage in a long process. In many regions terracing or irrigation, overgrazing, the cutting of forests by Romans to build ships to fight Carthaginians or by Crusaders to solve the logistics problems of their expeditions, have profoundly changed some ecologies. Observation that the French landscape falls into two basic types, the open fields of the north and the *bocage* of the south and west, inspired Marc Bloch to undertake his classic study of medieval agricultural methods. Quite unintentionally, changes in human ways often affect nonhuman nature. It has been noted, for example, that the advent of the automobile eliminated huge flocks of sparrows that once fed on the horse manure littering every street.

The history of ecologic change is still so rudimentary that we know little about what really happened, or what the results were. The extinction of the European aurochs as late as 1627 would seem to have been a simple case of overenthusiastic hunting. On more intricate matters it often is impossible to find solid information. For a thousand years or more the Frisians and Hollanders have been pushing back the North Sea, and the process is culminating in our own time in the reclamation of the Zuider Zee. What, if any, species of animals, birds, fish, shore life, or plants have died out in the process? In their epic combat with Neptune have the Netherlanders overlooked ecological values in such a way that the quality of human life in the Netherlands has suffered? I cannot discover that the questions have ever been asked, much less answered.

People, then, have often been a dynamic element in their own environment, but in the present state of historical scholarship we usually do not know exactly when, where, or with what effects man-induced changes came. As we enter the last third of the 20th century, however, concern for the problem of ecologic backlash is mounting feverishly. Natural science, conceived as the effort to understand the nature of things, had flourished in several eras and among several peoples. Similarly there had been an age-old accumulation of technological skills, sometimes growing rapidly, sometimes slowly. But it was not until about four generations ago that Western Europe and North America arranged a marriage between science and technology, a union of the theoretical and the empirical approaches to our natural environment. The emergence in widespread practice of the Baconian creed that scientific knowledge means technological power over nature can scarcely be dated before about 1850, save in the chemical industries, where it is anticipated in the 18th century. Its acceptance as a normal pattern of action may mark the greatest event in human history since the invention of agriculture, and perhaps in nonhuman terrestrial history as well.

Almost at once the new situation forced the crystallization of the novel concept of ecology; indeed, the word *ecology* first appeared in the English language in 1873. Today, less than a century later, the impact of our race upon the environment has so increased in force that it has changed in essence. When the first cannons were fired, in the early 14th century, they affected ecology by sending workers scrambling to the forests and mountains for more potash, sulfur, iron ore, and charcoal, with some resulting erosion and deforestation. Hydrogen bombs are of a different order: a war fought with them might alter the genetics of all life on this planet. By 1285 London had a smog problem arising from the burning of soft coal, but our present combustion of fossil fuels threatens to change the chemistry of the globe's atmosphere as a whole, with consequences which we are only beginning to guess. With the population explosion, the carcinoma of planless urbanism, the now geological deposits of sewage and garbage, surely no creature other than man has ever managed to foul its nest in such short order.

There are many calls to action, but specific proposals, however worthy as individual items, seem too partial, palliative, negative: ban the bomb, tear down the billboards, give the Hindus contraceptives and tell them to eat their sacred cows. The simplest solution to any suspect change is, of course, to stop it, or, better yet, to revert to a romanticized past: make those ugly gasoline stations look like Anne Hathaway's cottage or (in the Far West) like ghost-town saloons. The "wilderness area" mentality invariably advocates deep-freezing an ecology, whether San Gimignano or the High Sierrra, as it was before the first

Kleenex was dropped. But neither atavism nor prettification will cope with the ecologic crisis of our time.

What shall we do? No one yet knows. Unless we think about fundamentals, our specific measures may produce new backlashes more serious than those they are designed to remedy.

As a beginning we should try to clarify our thinking by looking, in some historical depth, at the presuppositions that underlie modern technology and science. Science was traditionally aristocratic, speculative, intellectual in intent; technology was lowerclass, empirical, action-oriented. The quite sudden fusion of these two, towards the middle of the 19th century, is surely related to the slightly prior and contemporary democratic revolutions which, by reducing social barriers, tended to assert a functional unity of brain and hand. Our ecologic crisis is the product of an emerging, entirely novel, democratic culture. The issue is whether a democratized world can survive its own implications. Presumably we cannot unless we rethink our axioms.

The Western Traditions of Technology and Science

One thing is so certain that it seems stupid to verbalize it: Both modern technology and modern science are distinctively *Occidental*. Our technology has absorbed elements from all over the world, notably from China; yet everywhere today, whether in Japan or in Nigeria, successful technology is Western. Our science is the heir to all the sciences of the past, especially perhaps to the work of the great Islamic scientists of the Middle Ages, who so often outdid the ancient Greeks in skill and perspicacity: al-Rāzī in medicine, for example; or Ibn-al-Haytham in optics; or Omar Khayyám in mathematics. Indeed, not a few works of such geniuses seem to have vanished in the original Arabic and to survive only in medieval Latin translations that helped to lay the foundations for later Western developments. Today, around the globe, all significant science is Western in style and method, whatever the pigmentation or language of the scientists.

A second pair of facts is less well recognized because they result from quite recent historical scholarship. The leadership of the West, both in technology and in science, is far older than the so-called Scientific Revolution of the 17th century or the so-called Industrial Revolution of the 18th century. These terms are in fact outmoded and obscure the true nature of what they try to describe—significant stages in two long and separate developments. By A.D. 1000 at the latest—and perhaps, feebly, as much as 200 years earlier—the West began to apply water power to industrial processes other than milling grain. This was followed in the late 12th century by the harnessing of wind power. From simple beginnings, but with remarkable consistency of style, the West rapidly expanded its skills in the development of power machinery, labor-saving devices, and automation. Those who doubt should contemplate that most monumental achievement in the history of automation: the weight-driven mechanical clock, which appeared in two forms in the early 14th century. Not in craftsmanship but in basic technological capacity, the Latin West of the later Middle Ages far outstripped its elaborate, sophisticated, and esthetically magnificent sister cultures, Byzantium and Islam. In 1444 a great Greek ecclesiastic, Bessarion, who had gone to Italy, wrote a letter to a prince in Greece. He is amazed by the superiority of Western ships, arms, textiles, glass. But above all he is astonished by the spectacle of waterwheels sawing timbers and pumping the bellows of blast furnaces. Clearly, he had seen nothing of the sort in the Near East.

By the end of the 15th century the technological superiority of Europe was such that its small, mutually hostile nations could spill out over all the rest of the world, conquering, looting, and colonizing. The symbol of this technological superiority is the fact that Portugal, one of the weakest states of the Occident, was able to become, and to remain for a century, mistress of the East Indies. And we must remember that the technology of Vasco da Gama and Albuquerque was built by pure empiricism, drawing remarkably little support or inspiration from science.

In the present-day vernacular understanding, modern science is supposed to have begun in 1543, when both Copernicus and Vesalius published their great works. It is no derogation of their accomplishments, however, to point out that such structures as the *Fabrica* and the *De revolutionibus* do not appear overnight. The distinctive Western tradition of science, in fact, began in the late 11th century with a massive movement of translation of Arabic and Greek scientific works into Latin. A few notable books —Theophrastus, for example—escaped the West's avid new appetite for science, but within less than 200 years effectively the entire corpus of Greek and Muslim science was available in Latin, and was being eagerly read and criticized in the new European universities. Out of criticism arose new observation, speculation, and increasing distrust of ancient authorities. By the late 13th century Europe had seized global scientific leadership from the faltering hands of Islam. It would be as absurd to deny the profound originality of Newton, Galileo, or Copernicus as to deny that of the 14th century scholastic scientists like Buridan or Oresme on whose work they built. Before the 11th century, science scarcely existed in the Latin West, even in Roman times. From the 11th century onward, the scientific sector of Occidental culture has increased in a steady crescendo.

Since both our technological and our scientific movements got their start acquired their character, and achieved world dominance in the Middle Ages, it would seem that we cannot understand their nature or their present impact upon ecology without examining fundamental medieval assumptions and developments.

Medieval View of Man and Nature

Until recently, agriculture has been the chief occupation even in "advanced" societies; hence, any change in methods of tillage has much importance. Early plows, drawn by two oxen, did not normally turn the sod but merely scratched it. Thus, cross-plowing was needed and fields tended to be squarish. In the fairly light soils and semiarid climates of the Near East and Mediterranean, this worked well. But such a plow was inappropriate to the wet climate and often sticky soils of northern Europe. By the latter part of the 7th century after Christ, however, following obscure beginnings, certain northern peasants were using an entirely new kind of plow, equipped with a vertical knife to cut the line of the furrow, a horizontal share to slice under the sod, and a moldboard to turn it over. The friction of this plow with the soil was so great that it normally required not two but eight oxen. It attacked the land with such violence that cross-plowing was not needed, and fields tended to be shaped in long strips.

In the days of the scratch-plow, fields were distributed generally in units capable of supporting a single family. Subsistence farming was the presupposition. But no peasant owned eight oxen: to use the new and more efficient plow, peasants pooled their oxen to form large plow-teams, originally receiving (it would appear) plowed strips in proportion to their contribution. Thus, distribution of land was based no longer on the needs of a family but, rather, on the capacity of a power machine to till the earth. Man's relation to the soil was profoundly changed. Formerly man had been part of nature; now he was the exploiter of nature. Nowhere else in the world did farmers develop any analogous agricultural implement. Is it coincidence that modern technology, with its ruthlessness toward nature, has so largely been produced by descendants of these peasants of northern Europe?

This same exploitive attitude appears slightly before A.D. 830 in Western illustrated calendars. In older calendars the months were shown as passive personifications. The new Frankish calendars, which set the style for the Middle Ages, are very different: they show men coercing the world around them—plowing, harvesting, chopping trees, butchering pigs. Man and nature are two things, and man is master.

These novelties seem to be in harmony with larger intellectual patterns. What people do about their ecology depends on what they think about themselves in relation to things around them. Human ecology is deeply conditioned by beliefs about our nature and destiny—that is, by religion. To Western eyes this is very evident in, say, India or Ceylon. It is equally true of ourselves and of our medieval ancestors.

The victory of Christianity over paganism was the greatest psychic revolution in the history of our culture. It has become fashionable today to say that, for better or worse, we live in "the post-Christian age." Certainly the forms of our thinking and language have largely ceased to be Christian, but to my eye the substance often remains amazingly akin to that of the past. Our daily habits of action, for example, are dominated by an implicit faith in perpetual progress which was unknown either to Greco-Roman antiquity or to the Orient. It is rooted in, and is indefensible apart from, Judeo-Christian teleology. The fact that Communists share it merely helps to show what can be demonstrated on many other grounds: that Marxism, like Islam, is a Judeo-Christian heresy. We continue today to live, as we have lived for about 1700 years, very largely in a context of Christian axioms.

What did Christianity tell people about their relations with the environment?

While many of the world's mythologies provide stories of creation, Greco-Roman mythology was singularly incoherent in this respect. Like Aristotle, the intellectuals of the ancient West denied that the visible world had had a beginning. Indeed, the idea of a beginning was impossible in the framework of their cyclical notion of time. In sharp contrast, Christianity inherited from Judaism not only a concept of time as nonrepetitive and linear but also a striking story of creation. By gradual stages a loving and all-powerful God had created light and darkness, the heavenly bodies, the earth and all its plants, animals, birds, and fishes. Finally, God had created Adam and, as an afterthought, Eve to keep man from being lonely. Man named all the animals, thus establishing his dominance over them. God planned all of this explicitly for man's benefit and rule: no item in the physical creation had any purpose save to serve man's purposes. And, although man's body is made of clay; he is not simply part of nature: he is made in God's image.

Especially in its Western form, Christianity is the most anthropocentric religion the world has seen. As early as the 2nd century both Tertullian and Saint Irenaeus of Lyons were insisting that when God shaped Adam he was foreshadowing the image of the incarnate Christ, the Second Adam. Man shares, in great measure, God's transcendence of nature. Christianity, in absolute contrast to ancient paganism and Asia's religions (except, perhaps, Zoroastrianism), not only established a dualism of man and nature but also insisted that it is God's will that man exploit nature for his proper ends.

At the level of the common people this worked out in an interesting way. In Antiquity every tree, every spring, every stream, every hill had its own *genius loci*, its guardian spirit. These spirits were accessible to men, but were very unlike men; centaurs, fauns, and mermaids show their ambivalence. Before one cut a tree, mined a mountain, or dammed a brook, it was important to placate the spirit in charge of that particular situation, and to keep it placated. By destroying pagan animism, Christianity made it possible to exploit nature in a mood of indifference to the feelings of natural objects.

It is often said that for animism the Church substituted the cult of saints. True; but the cult of saints is functionally quite different from animism. The saint is not *in* natural objects; he may have special shrines, but his citizenship is in heaven. Moreover, a saint is entirely a man; he can be approached in human terms. In addition to saints, Christianity of course also had angels and demons inherited from Judaism and perhaps, at one remove, from Zoroastrianism. But these were all as mobile as the saints themselves. The spirits *in* natural objects, which formerly had protected nature from man, evaporated. Man's effective monopoly on spirit in this world was confirmed, and the old inhibitions to the exploitation of nature crumbled.

When one speaks in such sweeping terms, a note of caution is in order. Christianity is a complex faith, and its consequences differ in differing contexts. What I have said may well apply to the medieval West, where in fact technology made spectacular advances. But the Greek East, a highly civilized realm of equal Christian devotion, seems to have produced no marked technological innovation after the late 7th century, when Greek fire was invented. The key to the contrast may perhaps be found in a difference in the tonality of piety and thought which students of comparative theology find between the Greek and the Latin Churches. The Greeks believed that sin was intellectual blindness, and that salvation was found in illumination, orthodoxy—that is, clear thinking. The Latins, on the other hand, felt that sin was moral evil, and that salvation was to be found in right conduct. Eastern theology has been intellectualist. Western theology has been voluntarist. The Greek saint contemplates; the Western saint acts. The implications of Christianity for the conquest of nature would emerge more easily in the Western atmosphere.

The Christian dogma of creation, which is found in the first clause of all the Creeds, has another meaning for our comprehension of today's ecologic crisis. By revelation, God had given man the Bible, the Book of Scripture. But since God had made nature, nature also must reveal the divine mentality. The religious study of nature for the better understanding of God was known as natural theology. In the early Church, and always in the Greek East, nature was conceived primarily as a symbolic system through which God speaks to men: the ant is a sermon to sluggards; rising flames are the symbol of the soul's aspiration. This view of nature was essentially artistic rather than scientific. While Byzantium preserved and copied great numbers of ancient Greek scientific texts, science as we conceive it could scarcely flourish in such an ambience.

However, in the Latin West by the early 13th century natural theology was following a very different bent. It was ceasing to be the decoding of the physical symbols of God's communication with man and was becoming the effort to understand God's mind by discovering how his creation operates. The rainbow was no longer simply a symbol of hope first sent to Noah after the Deluge: Robert Grosseteste, Friar Roger Bacon, and Theodoric of Freiberg produced startlingly sophisticated work on the optics of the rainbow, but they did it as a venture in religious understanding. From the 13th century onward, up to and including Leibnitz and Newton, every major scientist, in effect, explained his motivations in religious terms. Indeed, if Galileo had not been so expert an amateur theologian he would have got into far less trouble: the professionals resented his intrusion. And Newton seems to have regarded himself more as a theologian than as a scientist. It was not until the late 18th century that the hypothesis of God became unnecessary to many scientists.

It is often hard for the historian to judge, when men explain why they are doing what they want to do, whether they are offering real reasons or merely culturally acceptable reasons. The consistency with which scientists during the long formative centuries of Western science said that the task and the reward of the scientist was "to think God's thoughts after him" leads one to believe that this was their real motivation. If so, then modern Western science was cast in a matrix of Christian theology. The dynamism of religious devotion, shaped by the Judeo-Christian dogma of creation, gave it impetus.

An Alternative Christian View

We would seem to be headed toward conclusions unpalatable to many Christians. Since both *science* and *technology* are blessed words in our contemporary vocabulary, some may be happy at the notions, first, that, viewed historically, modern science is an extrapolation of natural theology and, second, that modern technology is at least partly to be explained as an Occidental, voluntarist realization of the Christian dogma of man's transcendence of, and rightful mastery over, nature. But, as we now recognize, somewhat over a century ago science and technology—hitherto quite separate activities—joined to give mankind powers which, to judge by many of the ecologic effects, are out of control. If so, Christianity bears a huge burden of guilt.

I personally doubt that disastrous ecologic backlash can be avoided simply by applying to our problems more science and more technology. Our science and technology have grown out of Christian attitudes toward man's relation to nature which are almost universally held not only by Christians and neo-Christians but also by those who fondly regard themselves as post-Christians. Despite Copernicus, all the cosmos rotates around our little globe. Despite Darwin, we are *not*, in our hearts, part of the natural process. We are superior to nature, contemptuous of it, willing to use it for our slightest whim. The newly elected Governor of California, like myself a churchman but less troubled than I, spoke for the Christian tradition when he said (as is alleged), "When you've seen one redwood tree, you've seen them all." To a Christian a tree can be no more than a physical fact. The whole concept of the sacred grove is alien to Christianity and to the ethos of the West. For nearly 2 millennia Christian missionaries have been chopping down sacred groves, which are idolatrous because they assume spirit in nature.

What we do about ecology depends on our ideas of the man-nature relationship. More science and more technology are not going to get us out of the present ecologic crisis until we find a new religion, or rethink our old one. The beatniks, who are the basic revolutionaries of our time, show a sound instinct in their affinity for Zen Buddhism, which conceives of the man-nature relationship as very nearly the mirror image of the Christian view. Zen, however, is as deeply conditioned by Asian history as Christianity is by the experience of the West, and I am dubious of its viability among us.

Possibly we should ponder the greatest radical in Christian history since Christ: Saint Francis of Assisi. The prime miracle of Saint Francis is the fact that he did not end at the stake, as many of his left-wing followers did. He was so clearly heretical that a General of the Franciscan Order, Saint Bonaventura, a great and perceptive Christian, tried to suppress the early accounts of Franciscanism. The key to an understanding of Francis is his belief in the virtue of humility—not merely for the individual but for man as a species. Francis tried to depose man from his monarchy over creation and set up a democracy of all God's creatures. With him the ant is no longer simply a homily for the lazy, flames a sign of the thrust of the soul toward union with God; now they are Brother Ant and Sister Fire, praising the Creator in their own ways as Brother Man does in his.

Later commentators have said that Francis preached to the birds as a rebuke to men who would not listen. The records do not read so: he urged the little birds to praise God, and in spiritual ecstasy they flapped their wings and chirped rejoicing. Legends of saints, especially the Irish saints, had long told of their dealings with animals but always, I believe, to show their human dominance over creatures. With Francis it is different. The land around Gubbio in the Apennines was being ravaged by a fierce wolf. Saint Francis, says the legend, talked to the wolf and persuaded him of the error of his ways. The wolf repented, died in the odor of sanctity, and was buried in consecrated ground.

What Sir Steven Ruciman calls "the Franciscan doctrine of the animal soul" was quickly stamped out. Quite possibly it was in part inspired, consciously or unconsciously, by the belief in reincarnation held by the Cathar heretics who at that time teemed in Italy and southern France, and who presumably had got it originally from India. It is significant that at just the same moment, about 1200, traces of metempsychosis are found also in western Judaism, in the Provençal *Cabbala*. But Francis held neither to transmigration of souls nor to pantheism. His view of nature and of man rested on a unique sort of pan-psychism of all things animate and inanimate, designed for the glorification of their transcendent Creator, who, in the ultimate gesture of cosmic humility, assumed flesh, lay helpless in a manger, and hung dying on a scaffold.

I am not suggesting that many contemporary Americans who are concerned about our ecologic crisis will be either able or willing to counsel with wolves or exhort birds. However, the present increasing disruption of the global environment is the product of a dynamic technology and science which were originating in the Western medieval world against which Saint Francis was rebelling in so original a way. Their growth cannot be understood historically apart from distinctive attitudes toward nature which are deeply grounded in Christian dogma. The fact that most people do not think of these attitudes as Christian is irrelevant. No new set of basic values has been accepted in our society to displace those of Christianity. Hence we shall continue to have a worsening ecologic crisis until we reject the Christian axiom that nature has no reason for existence save to serve man.

The greatest spiritual revolutionary in Western history, Saint Francis, proposed what he thought was an alternative Christian view of nature and man's relation to it: he tried to substitute the idea of the equality of all creatures, including man, for the idea of man's limitless rule of creation. He failed. Both our present science and our present technology are so tinctured with orthodox Christian arrogance toward nature that no solution for our ecologic crisis can be expected from them alone. Since the roots of our trouble are so largely religious, the remedy must also be essentially religious, whether we call it that or not. We must rethink and refeel our nature and destiny. The profoundly religious, but heretical, sense of the primitive Franciscans for the spiritual autonomy of all parts of nature may point a direction. I propose Francis as a patron saint for ecologists.

ECONOMICS OF ECOLOGY

If the ecology of the world is to include man, then it must include the systems of man such as economics. Perhaps most of the uses of environmental improvement or control can be stated in economic terms: how much are taxpayers willing to pay for cleaner air, unscathed land and purer water? Of course, economics is the study of the production, distribution and consumption of resources—especially the scarce resources. For example, with respect to the urban air, clean air is a question of what the members of the urban community will expend in capital resources to add pollution-control devices to their power plants and automobiles, and to purchase more costly fuel which contains fewer pollutants. The total cost of pollution control in the United States for the period 1971–1980 is estimated to be $287.1 billion, as shown in Figure 13–4. These estimates are by the United States Environmental Protection Agency.[3]

These costs can be seen as debits incurred over a hundred and fifty years of use of the American soil, air and water. Perhaps it is the present generation that will pay the cost of previous "free" use of the natural resources of America as well as paying the cost of maintaining the quality of our natural habitat with our increasing use and consumption.

POLITICAL RESPONSE TO ECOLOGICAL MOVEMENTS

Of course, not all proposals for environmental control are met with acceptance or pleasure. Significant debate on the wisdom of the ecological movement has ensued in the past few years. For example, an initiative from a group called The People's Lobby, which collected more than 325,000 signatures of registered voters, was placed on the California primary ballot of June 1972. That environmental intiative read: "ENVIRONMENT. Initiative. Specifies permissible composition and quality of gasoline and other fuel for internal combustion engines. Authorizes shutting down of businesses and factories violating air pollution standards. Imposes restrictions on leasing and extraction of oil and gas from tidelands or submerged lands, or onshore areas within one mile of mean high tide line. Prohibits construction of atomic powered electric generating plants for five years. Establishes restrictions on manufacture, sale, and

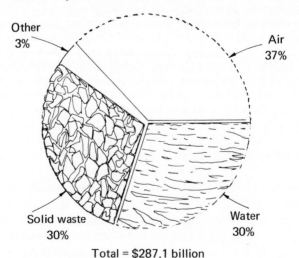

Total = $287.1 billion

FIGURE 13-4   Total cumulative environmental expenditures by category, 1971-1980.

use of pesticides. Prohibits enforcement officials from having conflicting interests. Provides for relief by injunction and mandate to prevent violations. Imposes penal sanctions and civil penalties."

This initiative was defeated by the voters of California.

Some observers believe we may experience an excess of new pollution controls by 1980 in the United States.[4] Others have noted that the debate over the proper balance of environmental measures will result in a backlash against environmental groups and an eventual working-out of compromises that will provide opportunity for economic growth along with reasonable environmental safeguards.[5]

With the potential for reasoned and planned compromise in environmental controls, man may achieve "a bountiful economy and a beautiful environment" in the words of Athelstan Spilhaus. Mr. Spilhaus discusses the possibility of an *ecolibrium*—a balance in our earthly home.[7]

Spilhaus believes that the proper uses of the materials of nature are for man's needs, but the resources must be renewed. Therefore, if air or water are to be used in industrial processes, the processes must include a renewal step in which the air or water is returned to its original state. The price of resource-renewal must be included in the cost of the product, and not in some general tax.

Spilhaus also believes that a permanent United States Planning Board could serve to formulate long-range directions for our economy so that a desirable ecological equilibrium could be maintained.

COASTLINES AS EXAMPLES

Parts of our ecology to which we have not paid great attention are the coastlines. They are an area, bordering our land and oceans, which require new understanding, new technology and new planning for a balanced ecology—an ecolibrium.[8] Although the coastal regions comprise only 5% of the area of the world, about two-thirds of the world's population lives near some coast. The fragile beaches and coastline therefore receive a large proportion of the wastes of civilization. The uses of beaches and coastal zones for resources such as building sand is also significant. For example, solid wastes dumped into United States coastal waters each year weigh 500,000 tons; the amount of sewage emptied into the coastal zone of southern California is 1.1 billion gallons *each day*.[8] There are 500,000 pleasure craft registered in California, primarily using the coast. Major oil spills from tankers and ships are having their effect on the coast. Drilling accidents can also occur. A photo of the oil leak off Santa Barbara, California is shown in Figure 13–5. This oil flowed from a break in the oil pipe connected to the floor of the ocean off the coast at Santa Barbara. It is imperative that new technologies be developed for the preservation of the beaches and harbors as well as for the creation of new beaches and shoreline structures in harmony with the ecology of the coast. As Inmen and Brush state, in the case of the coast zones,

"An informed and concerned public reaction of hitherto unknown magnitude may possibly lead to solutions for the problem of slowing and redirecting the machine, without wrecking the machinery: that is the coastal challenge."[8]

INTERNATIONAL ECOLOGICAL RESPONSIBILITIES

As in the case of our coastal challenge, our environmental problems are worldwide. One of the answers to global problems is interpreted global analysis.[9] A United Nations Conference on the Human Environment was held in Stockholm in June 1972 for the purpose of developing an international committment to environmental action. Principle 21 of the Stockholm Conference Declaration affirms, "States have ... the responsibility to ensure that activities

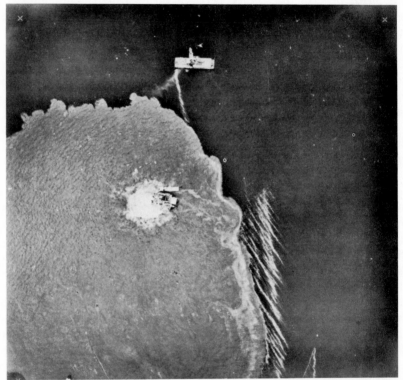

FIGURE 13-5 Oil slick spreading from a leaking oil well near Santa Barbara, California, 1969. This photograph was taken from 1,800 feet. At the top of the photograph is a barge which put down a pipe to the well to try to reduce the outpouring of oil. Thousands of gallons of oil covered the channel and floated to nearby shores and polluted them before the leakage stopped. The light gray area is oil slick. Courtesy United States Environmental Protection Agency.

within their jurisdiction or control do not cause damage to the environment of other States or of areas beyond the limits of national jurisdiction."

This common view is necessary because many countries use common resources. For example, France and Italy both pollute the Mediterranean Sea. Also, the damming of the Rio de la Plata should require a common agreement between Argentina and Brazil. At present the primary environmental problems relate to the polluting of the air of a neighboring nation, polluting the waters of a downstream neighbor and weather modification of neighboring nations. The application of mutually beneficial controls among nations is a task immediately before us.

The United States and Russia have established an agreement on cooperation in the field of environmental protection. The two countries are particularly interested in discussing the role of domestic laws and regulations in environmental protection.[10] Problems associated with natural resource utilization in the Soviet Union suggest many parallels with the experience of the United States, despite fundamental economic and political differences.[11,12] It appears that economic expediency and unexamined technological development is as prevalent in the U.S.S.R. as in Western countries. The primacy of increased production in the U.S.S.R. has overshadowed the environmental goals. In recent years, however, Moscow's air is clean because some 200 industries have been moved outside its city limits. Another city of 800,000 on the Volga River will completely purify its waste water and recirculate it within the city by 1980.[16]

In a recent book, *A God Within*, Professor René Dubos writes about the interplay between man and nature.[13] Man's creative response to this situation is a process of interdependence between the human spirit and nature, between culture and environment. The ecological crisis, in Dubos' view, came about because of failure to recognize this interdependence. The "god within" is those hidden forces of spontaneity and freedom, values and fulfillment. Dubos is optimistic about the uses of technology while he is also critical of the exploitive attitude toward nature. In contrast to Professor White's article, Dubos recommends stewardship based on the principles of St. Benedict, rather than St. Francis.[14] While St. Francis of Assisi preached and practiced absolute identification with nature, St. Benedict of Russia established a monastery in the sixth century on the basis that "to labor is to pray." The Benedictines acquired both practical and theoretical skills and created an atmosphere where technology could flourish. They developed windmills as sources of power, and they manufactured goods. While the first chapter of Genesis speaks of man's domination over nature, the second chapter speaks of man as steward in the Garden. Man as steward in wise management of the earth is a principle entirely compatible with the maintenance of environmental quality.

STEWARDSHIP

In the article "Humanizing the Earth," which follows, Dubos further explores the human use of technology and natural resources and their relationship to ecological health. In his view, nature does not always provide the best ecological equilibrium, but can do so in partnership with man.[15]*

Humanizing the Earth

René J. Dubos

How gray and drab, unappealing and unsignificant, our planet would be without the radiance of life. If it were not covered with living organisms the surface of the earth would resemble that of the moon. Its colorful and diversified appearance is largely the creation of microbes, plants, and animals which endlessly transform its inanimate rocks and gases into an immense variety of organic substances. Man augments still further this diversification by altering the physical characteristics of the land, changing the distribution of living things, and adding human order and fantasy to the ecological determinism of nature.

Many of man's interventions into nature have, of course, been catastrophic. History is replete with ecological disasters caused by agricultural and industrial mismanagement. The countries which were most flourishing in antiquity are now among the poorest in the world. Some of their most famous cities have been abandoned; lands which were once fertile are now barren deserts.

Disease, warfare, and civil strife have certainly played important roles in the collapse of ancient civilizations; but the primary cause was probably the damage caused to the quality of the soil and to water supplies by poor ecological practices. Similarly today, the environment is being spoiled in many parts of the world by agricultural misuse or overuse, by industrial poisoning, and of course by wars.

The primary purpose of the recent United Nations Conference on the Human Environment, held in Stockholm in June 1972, was to formulate global approaches to the correction and prevention of the environmental defects resulting from man's mismanagement of the earth. I shall not discuss the technical aspects of these problems, but rather shall try to look beyond them and present facts suggesting that man can actually improve on nature. In my opinion, the human use of natural resources and of technology is compatible with ecological health, and can indeed bring out potentialities of the earth which remain unexpressed in the state of wilderness.

The disastrous ecological consequences of many past and present human activities point to the need for greater knowledge and respect of natural laws. This view is succinctly expressed by Barry Commoner in his fourth law of ecology: "Nature knows best." I shall first discuss the limitations of this law.

When left undisturbed, all environments tend toward an equilibrium state, called the climax or mature state by ecologists. Under equilibrium conditions, the wastes of nature are constantly being recycled in the ecosystem, which becomes thereby more or less self-perpetuating. In a natural forest, for example, acorns fall to the ground and are eaten by squirrels, which in turn may be eaten by foxes or other predators; the dead leaves and branches, the excrements of animals, are utilized by microbes, which return their constituents to the soil in the form of humus and mineral nutrients. More vegetation grows out of the recycled materials, thus assuring the maintenance of the ecosystem.

When applied to such equilibrated systems, the phrase "Nature knows best" is justified, but is in fact little more than a tautology. As used in this phrase, the word *nature* simply denotes a state of affairs spontaneously brought about by evolutionary adaptation resulting from feedbacks which generate a coherent system. There are no problems in undisturbed nature; there are only solutions, precisely because the equilibrium state is an adaptive state. But in a given area, there is usually more than one possible equilibrium state, and there is no evidence that the *natural* solution is necessarily the best or the most interesting solution. In fact, it is likely, as I shall illustrate later, that the symbiotic interplay between man and nature can generate ecosystems more diversified and more interesting than those occurring in the state of wilderness.

What is surprising is not that natural environments are self-sustaining and generally appear efficient but, rather, that many of them constitute clumsy solutions to ecological problems. Many of these solutions appear inadequate, even where nature has not been disturbed by man or by cataclysms, and therefore could have been expected to reach the optimum ecological state.

That the wisdom of nature is often shortsighted is illustrated by the many disasters that repeatedly affect plants and animals in their undisturbed native habitats. The repeated population crashes among animal species such as lemmings, muskrats, or rabbits result from the defectiveness in the natural mechanisms which control population size. These crashes unquestionably constitute traumatic experiences for the animals, as indicated by the intense behavioral disturbances which often occur among them long before death. The crashes constitute, at best, clumsy ways of reestablishing an equilibrium between population size and local resources, Judging from the point of view of lemmings, muskrats, and rabbits—let alone human beings—only the most starry-eyed Panglossian optimist could claim that nature knows best how to achieve population control.

Most surprising is the fact that, even without environmental changes caused by human interference or accidental cataclysms, nature fails in many cases to complete the recycling processes that are considered the earmarks of ecological equilibrium. Examples of such failures are the accumulation of peat, coal, oil, shale, and other deposits of organic origin. These materials are largely derived from the bodies of plants and other living things that have become chemically stabilized after undergoing only partial decomposition. The very fact that they have accumulated in fantastic amounts implies that they have not been recycled. Paradoxically, man helps somewhat in the completion of the cycle when he burns peat, coal, or oil, because he thereby makes the carbon and minerals of these fuels once more available for plant growth. The trouble with this form of recycling is that the breakdown products of the fuels are so rapidly put back into circulation through air, water, and soil that they overload contemporary ecological systems.

The accumulation of guano provides another example of recycling failure on the part of nature. This material, now used as a fertilizer, consists of the excrements deposited by birds on certain islands and cliffs. For example, millions of seabirds use the Chincha Islands off the coast of Peru as a resting place and breeding ground; their droppings, accumulated through centuries or perhaps millennia, have formed layers of guano from 60 to 100 feet (18 to 30 meters) in thickness. Guano, being rich in nitrogen, phosphate, and potash, constitutes an ideal fertilizer, and its accumulation therefore represents a spectacular example of recycling failure. Here again, man completes the recycling process by collecting guano and transporting it to agricultural fields where it reenters the biological cycle in the form of plant nutrient.

Just as it is erroneous to claim that nature has no waste, so it is erroneous to claim that it has no junkyards. The science of paleontology is built from the wastes and artifacts casually abandoned by primitive man. Admittedly, the accumulation of solid wastes in technological societies is evidence of a massive failure of recycling for which man is responsible. But this ecological failure is the expression of behavioral characteristics that have always existed in human nature. Like the great apes, man in his primative state was wasteful and careless of his wastes, and we have remained so throughout history.

The solid waste problem has become grave in our times because we produce more wastes than in the past, and what we reject is commonly of a chemical composition not found in natural ecosystems. Nature does not know how to deal with situations that have no precedents in the evolutionary past. The solution to the problem of solid wastes, therefore, cannot be found in the ways of nature. It requires new technological methods and changes in the innate (natural!) behavior of man.

Hailstorms, droughts, hurricanes, earthquakes, and volcanic eruptions are common enough to make it obvious that the natural world is not the best possible world; man is not responsible for these disasters, but he suffers from them as do other living things. Of greater interest, perhaps, is the fact that nature is incapable, by itself, of fully expressing the diversified potentialities of the earth. Many richnesses of nature are brought to light only in the regions that have been humanized: agricultural lands, gardens, and parks have to be created and maintained by human toil.

Until man intervened, much of the earth was covered with forests and marshes. There was grandeur in this seemingly endless green mantle, but it was a monotonous grandeur chiefly derived from immensity and uniformity. The primeval forest almost concealed the underlying diversity of the earth. This diversity was revealed by man in the process of producing food and creating his civilizations. Since an extensive analysis of the creative transformations of the earth by man would be impossible here, I shall illustrate it with one single example, namely, that of the part of France where I was born and raised.

Before human occupation, the Île de France was a land without any notable characteristics. The hills have such low profiles that they would be of little interest without the venerable churches and clusters of houses that crown their summits. The rivers are sluggish and the ponds muddy, but their banks have been adapted to human use and their names have been celebrated so often in literature that they evoke the enchantment of peaceful rural scenes. The sky is rarely spectacular, but painters have created a rich spectrum of visual and emotional experiences from its soft luminosity.

Ever since the primeval forest was cleared by Neolithic settlers and medieval farmers, the province of the Île de France has retained a humanized charm that transcends its natural endowments. To this day, its land has remained very fertile, even though part of it has been in continuous use for more than 2000 years. Far from being exhausted by intensive agriculture over such long periods of time, the land still supports a large population and a great variety of human settlements.

What I have just stated about the Île de France is, of course, applicable to many other parts of the world. Ever since the beginning of the agricultural revolution during the Neolithic period, settlers and farmers have been engaged all over the world in a transformation of the wilderness. Their prodigious labors have progressively generated an astonishing diversity of man-made environments, which have constituted the settings for most of human life. A typical landscape consists of forested mountains and hills serving as a backdrop for pastures and arable lands, villages with their greens, their dwellings, their houses of worship, and their public buildings. People now refer to such a humanized landscape as "nature," even though most of its vegetation has been introduced by man and its environmental quality can be maintained only by individualized ecological management.

Just as nature has not been capable by itself of giving full expression to the potential diversity of our globe, likewise it is not capable of maintaining man-made environments in a healthy state. Now that so much of the world has been humanized, environmental health depends to a very large extent on human care. Forests must be managed, swampy areas which are under cultivation must continually be drained, the productivity of farmlands must be maintained by crop rotation, irrigation, fertilization, and destruction of weeds. From historical times, the Campagna Romana has been infested with mosquitoes and devastated by malaria every time men have lacked the stamina to control its marshes. Similarly, farmlands that have been economically projuctive and esthetically attractive for a thousand years are invaded by brush and weeds as soon as farmers neglect to cultivate them. The rapid degradation of abandoned gardens, farmlands, and pastures is evidence that humanized nature cannot long retain its quality without human care.

It is true that many ancient civilizations have ruined their environment and that a similar process is going on now in certain highly industrialized areas, but this is not inevitable. Intensive agriculture has been practiced for a thousand years in certain lands without decreasing their fertility or ruining their scenery. Man can create artificial environments from the wilderness and manage them in such a manner that they long remain ecologically stable, economically profitable, esthetically rewarding, and suited to his physical and mental health. The immense duration of certain man-made landscapes contributes a peculiar sense of tranquility to many parts of the Old World; it inspires confidence that mankind can act as steward of the earth for the sake of the future.

Lands could not remain fertile under intense cultivation unless managed according to sound ecological principles. In the past, these principles emerged empirically from practices that assured the maintenance of fairly high levels of humus in the soil. But scientific knowledge of soil composition and texture, of plant physiology, and of animal husbandry is providing a new basis for agricultural management. During the past century, the sound empirical practices of the past have been progressively replaced by more scientific ones, which include the use of artificial fertilizers and pesticides. Scientific agriculture has thus achieved enormous yields of plant and animal products. Furthermore, experimental studies have revealed that many types of lands can remain fertile for long periods of time without organic manure, provided that they are continuously enriched with chemical fertilizers in amounts and compositions scientifically determined.

Efficiency, however, cannot be measured only in terms of agricultural yields. Another criterion is the amount of energy (measured in calories) required for the production of a given amount of food. And when scientific agriculture is judged on this basis its efficiency is often found to be very low. Paradoxical as this may sound, there are many situations in which the modern farmer spends more industrial calories than the food calories he recovers in the form of food. His caloric expenditure consists chiefly of gasoline for powering his equipment and of electricity for producing chemical fertilizers and pesticides—let alone the caloric input required to irrigate the land and to manufacture tractors, trucks, and the multifarious kinds of machines used in modern farming.

Needless to say, modern civilizations would be inconceivable if the energy (calories) now required by agriculture had to come from human muscles instead of from gasoline and electricity. But it is a fact, nevertheless, that if fossil fuels were to remain the most important source of power, the sheer size of the world population would make it impossible to continue for long the energy deficit spending on which agriculture depends in prosperous industrialized countries. And there would be no hope of extending these modern agricultural practices to the developing countries, which constitute the largest part of the world.

No matter how the situation is rationalized, the present practices of scientific agriculture are possible only as long as cheap sources of energy are available. After the world supplies of fossil fuels have been exhausted, the modern farmer, like the modern technologist, will become ineffective unless energy derived from nuclear reactions, geothermal sources, or solar radiation can be supplied in immense amounts at low cost. Thus, the future of land management is intimately bound to the development of new sources of energy, as are all other aspects of human life.

Of the 70 to 100 billion people who have walked the surface of the earth since *Homo sapiens* acquired his biological identity, by very far the largest percentage have lived on the man-made lands that have been created since the agricultural revolution.

In every part of the world, the interplay between man and nature has commonly taken the form of a true symbiosis, namely, a biological relationship which alters somewhat the two components of the symbiotic system in a way that is beneficial to both. Such transformations, achieved through symbiosis, account in large part for the immense diversity of places on earth and for the fitness between man and environment so commonly observed in areas that have been settled and have remained stable for long periods of time.

Furthermore, the reciprocal transformations of man and environment have generated various new situations, each with its own human and environmental characteristics. For example, the agricultural techniques, social policies, and behavioral patterns in the various islands of the South Pacific are determined not only by geologic and climatic factors but even more by the cultural attitudes of the early settlers—Polynesians. Melanesians, or Indonesians—and then later of the Western and Oriental people who colonized the islands. Cultural attitudes, more than natural conditions, are responsible for the profound differences between Fiji, Tahiti, and the Hawaiian islands. The Pacific Islands were initially settled by different groups of people and, in addition to these early human influences, they exhibit today the more recent influences respectively of their English, French, or American colonizers.

The shaping of nature by culture can be recognized in many other parts of the world. As the process of humanization of the earth continues, however, it will increasingly be influenced by the fact that most of the globe will soon be completely occupied and utilized. This colonization process began, of course, long before the days of modern technology. But the difference is that men now occupy and utilize all land areas except those that are too cold, too hot, too dry, too wet, too inaccessible, or at too high an altitude for prolonged human habitation.

According to the United Nation's Food and Agricultural Organization (FAO), practically all the best lands are already farmed; future agricultural developments are more likely to result from intensification of management than from expansion into marginal lands. There probably will be some increase in forest utilization; but, otherwise, land use will soon be stabilized. In fact, expansion into new lands has already come to an end in most developed countries and is likely to be completed within a very few decades in the rest of the world. A recent FAO report states the probable final date as 1985.

The U.N. Conference on the Human Environment came therefore at a critical time in man's history. Now that the whole earth has been explored and occupied, the new problem is to manage its resources. Careful management need not mean stagnation. In many places, as already mentioned, the interplay between man and nature results in creative symbiotic relationships that facilitate evolutionary changes. Man continuously tries to derive from nature new satisfactions that go beyond his elementary biological needs—and he thereby gives expression to some of nature's potentialities that would remain unrecognized without his efforts.

Man has now succeeded in humanizing most of the earth's surface but, paradoxically, he is developing simultaneously a cult for wilderness. After having been for so long frightened by the primeval forest, he has come to realize that its eerie light evokes in him a mood of wonder that cannot be experienced in an orchard or a garden. Likewise, he recognizes in the vastness of the ocean and in the endless ebb and flow of its waves a mystic quality not found in humanized environments. His response to the thunderous silence of deep canyons, the solitude of high mountains, the luminosity of the deserts is the expression of an aspect of his fundamental being that is still in resonance with cosmic events.

As was mentioned earlier, nature is not always a good guide for the manipulation of the forces that affect the daily life of man; but undisturbed nature knows best—far better than ordinary human intelligence—how to make men aware of the cosmos and to create an atmosphere of harmony between him and the rest of creation.

Humanizing the earth thus implies much more than transforming the wilderness into agricultural lands, pleasure grounds, and healthy areas suitable for the growth of civilization. It also means preserving the kinds of wilderness where man can experience mysteries transcending his daily life, and also recapture direct awareness of the cosmic forces from which he emerged. It is obvious, however, that man spends his daily life not in the wilderness but in environments that he creates—in a man-made nature. Let me restate in conclusion my belief that by using scientific knowledge and ecological wisdom we can manage the earth so as to create environments which are ecologically stable, economically profitable, esthetically rewarding, and favorable to the continued growth of civilization.

SUMMARY

Man's proper place in the world is as a steward. His technology must merge wisely with nature so that the ecological equilibrium achieved is one that leads to ecological health and well-being. Despite his unique status, man is a part of nature as much as trees and other animals are. The natural world is an interdependent organism in process and actions taken by man must recognize this continuing process and interdependency. Nature is a complex and intricate system of balances in which all living things share and contribute. With wise stewardship of technology, natural resources and the earth man can prosper and survive.

CHAPTER 13 REFERENCES

1. *Great Speeches of Col. R.G. Ingersoll*, Rhodes and McClure, Inc., Chicago, 1885.
2. L. White, Jr., "The Historical Roots of Our Ecologic Crisis," *Science*, March 10, 1967, pp. 1203–1207.

3. *Environmental Quality*, Council on Environmental Quality, Third Annual Report, August 1972, p.278, United States Government Printing Office.
4. J. Diebold, "Industrial Elasticity, Astute Prophecy," *Chemical and Engineering News*, May 21, 1973, p.3.
5. A. L. Oten, "The Pendulum Swings," *Wall Street Journal*, Aug. 2, 1973, p. 12.
6. W. Leiss, *The Domination of Nature*, Braziller, Inc., New York, 1973.
7. A. Spilhaus, "Ecolibrium," *Science*, Feb. 18, 1972, pp. 711–715.
8. D.L. Inman and B.M. Brush, "The Coastal Challenge," *Science*, July 6, 1973, pp. 20–31.
9. L.R. Brown, *World Without Borders*, Random House, New York, 1972.
10. P.W. Quizz, "Environment Opens a Door," *World*, July 17, 1973, p. 36.
11. P.R. Pryde, "The Quest for Environmental Quality in the U.S.S.R.," American Scientist, Dec. 1972, pp. 739–745.
12. P.R. Pryde, *Conservation in the Soviet Union*, Cambridge University Press, New York, 1972.
13. R. Dubos, *A God Within: A Positive Philosophy for a More Complete Fulfillment of Human Potentials*, Scribners and Sons, New York, 1972.
14. R. Dubos, "St. Francis *versus* St. Benedict," Psychology Today, May 1973, pp. 54–60.
15. R. Dubos, "Humanizing the Earth," *Science*, Feb. 23, 1973, pp. 769–772.
16. "Russia on the Environment: Concern," *Technology Review*, July/August 1973, p. 67.
17. D. Brand, "Question of Who Owns Oceans Becomes Vital as Exploitation Grows," *Wall Street Journal*, Sept. 13, 1973, p. 1,12.
18. B. Newman, "Mysterious Nodules at Bottom of Oceans May Yield a Treasure," *Wall Street Journal*, Sept. 21, 1973, p. 1,6.
19. M. Slesser, "Energy Analysis in Policy-Making," *New Scientist*, Nov. 1, 1973, pp. 328–333.
20. J. Maddox, *The Doomsday Syndrome*, McGraw-Hill Book Co., 1972.

CHAPTER 13 EXERCISES

13–1. The resources of the coastal areas of the oceans are vast, and they are rapidly being exploited. The ownership of offshore areas of the oceans is often the result of national territorial claims rather than international law. The concept of the past centuries was that the seas must be free for navigation and shipping. Now laws must be developed for the appropriate use of the seas.

　　The United States has claimed a 12 mile-limit off its shores for fishing restricted to United States boats. Now, however, supertrawlers without onboard factories make possible the harvesting and processing of fish by the thousands of tons. These ships are capable of hauling in 10 tons of fish at a time.[17] Many countries want protected fishing limits to 100 miles off their shores. Develop a set of laws and appropriate controls to be submitted to the United Nations for deliberation and possible acceptance for the appropriate regulation and use of fishing rights in the oceans.

13–2. In 1873 the ship *HMS Challenger* sailed on a three-year, around-the-world scientific voyage. This ship returned with some innocuous rocks from the sea along with the vast scientific materials of Charles Darwin. Those dull rocks turned out to conceal a treasure. The nodules, as they are called, contain large quantities of metals such as Manganese, Copper, Nickel and Cobalt. They are scattered over vast areas of the ocean floor at depths of 12,000 to 20,000 feet. There could be as much as 1.5 trillion tons of nodules in the Pacific alone and they are growing at the rate of 16 million tons per year.[18] As many as 30 companies have already invested up to $300 million to develop technologies to mine the ocean. By 1976, it is estimated that perhaps 10,000 tons of nodules a day would be taken from the ocean floor. The United States relies on imports for much of its Copper, Nickel, Cobalt and Manganese. Thus, deep—sea mining is an important venture.

To whom do these minerals belong? What are the ecological factors to be considered in the mining of the seas? The processing of the nodules will require plants yielding waste. Where should these plants be constructed? On the coastline? What international controls for deep-sea mining should be instituted?

13–3. Average annual rainfall varies from three inches in dry deserts to over 100 inches in other locations. From early times man has tried to increase rainfall in times of drought. Currently silver iodide crystals are used to seed clouds and to try to stimulate rainfall. Nevertheless, while farmers in need of rain welcome the cloud seeding, others believe it is ill advised. Examine the adverse effects on the environment. Also, explore the following question: With the prevailing weather motion west to east in the United States, what happens to precipitation eastward of the seeded area? Have we just moved the location of the rainfall?

13–4. The major source of lead poisoning for most city dwellers comes from the air. The largest contributor to lead in city air is Tetraethyl Lead, used in auto gasoline since 1923. Autos discharge about 500 million pounds of lead into the air each year in the United States. Unleaded or low-lead gas is now regularly available in the United States. Explore the effect of lead on the health of persons breathing city air, and propose steps to ameliorate the problem.

13–5. The catalytic converter is one way that the strongest exhaust-control standards may be met for autos. Yet catalyst controls on autos may increase the gas consumption of new autos. Also, while decreasing some auto emissions, catalysts may promote the formation of other harmful emissions—namely sulphuric acid mist and sulphur particulates. Explore the advantages and disadvantages of catalytic converters for reducing emissions.

13–6. The use of agricultural land for suburban growth has grown rapidly over the past two decades. Also, urban sprawl has added to aesthetic and transportation problems as shown in Figure E13–6. Examine the alternatives to urban sprawl and identify the environmental costs of suburban development.

FIGURE E13–6   A northern California suburb near San Francisco. Courtesy Pacific Resources Incorporated.

13–7. The use of water for industrial and residential purposes has grown steadily over the past 70 years. The United States used 400 billion gallons daily in 1970 and it is expected that we shall be using 700 billion gallons daily by 2000. Our use of water has been growing at a rate of about 5% during this century. The use of water for irrigation, electric power plants, industrial use and the public water supply is critical to the welfare of the country. Discuss what effects this use of water can have on the ecology of the United States.

13–8. The use of materials has grown extensively since 1900 in the United States. Figure E13–8 shows the growth in the use of different materials. Discuss the effects which this rapid growth of the use of materials has had on the ecology of the United States. Note that the use of plastics has grown 10,000 times since 1900.

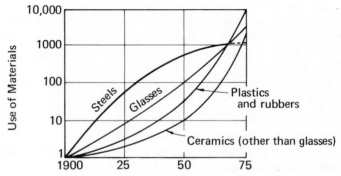

FIGURE E13–8   Growth of use of materials in the United States since 1900.

13-9. The amount of energy expended on a given territory reflects the degree of activity going on there. In Los Angeles it reflects affluence and in New York City it reflects population density. The table below shows the energy density in tons-of-coal equivalent per hectare in 1967 and when the countries might reach an energy density equivalent of 1% of the solar radiation level.[19]

This index of energy density may be considered an index of environmental pressure. Examine the effects of energy density use in countries and regions and the effects on the environment. Calculate, if you can, the energy density for the metropolitan region nearest to you or in which you reside.

Country	Energy Density	Date for 1% of Solar Energy
Japan	5.2	1987
Germany	11.1	1989
France	2.8	1999
United Kingdom	11.9	2012
China	0.325	2029
U.S.S.R.	0.4	2042
U.S.A.	2.0	2054

13-10. In *The Doomsday Syndrome*, John Maddox, editor of *Nature* magazine, criticizes many environmentalists for stating a case for conservation that could lead to stagnation. Mr. Maddox says, "After all, the best way to prevent the accumulation of carbon dioxide in the atmosphere is not to build power stations. The best way to insure against the remote risk that DDT may be damaging to people is not to use DDT. In this sense, the environmental movement tends toward passivity, true conservatism. It is understandable that policies of doing nothing should seem easier than policies which require vigorous and expensive action, but widespread acceptance of what the doomsday men are asking for could so undermine the pattern of modern economic life as to create social stagnation."[20]*

Do you agree that the environmental movement could lead to economic and social stagnation? What counterbalancing effects are present to avoid any tendency to stagnation in our economy?

---

*From *The Doomsday Syndrome* by John Maddox. Copyright 1972 by Maddox Editorial Ltd. Used with permission of McGraw-Hill Book Co. and the author.

14

TECHNOLOGY, AFFLUENCE AND POPULATION

POPULATION, AFFLUENCE AND TECHNOLOGY ITSELF all contribute to the growth of technology and its impact on the environment. Economic growth was the American ethical standard for many for two centuries. Although Thomas Jefferson believed that industrialization was dangerous to democracy and Theodore Roosevelt opposed the greedy interests despoiling the nation, growth has been the prevailing principle. It is the purpose of this chapter to explore the roles of the growth of technology, affluence and population and their joint impact on the environment.

NO-GROWTH

The no-growth attitude may be best exemplified by a statement by Oregon's Governor Tom McCall, who likes to invite tourists to his state but adds, "Just don't come here to live." Many cities and towns are talking about population limits. These attempts are in response to the view that population growth is the primary force of environmental degradation. Or is it technological development? In San Francisco many wish to ban the new skyscrapers being built on the lovely hills overlooking the bay. One proposal in Boulder, Colorado is to rezone the remaining undeveloped industrial land so as to discourage industrial development.

Antigrowth forces are seen by some observers as faintly reminiscent of the border guards that tried to keep the Okies of the 1930s out of California.[7] A citizens' group in California has developed a plan through the sponsorship of the organization California Tomorrow.[1] The California Tomorrow Plan recommends that the state be authorized to establish a capital investment fee of $1,000 that would be charged to each new resident.[7] The California Tomorrow Plan also calls for a policy of zero population growth, the development of more balanced transportation systems, wiser use and re-use of resources and enlightened zoning. Also, the Plan calls for $6,000 minimum income (1972 dollars) for a family of four, linked to a work-incentive provision. The proposed programs are presented within the context of a planning process to be carried out by a new organization called the State Planning Council, which would prepare and annually update a Comprehensive California State Plan. Regional governments are proposed to be established to prepare and administer regional conservation and development programs. Furthermore, a state zoning plan is part of the overall plan to preserve agricultural land, open space and natural resources. The proposed plan is based on a vision of a better state; it gives one way that the citizens of that state might achieve the vision.

You will recall that Malthus believed population growth would result in a population beyond the means of man to provide food.[2] In his essay on population, he said, "Population, when unchecked, increases in a geometrical ratio. Subsistence increases only in an arithmetical ratio. A slight acquaintance with numbers will shew the immensity of the first power in comparison of the second."

THREE CAUSES OF DETERIORATION

We find that at present there are three primary forces to which are attributed the effects of environmental deterioration. They are the growth of population, technology and affluence. Affluence in this case is defined as having great wealth and abundance beyond man's simple needs.

Barry Commoner has published a book entitled *The Closing Circle* which considers the effects of technology, population and affluence and their contributions to our environmental problems.[3] A stable ecosystem has always been dependent upon its own self-renewing quality, according to Commoner. Recently, technology has been upsetting this stable equilibrium and tampering with the natural processes. Eventually, the closed circle will open and the environment may collapse.

What is uniquely the challenge today is, according to Commoner, the growth of technology that has resulted in a technology of a scale and intensity to match the ecosystem itself. A singular example is provided in the following excerpt:*

When the world learned, on that fateful day in 1945, of the successful construction of an atomic bomb, it was clear that a new period in human history had begun. Those who have marked the day by remembering the Hiroshima dead foresaw an era of deadly peril for humanity and feared an inexorable march toward the holocaust of a Third, and final, World War. Those who sensed, instead, in the brilliant flash of the bomb that man had at last "harnessed the power of the stars," dreamed of an era in which, with unlimited power, mankind—or some lesser portion of it—could achieve all the goals that power commands.

As the atomic era has unfolded since 1945, the contrast between these two visions has sharpened and the gulf between those who follow them has grown wider. On one side are those who fear that humanity will be crushed beneath the ungovernable power of nuclear technology; many of them are the young, who were born with the bomb and have lived a life in which, because of it, doomsday may come tomorrow. On the other side are some of their elders, who possess or hope to possess some of the new power, if need be at the cost of human lives.

Despite this confrontation, there is a widespread conviction that the new knowledge is sound, that the new technology is therefore competent, and that the new power is thereby irresistible. The first 25 years of the atomic age tell us that this belief is deeply, tragically, wrong. Isolated on a Pacific island or confined to the grounds of a power plant, nuclear energy is a success. It works: it vaporizes the island; it sends electricity surging out of the power plant. But neither the island nor the power plant—nor anything else on the earth's surface—exists apart from the thin, dynamic fabric that envelopes the planet: its environment, the ecosphere. And once power from the split atom impinges on the environment, as it must, we discover that our knowledge is incomplete, that the new technology is therefore incompetent and that the new power is thereby something that *must* be governed if we are to survive.

This, it seems to me, is the meaning of the first environmental encounter of the new age of technology. Our experience with nuclear power tells us that modern technology has achieved a scale and intensity that begin to match that of the global system in which we live. It reminds us that we cannot wield this power without deeply intruding on the delicate environmental fabric that supports us. It warns us that our capability to intrude on the environment far outstrips our knowledge of the consequences. It tells us that every environmental incursion, whatever its benefits, has a cost—which, from the still silent testimony of the world's nuclear weapons, may be survival.

Yet this same experience with the first 25 years of the nuclear age has a more hopeful message: seen in its true, environmental context, the power of nuclear technology is subject less to the control of the technologist than to the governance of the public will.

While nuclear technology has the ability to overwhelm the environment, one hopeful sign emerges: the power of technology is less in the hands of the technologist than (properly) in the hands of the public will.

Nevertheless, the ecosphere remains profoundly threatened by a growing use of our natural resources which may lead to a potential break in the circle or the nature of the ecological system. As Commoner observes:

*From *The Closing Circle* by Barry Commoner. Copyright © 1971 by Barry Commoner. Reprinted by permission of Alfred A. Knopf, Inc. A substantial portion of this book originally appeared in *The New Yorker*.

Kept in proper balance, the earth's ecological cycle is self-renewable, at least over the time-scale involved in human history. In this time-scale, it can operate and support some number of human beings as one of its constitutents more or less indefinitely. However, mineral resources, if used, can only move in one direction—downward in amount. Unlike the constituuents of the ecosphere, mineral resources are *nonrenewable*. Fossil fuels, such as coal, oil, and natural gas, were deposited in the earth during a special period of its evolution that has not since been repeated—with the exception of the slow accumulation of very slight modern fuel deposits such as peat. Once fossil fuels are used, solar energy trapped within them millions of years ago is dissipated irrevocably.

The earth's store of metals, laid down by not-to-be-repeated geological events, is also nonrenewable. Of course, since matter is never destroyed, metals taken from the earth's ores remain on the earth after use and in theory could be used again. However, when iron, for example, is taken from the earth as a concentrated ore, converted into useful products that later are scattered, as rust, across the face of the globe, what is lost, irrevocably, is energy. Whenever any material is scattered from a concentrated origin and mingles with other substances, a property known as "entropy" is increased. And an increase in entropy always involves a loss in available energy. This is perhaps more easily seen in reverse: that the gathering together of scattered material into an ordered arrangement requires the addition of energy. (Anyone who has tried to reassemble a jigsaw puzzle from its scattered parts has experienced this law of nature.) Since any use of a metallic resource inevitably involves some scattering of the material, if only from the effects of friction, the *availability* of the resource tends constantly downward and can be reversed only at the expense of added energy—which is itself a limited resource.

We have here a statement of the Second Law of Thermodynamics and the concept of entropy. As materials are used, the level of usable energy is degraded and the availability of energy moves downward to a less useful state. However, there is nothing inevitable about the high rate at which many resources are lost to reuse. If we wished, and it were economic, nearly all the copper, for example, could be recovered from products and used again when the product outlives its usefulness. Obviously this becomes economic when the supply of copper ore is depleted to the point where it costs less to recycle copper than to obtain it from ore. This is exactly what has happened with precious metals such as gold and platinum.

In Dr. Commoner's view, we have "come, then, to a fundamental paradox of man's life on the earth: that human civilization involves a series of cyclically interdependent processes, most of which have a built-in tendency to grow, except one—the natural, irreplaceable, absolutely essential resources represented by the earth's minerals and the ecosphere. A clash between the propensity of man-dependent sectors of the cycle to grow and the intractable limits of the natural sector of the cycle is inevitable. Clearly, if human activity on the earth —civilization—is to remain in harmony with the whole global system, and survive, it *must* accommodate to the demands of the natural sector, the ecosphere. Environmental deterioration is a signal that we have failed, thus far, to achieve this essential accommodation."

A picture illustrative of a region that has experienced a large population and technological growth is shown in Figure 14–1. This area in northern New Jersey serves a highly industrialized and affluent area near New York City.

The effect of population growth, while not negligible, is secondary in Commoner's view. He says, "the increase in population accounts for 12 to 20 percent of the various increases in total pollutant output since 1946." However, not all calculations bear out this conclusion. Consider the growth of energy production in the United States in the period 1940–1969.[4] During this period population increased by 53% and energy production per capita increased 57%. Since the effects are multiplicative, not additive, we calculate the product of the two factors as follows:

$$I = 1.53 \times 1.57 = 2.4 \qquad (14\text{-}1)$$

The increase by a factor of 2.4 (or 140%) agrees with the data for the increase in energy production during the period. Commoner defines the production per capita as the increase in production due to technology, or the technology factor T. The increase, I, is due to the population factor P and the technology factor T in this calculation. Clearly, in the case of the increase in energy production, technology and population have made essentially equal contributions.

EFFECTS OF AFFLUENCE

How does affluence contribute to the failure of industrialized nations to accommodate thus far to the demands of the ecosphere? Commoner notes that in industrialized nations there appears to be no sign of increasing affluence with respect to food consumption or clothing. Affluence in this sense is defined as the increased amount of consumption per capita. For example, the annual production of shoes per capita remained constant at about three pairs per person over the period 1946–1966. With respect to shelter, Commoner notes that 0.272 housing units per capita were occupied in 1946 and in 1966 there were 0.295 housing units per capita occupied. (These figures do not account for quality of housing.) On this basis, Commoner concludes:

We can sum up the possible contribution of increased affluence to the United States pollution problem as follows: per capita production of goods to meet major human needs—food, clothing, and shelter—have not increased significantly between 1946 and 1968 and have even declined in some respects. There has been an increase in the per capita utilization of electric power, fuels, and paper products, but these changes cannot fully account for the striking rise in pollution levels. If affluence is measured in terms of certain household amenities, such as television sets, radios, and electric can-openers and corn-poppers, and in leisure items such as snowmobiles and boats, then there have been certain striking increases. But again, these items are simply too small a part of the nation's over-all production to account for the observed increase in pollution level.

What these figures tell us is that, in the most general terms—apart from certain items mentioned above—United States production has about kept pace with the growth of the United States population in the period from 1946 to 1968. This means that over-all

FIGURE 14-1 A highly industrialized region in northern New Jersey. The photo shows the Passaic River below the Falls near Paterson. Smoke means employment in this region. Courtesy The Bettmann Archive, Inc.

production of basic items, such as food, steel, and fabrics has increased in proportion to the rise in population, let us say from 40 to 50 per cent. This over-all increase in total United States production falls far short of the concurrent rise in pollution levels, which is in the range of 200 to 2,000 per cent, to suffice as an explanation of the latter. It seems clear, then, that despite the frequent assertions that blame the environmental crisis on "overpopulation," "affluence," or both, we must seek elsewhere for an explanation.

Nevertheless, the production of air conditioner compressors, as shown in Table 4-1, is up 30 fold (2850%). Also, the production of electric housewares during the period 1946–1971 was up by a factor of 12 (1100%). Commoner defines affluence to account for the contribution of per capita production of goods to meet the needs for food, clothing and shelter—the necessities. But today, many would call air conditioners or electric houseware appliances necessities. In any case the growth in use of these items would be ascribed by most people to affluence.

TECHNOLOGICAL EFFECTS

Since there are three factors which contribute to the environmental impact of technology, Commoner turns to the technological factor for consideration after asserting that the population increases and the growth in affluence are of a secondary nature. He says, "The over-all evidence seems clear. The chief reason

TABLE 14-1

The growth of the use of technological items over a 25 year period (1946-1971)
in the United States.

Item	Percentage Increase
Synthetic fibers	5,980
Air conditioner compressor units	2,850
Plastics	1,960
Fertilizer nitrogen	1,050
Aluminum	680
Electric power	530
Pesticides	390
Consumer electronics (TV sets, tape recorders, etc.)	217
Motor fuel consumption	190
Automobile	197
Automobile engine horsepower	150

for the environmental crisis that has engulfed the United States in recent years is the sweeping transformation of productive technology since World War II. The economy has grown enough to give the United States population about the same amount of basic goods, per capita, as it did in 1946. However, productive technologies with intense impacts on the environment have displaced less destructive ones. The environmental crisis is the inevitable result of this counterecological pattern of growth."

It is clear that Commoner believes faulty technology to be the cause, in his view, of increased deleterious environmental inpact and other disabilities.

An example of a faulty technology and the effect of affluence is illustrated by the street scene in Figure 14–2. On this street in Palo Alto, California we have the effects of strip zoning and uncontrolled advertising. Ironically, the name of the street is El Camino Real—The King's Highway.

CALCULATING MULTIPLE FACTORS

Because it is generally agreed that technology, population and affluence have contributed to the increased impact on our environment during the past 25 years, let us consider the calculations behind Commoner's assertions and then return again to an analysis of the factors.

As we have seen above, three factors which contribute to the impact of technological development are increases in population, affluence and technology. These factors are multiplicative in their effect. That is, it is the *product* of their

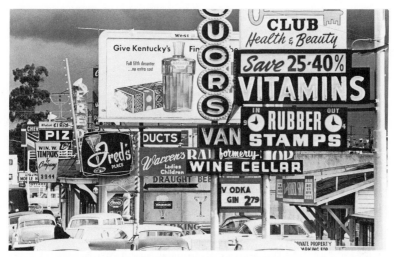

FIGURE 14-2   A photo of a highway that passes through Palo Alto, California. The
city is a center of advanced technology and the population is generally
affluent. Strip zoning such as is shown is an example of a faulty
technology. (Other parts of Palo Alto are designed for optimum use of
residential and recreational space, it should be noted.) Courtesy
Rondal Partidge.

increases that yields the environmental impact. The impact of a technological
development can be represented by equation

$$(I_0 + \Delta I) = (P_0 + \Delta P) \times (A_0 + \Delta A) \times (T_0 + \Delta T) \qquad (14\text{--}2)$$

where I_o is the initial value of the impact at some specified period in time and ΔI
is the change in the impact during a specified number of years. Similarly, we have
the initial values and incremental changes for the variables $P =$ population, $A =$
affluence and $T =$ technology.

We can rewrite Equation 14–2 as follows:

$$I_0\left(1 + \frac{\Delta I}{I_0}\right) = P_0\left(1 + \frac{\Delta P}{P_0}\right) \times A_0\left(1 + \frac{\Delta A}{A_0}\right) \times T_0\left(1 + \frac{\Delta T}{T_0}\right) \qquad (14\text{--}3)$$

Since $I_o = P_o A_o T_o$, we then have

$$\left(1 + \frac{\Delta I}{I_0}\right) = \left(1 + \frac{\Delta P}{P_0}\right)\left(1 + \frac{\Delta A}{A_0}\right)\left(1 + \frac{\Delta T}{T_0}\right) \qquad (14\text{--}4)$$

The ratio $\Delta P/P_o$ is the relative change in the population during the period of
consideration. Finally, we may, for convenience, rewrite Equation 14–4 as
follows:

$$I = P\,A\,T \qquad (14\text{--}5)$$

where $A = (1 + \Delta A/A_o)$, for example. Then we have a summary relationship which states that the product of the increases in population, affluence and technology yields the impact factor.

We may restate the relationship as follows:

$$I = P\,A\,T$$

or, Amount of Pollutant = Population $\times$ $\dfrac{\text{Production}}{\text{Capita}}$ $\times$ $\dfrac{\text{Amount of Pollutant}}{\text{Unit of Production}}$

As an example, let us determine the impact of an increased number of beer bottles on the environment in the United States during the period 1950–1967.

The relative increase in the impact of beer bottles on the environment in the form of air pollution and litter, etc. is expressed by $\Delta I/I_o$. During the seventeen year period the United States population rose from 151,868,000 to 197,859,000. Thus,

$$P = 1 + \frac{\Delta P}{P_0} = 1 + \frac{45,991,000}{151,868,000} = 1.303 \qquad (14\text{--}6)$$

The beer consumption *per capita* is a measure of affluence; for this seventeen year period it yields

$$A = 1 + \frac{1.28}{24.99} = 1.051 \qquad (14\text{--}7)$$

where $A_o = 24.99$ gallons *per capita* in 1950. The increase in technology is measured by the number of beer bottles used per gallon of beer consumed. This factor increased greatly during this period because of the introduction of nonreturnable bottles. We then find

$$T = 1 + \frac{1.01}{.25} = 5.04 \qquad (14\text{--}8)$$

Therefore, in the case of bottle production we have

$$I = P\,A\,T$$
$$I = 1.303 \times 1.051 \times 5.04 = 6.90 \qquad (14\text{--}9)$$

We have calculated a 590% increase* (or a factor of 6.90) in beer bottle dissemination arising from a result of 30% increase in population, a 5% increase in consumption (the affluence factor) and a 400% increase in the technology factor. In this case, the changing technology of nonreturnable bottles accounts for the greatest part of the environmental impact.

---

*The amount of pollution I increases for the time period by a factor of 6.90 because the original value of $I = 1$ and the value of $I = 6.90$ at the end of the time period. The percentage increase is defined as the ratio of the difference ΔI to the original value I_o or as

$$\Delta I/I_o \times 100\% = 5.9/1 \times 100\% = 590\%$$

Let us consider the case of the introduction of the air pollutant Nitrous Oxide generated by emission from automobiles during the period 1946–1967. The impact results from the increase in the population, the increasing affluence resulting in increased automobile travel and increased emissions due to the introduction of larger, more powerful automobile engines. Population increased 41% during this period and vehicle miles *per capita* increased 100%.

$$P = 1 + \frac{\Delta P}{P_0} = 1.41, \quad \text{and} \quad A = 1 + \frac{\Delta A}{A_0} = 2.00$$

Nitrous Oxide emissions increased by 140% from an increase in engine compression ratios of 140%.

$$T = 1 + \frac{\Delta T}{T_0} = 2.40$$

Therefore, we find that since $I = P\, A\, T$,

$$I = 1.41 \times 2.00 \times 2.40 = 7.28 \tag{14–10}$$

In this case the relative influence of affluence and technology are primary, but the influence of population increase cannot be discounted. The total resulting increase in Nitrous Oxide pollution is a significant 628% over 21 years. This growth is at a compound growth rate of approximately 10% per year.

As a final exercise, let us consider the environmental impact of electric power used for automobile production. The pollution resulting from this portion of electric power usage is low relative to the other uses of electric power, but it is an interesting, illustrative example. Let us consider the period 1937–1967, in which the population of the United States increased by 54%. The electric power used to produce one automobile was 179 Kilowatts in 1937; it increased to 708 Kilowatts in 1967. The electric power usage will be considered to yield a direct pollution output. The vehicles produced *per capita* increased by 22% during the period. Calculate the impact, I, and compare the influence of each factor. The answer appears as a footnote.

SUMMARY OF IMPACT STUDIES

We have seen in the preceeding examples that technology contributed heavily to the growing impact of the consumption of a specific product. In the case of the impact of beer bottles on the environment, the impact grew by 590%, primarily because of a growth in technology, measured as number of bottles per

---

For the increased automobile production and usage of electric power, we have $I = P\, A\, T$ such that

$$I = 1.54 \times 1.22 \times 3.96 = 7.4$$

where $T = 1 + 529/179 = 3.96$. In this case, there is a large impact caused by the technology factor T.

gallon of beer consumed. However, this impressive percentage increase need not mean the impact is large in absolute terms or that it is an important pollutant. The increase in the technology factor is primarily the result of negligible production of nonreturnable bottles in 1946. A large percentage increase from a small beginning can yield a small absolute value at the end of the period.

In the case of the Nitrogen pollutant resulting from auto emissions, we found that the influences of technology and affluence were primary and essentially equal, while the increase from population could not be discounted.

In summary, it can be stated that the increasing impact on the environment from increasing population, increasing technology and increasing affluence is essentially equal. Whatever the case in specific examples, if man is to exert control over the impact on his environment, he must reduce the rate of population growth, moderate the rate of growth of affluence and moderate the effects of misused and faulty technology.

The effects of increased population cannot be discounted, because it is not simply the percent of increase in population that is relevant; the population is not distributed evenly over the nation. Recall that farm employment dropped by almost a factor of two during the period 1946–1971 (See Figure 4–4) with a concomitant increase in the urban population. Would Los Angeles smog be as bad if only one fifth as many people lived there? The increase in the urban population of Los Angeles was essentially a factor of five during that period.

The environmental impact of a new development or product will grow exponentially from a very low value, as a product is introduced, but at some point it reaches saturation as the market for the product saturates. Thus the question of the environmental impact also depends somewhat upon where the product is in its life cycle.[5] Also, the shifting social patterns of life in the United States must have some effect on the environment, such as the shift to suburban homes from urban apartments. Is this not a measure of rising affluence?

The fact is that a simple algebraic equation for the environmental impact of technology, population and affluence is revealing and interesting, but not finally incisive. The American environment has been influenced deleteriously over the past several decades because man has found a way to meet his needs and wants economically. It is only in recent years that the social consequences of such a technological development have been fully realized and documented. Nevertheless, the intricate web of causes of environmental degradation are many, including, among others, social rearrangements, political adjustments and restructuring of justice.

Faulty technology, population growth and rising affluence are equally important causes. Any one cause is too important to neglect and too much an equal contributor to label as primary. Commoner's Fourth Law of Ecology is "There is no such thing as a free lunch."[3] Obviously some technologies are better than others with respect to environmental impact, but barring a repeal of the laws of thermodynamics, no technology can reduce to zero the impact of the technology itself, the population or the rising affluence.[6] What man wants for support of the bountiful life, he must pay for as a member of the ecological system. We must overcome the effects of overpopulation, excessive affluence and faulty technology.

CHAPTER 14 REFERENCES

1. A. Heller, Editor, *The California Tomorrow Plan*, William Kaufman, Inc., Los Altos, California, 1972.
2. T.R. Malthus, *An Essay on the Principle of Population, As It Affects the Future Improvement of Society*, London, 1798.
3. B. Commoner, *The Closing Circle*, Knopf, New York, 1972.
4. P.R. Ehrlich and J.P. Holdren, "People in the Machinery," *Saturday Review*, Jan. 1, 1972, p. 71.
5. G. Hardin, "Population Skeletons in the Environmental Closet," *Bulletin of the Atomic Scientists*, June, 1972, pp. 37–41.
6. P.R. Ehrlich and J.P. Holdren, "One–Dimensional Ecology," *Bulletin of the Atomic Scientists*, May 1972, pp. 16–27.
7. "Fellow Americans Keep Out!" *Forbes*, June 15, 1971, pp. 22–30.
8. "Oregon: A Test Case for Returnable Containers," *Business Week*, July 28, 1073, pp. 75–77.

CHAPTER 14 EXERCISES

14–1. Even as the population of the United States reaches a constant level, population shifts cause growth in certain regions. Discuss in a few sentences the no-growth policy of Oregon. Also, read the California Tomorrow Plan and determine if it is a reasonable approach to limiting growth.

14–2. Explore the concept of entropy. Follow the use of paper through several paths of recycling and visualize the increasing entropy in the cycles.

14–3. Consider the level of smog in your city or in a nearby city. Would you ascribe the increase in smog over the past two decades in your city to rising affluence, population or faulty technology? Can you obtain the necessary data for your city to calculate the effect of each factor on the environmental impact I?

14–4. During the period 1940–1969, the energy production *per capita* (or conversion to useful forms) increased by 140%. The population increased by 53% during this period. Determine an estimate of the increase of pollutant per unit of fuel for this period and calculate the environmental impact of this use of energy.

14–5. There has been a marked increase in the size of homes over the past 30 years. Determine the population increase from 1944 to 1974 and the attendant increase in residential space in square feet *per capita*. Finally, estimate the environmental impact by calculating the increase in the amount of board feet of lumber used per square foot of house.

14–6. In 1972 the State of Oregon passed a law banning the sale of beer and soft drinks in nonreturnable bottles and cans. The aim of the laws is to curb litter.[8] A similar bill took effect in Vermont in 1973. Litter of beverage containers alongside roads has reduced 81% from a year earlier in Oregon. The increase in the use of nonreturnable beverage containers was 53,000% over the period 1946–1974. (Of course this large percentage increase came from the small initial number in 1946.) Examine the ecological effects of the Oregon Law and discuss its applicability in your state.

14–7. In the "throwaway" society of the United States, the nation's citizens each year use 48 billion metal cans, 26 billion bottles and 65 billion metal bottle caps; they junk 7 million automobiles. In addition, over 400 pounds of materials for wrapping, bottling and canning are used per person each year. These figures are rising at a rate of about 3% per year. Collection of this waste is quite expensive to our society.

Work out a waste disposal plan for your town or college which would retrieve and recycle the appropriate materials while using appropriate materials for compost and fertilizer.

14–8. During the period 1945–1968 the total annual use of fertilizer Nitrogen, and the environmental impact related to its use, increased by 648%. During this period the population increased by 34% and crop production *per capita* increased by 11%. Calculate the technology factor and determine which factor is dominant. The technology factor in this case is the amount of fertilizer used per unit of crop production. How does fertilizer Nitrogen impact the environment?

14–9. The use of Phosphate detergents is an example of the introduction of a faulty technology. The environmental impact index of Phosphate increased in cleaners between 1946 and 1968 by 1,845%. During this period the increase in population was 42%, while the effect of *per capita* use of cleaners did not change. Calculate the technology factor. What is the environmental effect of phosphates in the effluents into lakes and rivers?

15

ECONOMICS AND ENVIRONMENTAL CONTROL

THE ENVIRONMENTAL DEGRADATION of our planet is attributed to population increases, the rise of affluence, the growth of technology and the growth or change in our social institutions and political arrangements. Pollution of our air, water and environment has been seen in the past as inconsequential because resources were thought to be limitless and free of cost to society. As resources become scarce or the direct costs of environmental degradation are recognized, a new analysis is necessary.

AIR POLLUTION

Air pollution is not a recent phenomenon. London smog was recorded a hundred years ago. The most serious smog episodes are those in which atmospheric conditions permit a build-up of pollutant levels. The worst episode in London occurred December 4–12, 1952, when the number of deaths caused by smog-induced respiratory disease rose to 4,000. At that time the smoke and SO_2 levels rose by a factor of three and four respectively.

The most recent Federal estimate of the direct annual cost of air pollution damage in the United States is $16 billion. The sources of air pollution are emissions from industrial plants, automobiles and refuse-burning, among others. The effects are reflected in deterioration of buildings, injury to crops and damage to human health. For example, the incidence of the respiratory disease of emphysema increased by a factor of ten in the United States during the period 1950–1970.

The control of this air pollution, as well as other forms of pollution, is required under recent laws of the United States. The Clean Air Act of 1970 calls for a substantial reduction in air pollution by 1975. Industries increased their spending for control by a factor of four during the period 1966–1972 and they spent $6.2 billion in 1973 to curb both air and water pollution. Capital spending for pollution control in 1973 amounted to 5.9% of all capital investment. It is estimated that an expenditure of $22.3 billion would be necessary over the next several years for industry to accommodate to the new standards.

Air pollution in Los Angeles is particularly heavy because of the collection of pollutants under atmospheric thermal inversion layers. Downtown Los Angeles is shown in Figure 15–1 under three different conditions of smog. In Figure 15–1a, a clear condition prevails, while 15–1b and 15–1c illustrate the smog often present in Los Angeles.

MEASURING COSTS OF CONTROL

As the cost of pollution control increases and the cost of the environmental effects is reduced, there will be a point where the two costs are equal. At this point the total cost to society will have reached a minimum; this will be an optimum operating condition, providing it can be measured and agreed upon. A hypothetical plot showing the costs of control and effects is shown in Figure 15–2. Point A is the estimate of $16 billion per year of direct effects. Evidently we have a significant degree of control to implement before we reach the optimum point in pollution control.

In recent years there has been some discussion of a law to require zero pollution from industrial wastes by 1985.[2] Assuming that zero pollution implies the water discharged from a plant is to be of drinking-water standards, the degree of control necessary, as indicated on a chart similar to Figure 15–2, would be essentially 100%. In the case of 100% control, the control costs far exceed the societal costs; this postulates an uneconomic operating point. Governor Nelson Rockefeller of New York estimated this approach could cost his state $275 billion.[2] Dr. Joseph Ling, Director of Environmental Engineering of Minnesota Mining and Manufacturing Company, in his recent testimony to a United States House of Representatives committee, outlined the excessive total social cost of an attempt to achieve zero pollution level, or a 100% degree of control.[2] His calcuations were for a system to achieve an effluent that met drinking-water standards. In order to meet these standards he found that his company would need to spend $25 million for capital investment and $3.5 million per year for

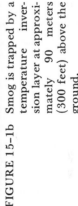

FIGURE 15-1a A clear day with essentially no air pollution present.

FIGURE 15-1b Smog is trapped by a temperature inversion layer at approximately 90 meters (300 feet) above the ground.

FIGURE 15-1c Smog engulfs the Civic Center area on a day when the inversion layer is approximately 460 meters (1,500 feet) above the surface.

FIGURE 15-1 Downtown Los Angeles viewed under three different air conditions. Courtesy of the Los Angeles Air Pollution Control District.

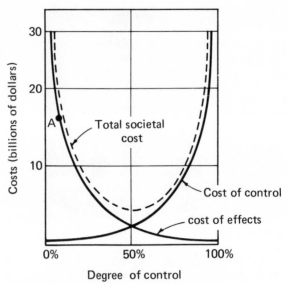

FIGURE 15-2   The cost of control and the cost of the environmental effects of air
pollution in the United States. Point A is the operating point as of
1972.

maintanence and operation. In addition, they would have to purchase 9,000 tons
of chemicals each year for the pollution-control devices and 1,500 kilowatts of
electricity for operation. Taking into account all the necessary resource inputs for
the pollution-control devices to achieve zero pollution level, he stated, "I draw
the conclusion: In order to remove approximately 4,000 tons of pollutants from
this particular plant, we would have to use more than 40,000 tons of natural
resources. My conclusion is that the zero discharge based on this particular
operation would produce a negative environmental impact. If you looked into
3M's effluent pipe, yes you would get a clear effluent. But you go up a little bit
higher to look at the overall environment for the country and you would find that
we created a lot more pollution than we have removed from this plant."

Operating a plant at an absolute zero level of pollution would normally be
uneconomic and, as in the case of this example, also can result in environmental
degradation, at other locations, of amounts equal to or greater than that from
pollutants removed by the control devices!

The economic analysis of pollution control is an important area of
consideration. If the uses of technology produce environmental effects, where
and how should these costs be assigned? As natural resources become scarce, or
we recognize the limited nature of their availability, a set of costs is developed. If
electricity is too cheap to the customer because he is not charged with the full
social cost, then industries and residences using electricity will really be
subsidized by society. Pollution control can be accomplished by the setting of
minimum standards for air or water, for example. Another approach is to charge

the producer of effluents for his waste production through user charges or taxes. Often financial incentives in the form of tax relief or subsidies are used to encourage industries to follow a desired course of action. In the following article, a professor of economics, Robert Solow, outlines an economist's approach to the control of pollution.[3]*

The Economist's Approach to Pollution and Its Control

Robert M. Solow

My object is not to tell you about pollution. In the nature of the case, many of you will know much more than I do about the physical and chemical causes and the biological consequences of environmental pollution. But pollution is also an economic problem, and economists have rather special ways of thinking about it, with implications for the design of environmental policy. What I can hope to do is to put the problem of pollution into the economist's framework, to see if that way of regarding it leads to any basic principles of regulation and control.

The ancient economists used to classify productive resources as Land, Labor, and Capital. In this classification "Land" stood for all those natural resources that are given in amount and cannot be augmented by human decision. Some natural resources are exhaustible: There is just a certain amount to begin with, like iron ore or oil, and when we have used it up there will be no more. If more iron ore or oil is being formed in nature, it happens much too slowly to matter to human society. Other natural resources, like water and forests, are being renewed all the time, though of course human action can interfere with the process.

In the early stages of economic development, some natural resources were "free," in the sense that there was more than enough available to saturate the demand. This was true even of agricultural land in the early history of the United States. It is still true of air and water in many places. Even if these natural resources are appropriated and a monopoly price is charged for them, they are still "free goods" to the society as the economist looks at the matter. No possible use needs to be suppressed so that some other possible use can take place.

As economic development proceeds, many resources become scarce. This may be either because a fixed supply is exhausted or, more usually, because growing population and increasing production of commodities put more pressure on the limited supply provided by nature. The use of scarce resources has to be rationed. One possibility is that the scarce resource becomes private property and is rationed by a market procedure; another possibility is that the scarce resource becomes public property and is rationed either by a market or by some other more political process. Either system may work well, or badly, depending on circumstances.

External Effects Can Distort Resource Use

Even with such a simple resource as residential land, we often find that a system of individual ownership needs public regulation because of "external effects." One person's use of a natural resource can inflict damage on other people who have no way of securing compensation, and who may not even know that they are being damaged. We would like to insure that each resource is allocated to that use in which its net social value is highest.

*From *Science*, Vol. 173, August 6, 1971, pp. 498–503. Copyright 1971 by The American Association for the Advancement of Science.

But if the full costs of some use of a resource do not fall upon the private owner or public decision-maker, but upon someone else, then the resource is unlikely to find its way into its socially best use. We could not allow pig farmers to bid freely for residential land, for example. (There is a symmetrical case where use of a resource confers external benefits that cannot be captured by the owner or decision-maker.)

There has been much economic analysis of these "external effects," and of possible corrective measures. But first I want to get further along in the story. Eventually, as an economy grows, even air and water become scarce. Air and water have only a limited capacity to assimilate wastes or to carry them away. Any modern industrial economy apparently generates so much waste—in the form of both matter and energy—that its disposal taxes the capacity of the atmosphere, the rivers, and eventually even the ocean. We used to think that these external or environmental effects were exceptions, but in modern industrial society they may become the rule.

In the situation we have now, the assimilative capacity of air and water has become a scarce resource, but it is provided free of charge as common property to anyone with some waste to dispose of. It is easy to see that, in these circumstances, the scarce resource will be overused. The normal system of incentives is biased. A costly (that is, scarce) resource does not carry a price to reflect its scarcity. If high-sulfur fuel is cheaper to produce than low-sulfur fuel, it will be burned and sulfur dioxide wastes will be dumped into the air. Society pays a price in terms of damage to paint, to metal surfaces, to plants, and to human health. But that cost is not normally attached to the burning of high-sulfur fuel; only a part of the full social costs become private costs and influence private decisions.

If an upstream factory deposits organic waste in a river—or merely raises its temperature by using the water for cooling purposes—the costs of water purification may rise for a downstream city that wants to use the water for drinking or recreation. But these costs do not fall on the party whose decision generated them, so he has no reason to take them into account. (By the way, comparative cost analysis will often show that it is cheaper to purify the industrial waste at the source. But the choice of a method of treatment is separable from the allocation of costs. Even if it is cheaper for the downstream city to treat its water, the allocation of resources will be distorted unless the extra costs of downstream water purification are treated as part of the full costs of operating the upstream factory.)

The private automobile is a similar case, but even more complicated, because each of a million cars contributes its small bit of the Los Angeles smog; each driver pays in coughs and tears, and perhaps lung disease, for everyone else's exhaust emissions, but his own responsibility is negligibly small.

We are used to these consequences of "external effects." I mean that we are accustomed to them as citizens; and we understand them as economists. But economists realize, as citizens sometimes do not, that the implications of external effects must be traced further. They have secondary effects on the system of resource allocation. If electric power is "too cheap" to the customer, because he is not charged with its full social cost, then other things will happen. Other commodities that are produced with the help of large amounts of electric power will also be cheap, and they will be overproduced. Other industries will be tempted to adopt techniques of production that use more electric power than they would if the price of electric power were higher. The rest of the society will find itself subsidizing those people—if they are an identifiable group—who consume a lot of electricity or a lot of goods made with a lot of electricity.

Similarly, if society is in fact subsidizing the private automobile—by not charging it for all the damage it does—then the location patterns of suburbs and of industry will be affected. A change in the private costs of automobile travel will have effects on house rents, residential choices, and eventually on the location of industry. If the use of DDT and other toxic chlorinated hydrocarbons were prohibited, or merely made more expensive, one would naturally expect certain changes in food prices and availability. But there might also be corresponding effects on the regional distribution of income and population, and these in turn might have further consequences difficult to calculate.

Piecemeal Regulation May Be Inefficient

The existence of all these systematic interdependencies means that piecemeal remedies for environmental pollution, by direct prohibitions and by setting specific standards for emission, may be inefficient and even harmful.

Piecemeal regulation may simply transfer pollution from one medium to another. For example, if the people of a densely populated city are denied the right to burn their trash in ordinary furnaces and incinerators, the city will have to find some other way to dispose of solid wastes; and this may be costly and difficult to do without spoiling the environment in other ways. New York and Philadelphia have tried to improve water quality by intensive municipal sewage treatment. One result has been the generation of a large volume of biologically active sludge. This is currently being carried out to sea in barges and dumped there, with unknown effects. To take another example, the scrubbing of stack gases can certainly reduce the emission of particle matter into the air; but it creates another liquid waste for disposal. One cannot know without a calculation whether the regulation of stack gases is in any particular case the right way to proceed.

The simplest way to deal with an acute pollution problem is to set minimum quality standards for air or water and enforce them on each polluter. But this ignores the fact that some sources of pollution are more readily remedied than others. If two factories producing different commodities both contaminate the same stream to the same extent, it might seem natural to require each of them to reduce its contamination by, say, 50 percent. If that were done, it would be almost certain that the incremental cost of a small further reduction would be different for the two factories; after all, they use different production techniques. But then it would be better if one of the factories—the one with the smaller incremental cost—were required to pollute still a little less, and the other permitted to pollute a little more. The total amount of pollution would be the same, but the total cost of accomplishing the 50 percent reduction would be smaller. Since it is the total amount of pollution that matters, the cheaper possibilities of reduction should be exploited first.

This would be accomplished if, instead of a direct imposition of standards, the two factories were charged or taxed an amount proportional to their emission of pollutants. The height of the tax could be varied until the desired total reduction in pollution occurred; the factories themselves would see to it that it occurred in the cheapest possible way. It is perfectly true that this way of doing things affects the distribution of income; the cost of preserving the environment is borne in a certain way. But that is true of any method, including simple prohibitions. The redistribution is only more visible in the case of a tax or effluent charge. The tax also provides some revenue which can be used either to further improve the environment or to assist genuine hardship cases or to accomplish socially desirable ends of any kind.

Here is a third example of the sort of mistake that can result from a piecemeal approach to policy. In the United States, it is often proposed that the government subsidize the construction of waste-treatment facilities. Only a few such proposals have worked their way into the law, usually indirectly. But why should the government promote the purchase of special equipment when other methods might be superior: the substitution of a cleaner fuel or other material for a dirtier one, or other changes in production methods, or the recirculation of cooling water, or the recovery of by-products for further use, or even the relocation of production altogether? I suppose the answer is that it is easier to subsidize treatment facilities: The amount is simply determined as a fraction of cost, and the industry that produces waste-treatment equipment is naturally anxious to have its product subsidized. Unfortunately, this may be an expensive way to accomplish the result, especially because, if the alternative to waste treatment is continued free dumping into the atmosphere or watercourse, the subsidy may have to be almost complete to induce polluters to take it.

Still another difficulty with piecemeal regulation is that it may be localized, in which case the environmental damage may simply be transferred to another town or district or region or—in Europe—another country. A sensible approach circumvents this possibility by having a unified regional policy, perhaps an international one.

A final difficulty with piecemeal policy is that it works automatically against large public investment projects in such fields as solid-waste disposal, low-flow augmentation in rivers, and perhaps others. It is not always the case that such large public investments are the best solution to a problem; but they are sometimes the best, and, when they are, this is unlikely to be discovered except as part of a system-wide analysis.

Use of Effluent Charges

In many cases, probably most cases, where direct regulation seems the natural approach to the concerned citizen, the economist will prefer to use taxes or effluent charges or user charges of some kind. I have mentioned one obvious reason for this: It is in the social interest that the cheapest method should be adopted to achieve any given reduction in pollution. A system of taxes and charges is more likely to accomplish this than direct regulation, given that we cannot possibly have all the desired facts.

This economizing on information is a second reason for favoring taxes over direct regulation. The construction of a good schedule of taxes or fees also requires information, but rather less information. And the process of collection itself produces new information that can be used to improve the schedule in use.

Third, financial incentives are usually easier to administer than direct regulation. They preserve decentralized decision-making, which is often good in itself, and in so doing they induce everyone directly concerned to seek for trade-offs and substitutions and improved techniques that could not be known to any central office.

I think it is also a good general principle that fees or taxes are better than subsidies. It is probably an unpopular principle—nobody likes a tax, but there is always at least one person who likes a subsidy. Subsidies, however, are more difficult to administer. A tax is levied against the amount of pollution actually discharged, an observable quantity. A correct subsidy depends on how much pollution has been reduced from what it would have been in the absence of the subsidy, a hypothetical quantity. If one subsidizes *actual* waste treatment, this may lead to the perverse result that techniques may be adopted that lead to the production of waste on an unnecessarily large scale, simply to collect the subsidy for treating it. Moreover, subsidies will lead to higher net profits in pollution-intensive industries, and perhaps attract a socially undesirable expansion of those industries.

Taxes are generally preferable to subsidies on grounds of equity, too. If some part of the population likes to do things or consume things whose production damages the national environment, it seems fairer that they should pay for the damage than that we should have to bribe them to stop.

This general principle may need to be modified, however, to the extent that the initial distribution of income in society is not equitable. If everyone had the same income, or if the distribution of income met some other standards of equity and justice, it would be right to say that anyone who indulges a taste that leads to the pollution of the environment should be required to pay the costs of restoring environmental quality, or at least to compensate others for the damage caused them.

But incomes are not equally distributed, nor do they meet any other test of equity or justice. The principle of assessing environmental costs on the activities that cause them would lead to material goods (which play a relatively bigger role in the budgets of the poor) becoming dearer relative to services (which play a proportionately bigger role in the budgets of the rich). It might lead to the taxation of the necessities of life of the poor to pay for the protection of the recreational amenities of the rich. There is not much to be said for that.

The economist's answer is that two wrongs don't usually make a right. It is irrational to befoul the environment by fudging a desirable system of taxes and user charges in order to accomplish redistributional aims. It would be far better to achieve an efficient allocation of resources by a proper system of effluent charges; and to correct an inequitable distribution of income and wealth directly by taxing the income and the wealth of the rich to subsidize the poor.

It hardly needs saying that there are situations in which immediate and decisive regulatory action is the only sensible thing. It seems very unlikely that we would ever regret simply having forbidden the disposal of heavy toxic metals like mercury or arsenic where they can be consumed by animals or humans. We may also want to stop irreversible deterioration of certain national resources at once. Formally, these are cases where the optimum taxes or user charges are prohibitive.

Even if it is granted that fees and taxes are generally preferable to specific regulations, that does not settle the matter. There is still need for a new and inclusive way of looking at the management of the physical environment as a natural resource which is common property. The usual textbook illustrations of external effects and their correction through bilateral negotiation or isolated fees and taxes are too simple, too isolated, too often artificially concerned with two parties who know each other, who realize how they interact, and who can negotiate a proper solution. At the present stage of economic development, something more serious is happening.

The Universal Problem of Materials Disposal

The leading specialist on these matters has remarked that we need to take account of the fundamental physical law of the conservation of mass. Our language speaks of the "consumption" of goods as if nothing is left of them after they are consumed. But of course everything is left of them. Every ton of material that is removed from the earth and transformed into goods still remains to be disposed of when the goods in question are finally used. Sometimes, as in the case of a building or a dam, disposal is postponed for a very long time. But the fact that thousands of automobiles are abandoned on city streets, or even deposited in country streams to rust reminds us that durable goods are not permanent. It is also true that much of the weight of each year's production is transformed into gas and disposed of into the atmosphere without any special handling. This is especially true of fuels. But that is part of the problem; the capacity of the air to absorb waste gases is not limitless.

In principle, then, the residuals from production weigh as much as (or slightly more) than the original weight of materials. All this has to be returned to the environment in one way or another, unless it is recycled. Even what we call "waste treatment" merely changes the form of waste material, presumably to something less unpleasant, but the disposal problem remains. This problem is growing in size along with the production of goods. It has been estimated that the total weight of basic materials produced in (or imported into) the United States in 1963 was 2261 million tons (excluding construction materials, mine wastes, and other materials that are just moved from one place to another without undergoing any real chemical change). By 1965 the figure was 2492 million tons, 10 percent higher after 2 years. There is no reason to doubt that the figure is considerably higher now. Over half of this total weight consists of mineral fuels, which are discharged, more or less unnoticed, into the atmosphere as carbon dioxide and water vapor. This seems to have minor short-run effects though, as you know, there is now beginning to be some worry about the possible effects on climate of the accumulation of carbon dioxide in the atmosphere.

The point of my reference to the conservation of mass is that there is not an air pollution problem and a water pollution problem; there is a materials disposal problem. Some ways of disposing of materials are less objectionable than others, just as some materials are less objectionable than others. But one must keep in mind that to "eliminate" air and water pollution means to transform them into the problem of disposing of solid waste, another pollution problem. The only "solution" to the combined problem is the recycling of materials (or the greater durability of material things or the increased efficiency of conversion of fuels into energy). The rest is a choice of the socially best way to dispose of a given weight of residual material.

Since this combined problem is getting bigger, it must be planned for, and it is probably best planned for as a large-scale problem in managing the flow of materials. This suggests that planning must be at least regional and, in principle, concerned with all the media of waste disposal.

Nevertheless, there are certain real physical differences between water pollution and air pollution, and the physical differences mean that the best available policies are likely to be different. Water flows downhill, and there is a natural asymmetry between upstream and downstream. In most cases it is possible to identify natural boundaries to river basins or coastal estuaries, and to deal with them as more or less isolated units. One can more clearly identify individual polluters and individual victims of pollution, and they tend to be distinct individuals or groups (though, of course, B may be a victim of A's pollution and a polluter of C). I do not mean to make it sound excessively simple, as an outsider often tends to do; planning for a water basin is not simple. There are networks of streams and mixed possibilities of treatment of waste at source, treatment of waste still further downstream, mechanical reaeration, and low-flow augmentation. Still, I understand that there have been several successful attempts to build mathematical models of water systems with a view to planning for improved water quality, such as the Delaware River Basin in the eastern United States, the Miami River Basin in Ohio, and recently Jamaica Bay at New York City. I gather that these models are oversimplified, usually account for only one pollutant (biochemical oxygen demand) and only one measure of water quality (dissolved oxygen). An economist is used to oversimplified models, and is even encouraged by seeing that others have to use them too.

These models can be used to find solutions to planning problems of the following kind. What combination of waste treatment at each source and low flow augmentation from a reservoir will maintain specified minimum water quality standards throughout the river basin at least total cost? And what system of effluent charges or taxes on untreated wastes will induce the polluters themselves to achieve that socially best degree of waste treatment? The best system of effluent charges would vary by time of year and location on the river basin. In principle, it ought to be possible to use the model to determine the appropriate minimum water-quality standards themselves, but for that one must know something about the actual damage caused by specified amounts of stream pollution. I will come back to that question.

I have mentioned the economist's tendency to prefer taxes or effluent charges to direct regulation or subsidies as a device for environmental planning. Let me emphasize again the reasons for this choice in this water-quality context. In the first place, effluent charges concentrate automatically on the cheap abatement of pollution, rather than on any artificial allocation of the abatement burden on polluters. For the same reason, effluent charges provide an incentive for the polluters themselves to search for new and cheaper methods of waste treatment and waste reduction including changes in their own production methods. Finally, effluent charges allow for a certain amount of decentralized decision-making. This is valuable for its own sake, and because it economizes on information, especially on information in the hands of the central authority controlling the river basin. It does not economize completely: intelligent management of water quality requires that the central authority have a lot of information about characteristics of stream flow and about the social costs of poor water quality. Individual polluters are likely to know most about the costs of reducing pollution at their own locations.

It seems to be possible to adapt the principles of two-level iterative planning that have been developed in Hungary and elsewhere for general purposes to the specific problem of water management. One such procedure requires the central authority to propose a scale of effluent charges to each polluter. Each polluter then makes his own cost calculation and responds to the central authority by reporting the amount of pollution he will discharge into the river and his total spending on purification. Using this information, the central authority calculates a new schedule of effluent charges. The procedure continues in this way until it converges to the optimum schedule of charges and the least-cost combination of treatments satisfying the minimum standards.

Such a system collects revenue, almost as a by-product of environmental policy, because the main function of the taxes is to induce polluters to do the socially optimum thing. But the revenues can be used for any good cause. In particular, the best policy for managing a river basin may well involve the construction of some large-scale public investments, like reservoirs or downstream treatment facilities. The revenue from the effluent charges can be applied toward the cost of these public investments. Some part might also be used to assist workers in marginal enterprises made unprofitable by the taxes to find new jobs elsewhere.

Air pollution, especially in large cities, is in some important physical respects different from water pollution. Obviously the air moves less predictably as compared to water. Meteorologists can and do make mathematical models of the atmosphere, but they cannot capture local and short-run events. Moreover, the number of actual and potential polluters of the air is usually much larger than the number of waste sources along a watercourse. Air pollution is rather like automobile congestion. Just as each driver in a traffic jam is inflicting delay costs on every other driver (as well as himself), so is every polluter of the air in a city polluting everyone else (including himself), and inflicting costs on the property and the health of everyone else. The analogy to automobile congestion is interesting also because automobile exhaust emissions are such an important contributor to the pollution problem in large cities.

For this reason, it is much more difficult to imagine optimum planning of pollution abatement in a city than to imagine it in a river basin. A system of effluent charges would involve metering altogether too many emissions: from industrial stacks, from domestic heating, from automobiles, from office buildings, and from public utilities, Moreover, the seriousness of air pollution often depends on photochemical reactions in the atmosphere which cannot be directly connected with any particular polluter. The best solution may involve metering the few very large polluters and treating the many small ones differently.

Even apart from these difficulties of measurement, urban air pollution presents difficulties because it could be inefficient to treat it in isolation from other modes of waste disposal. A city could easily clean its air by disposing of most of its wastes in water; or it could just as easily protect its water by concentrating its wastes and incinerating the residues. Nor can the burden be thrown entirely on solid-waste disposal, because that has become an equally costly and difficult process in most areas of dense population. Rational management of waste materials in a city will require something more complicated than a model of a single river basin.

Shortcuts and a Possible Systematic Scheme

It is probably possible to make some progress by using shortcuts, though this may often cause a certain amount of inefficiency and inequity. For example, major polluters of the air, like electricity-generating stations, can be metered and regulated or taxed. It would seem much easier to control sulfur emission by the mass of small users by regulating instead the small number of refiners of oil, preferably by a system of excise taxes based on the sulfur content of oil sold. Similarly, although it might be prohibitively expensive to meter the exhausts of individual automobiles, it is obviously possible to tax gasoline (perhaps at different rates according to lead content, octane rating, and such) or to require pollution-control devices on newly produced cars, as we now intend. These shortcuts have disadvantages. They tax a particular fuel or device rather than the thing that ought ideally to be taxed, namely, pollution itself, but they are clearly much better than doing nothing at all.

The interrelations of air pollution, water pollution, solid-waste disposal, sewage, old automobiles, plastic containers, and all the other paraphernalia of life in a high-income city remind us again that the whole problem really boils down to the general one of managing the material residuals of production. Mills has recently proposed a scheme that is worth describing, if not as an immediately practical proposal then as one leading in the right direction. In principle, this proposal is that the government collect a materials use fee on specified materials removed from the environment. The fee would have to be paid by the original producer or importer of raw materials. It would be set for each material to equal the social cost to the environment if the material were eventually returned to the environment in the most harmful way possible. The fees would be refunded to anyone who could certify that he had disposed of the material, with the size of the refund depending on the method of disposal. Recycled materials would be exempt from the materials use fee, which is equivalent to a full refund; disposal in a preferred way, relatively harmless to users of the environment, would earn a large refund; disposal in some moderately harmful way would earn a moderate refund; and disposal in the most harmful way would earn no refund at all.

The economic advantage of such a scheme is that whenever two or more materials can serve the same purpose—for example, biodegradable and nonbiodegradable materials for containers or detergents—the fees would make their prices reflect social costs, including disposal, rather than merely private costs. The original choices of materials would come nearer to being socially optimum. To the extent that the schedule of refunds were an accurate reflection of the social costs of various methods of disposal, they would provide a correct guide to individuals and private and public agencies in choosing a method of disposal in view of the direct costs and the accompanying refund. There is also an administrative advantage in such a scheme: it avoids the worst measurement problems. For many important materials, fuels, for instance, it is fairly easy to measure the amount removed from the earth by the first producer. It is much harder to measure disposal by various methods, but here the burden of proof is placed on the individual, not on the pollution control agency. In order to receive a refund, the individual must demonstrate that he has disposed of so much of the material by a relatively harmless method. One can easily imagine specialized firms springing up to perform disposal services and to provide certification of the method of disposal.

There are, of course, difficulties with any such scheme. It would have to apply over a wide geographic area, or else one place would be making refunds to those who disposed of materials that had paid the fee elsewhere—this adds insult to injury for the area serving as a dump. There would have to be some sort of price correction for materials incorporated in very durable objects; and the scheme would hardly work at all for materials incorporated in essentially permanent objects like buildings, which, perhaps, should be thought of as harmless disposal, entitled to a full refund. There would be problems of equity. Owners of deposits of certain materials would suffer an immediate capital loss if such a fee were legislated. For some materials it would be nearly impossible to detect their first removal from the environment or to verify the method of disposal.

Practical or not, the scheme has great merit, I think, if only because it puts the problem in the right setting—namely, the global materials balance—and characterizes it in the right way, in that it depends on a price system with a centralized correction for the divergence between private and social costs. In some form, it might be the only way of making a generalized attack on air pollution. Even if it cannot be done, it is a good guide to thinking.

Like any good guide to thinking, it points to gaps in our knowledge. I mentioned earlier that one of the advantages in using the price system to control pollution is that it economizes on centralized information. I also mentioned that any approach to an optimum environmental policy necessarily requires a certain amount of centralized information, and more than we are accustomed to having. It appears to be the case that we actually know very little about the damage costs of stream pollution and air pollution, and thus know very little about the standards of environmental quality at which our society should aim. If we are to begin routine pricing of our common-property environmental resources, which is probably a necessary development, we need to know much more than we do about the effects on health of various common pollutants. At the moment the main source of information is from statistical analysis of epidemiological data—scattered data at that. We need to know more than we do about the effects of air pollutants on the performance and lifetime of metal and other surfaces exposed to the air. We need to have some way of estimating the damage costs of stream pollution, including the value of lost recreational opportunities. We may even need to have some agreed way of putting a monteary value on clean buildings and unspoiled landscape. We must even estimate how many more people would wish to look at unspoiled landscape if we had more of it to look at. These sound like vague and almost foolish tasks, but we must take them seriously if we take our physical environment seriously.

There is also, I gather, much room for improvement in models of the circulation of water in river basins and coastal estuaries, and especially in models of atmospheric diffusion. Economists have little or nothing to contribute directly to this effort; but they may be indirectly helpful to the extent that the object is to construct models that illuminate the strategically important interactions of the physical environment and the economic system itself. What is meteorologically or hydrologically interesting need not coincide with what is economically important.

It is possible that here, at last, is a natural place for interdisciplinary work between the natural and social sciences. It would be very nice if, together, we could contribute a rational solution to a problem that concerns us all.

CHAPTER 15 REFERENCES

1. R.J. Bibben, "System Approach Toward Nationwide Air Pollution Control," *IEEE Spectrum*, Oct. 1971, pp. 20–31.
2. "The High Cost of Getting Water Too Clean," *Wall Street Journal*, April 1, 1972, p. 8.
3. R.M. Solow, "The Economist's Approach to Pollution Control," *Science*, Vol. 173, August 6, 1971, pp. 498–503.
4. "World Environment Newsletter," *World*, August 28, 1973, p. 50.
5. "New Life for Venice If Industry Cleans Up," *Business Week*, April 28, 1973 pp. 72–73.
6. M. Edel, "Autos, Energy and Pollution," *Environment*, Oct. 1973, pp. 10–17.
7. *National Commission on Materials Policy, Material Needs and the Environment Today and Tomorrow*, United States Government Printing Office, Washington, D.C., June 1973.
8. J. Cannon, "Steel: The Recycleable Material," *Environment*, Nov. 1973, pp. 11–12.
9. "Discounting the Engineers," *The Sierra Club Bulletin*, Nov./Dec. 1973, pp. 24–25.

CHAPTER 15 EXERCISES

15-1. If action is not taken, it is estimated that 66 billion gallons of oil will have been spilled and discharged worldwide between the years 1974 and 1995.[4] Each year there are an estimated 10,000 discharges of oil into the navigable waters of the United States. In 1972, the United States was subjected to oil spills amounting to 30 million gallons. Discuss the use of effluent charges, taxes, subsidies or other economic measures to control the discharge of oil in the United States.

15-2. "Right now, we plan to stop any new major investing in the area, clean up what we have, and move out in about 20 years," says an executive of Montedison, the Italian chemical giant. He refers to the massive and costly cleanup of his company's $800 million petrochemical plant complex outside the environmentally beleaguered, beautiful canal-city of Venice. [5] The cleanup of the canals of Venice is necessary to save the beauty of Venice. The Italian government has allocated $516 million for massive flood-control projects. Also, a three-year deadline for cleanup has been set for industry. Examine the various economic measures useful for improving the quality of the water in the Venetian canals.

15-3. As economic development proceeds, many resources become scarce. Scarce resources must be rationed. One possibility is that the resource becomes private property and is rationed in the market. Another possibility is that the scarce resource becomes public property and is rationed by a process of government. Consider the following scarce resources and indicate whether you would use the market or government to ration them.

1. Areas of great beauty and interest to tourists, such as Yosemite; Aspen, Colorado; and the beaches of California and North Carolina.
2. Electric power.
3. Land in downtown areas of your city.
4. The ability to dump effluents into the river or lake near your town.

15-4. Develop a plan for effluent charges for those parties who use a river or lake near your place of residence.

15-5. The increased use of mass transit systems rather than private autos for commuting to work would substantially reduce traffic congestion and air pollution in cities. One recent study of Boston residents examined those who face differing fares, riding times and driving times to reach the central business district by subway and auto.[6] It was found that the number of trips to work by public transit was not affected greatly by differences in fares. Cutting fares in half would increase transit use by only about 10%.* Use of public transit was more a response to relative times required to travel downtown by public and private modes of travel. If there is a substantial amount of subsidy available in your town for mass transit (subway or buses as appropriate for your city), consider its use for reducing fares, increasing the frequency or number of trains or buses, or for instituting express service. Determine the effects of each approach in your town by survey or analysis and consider alternative uses of the available subsidy.

* Fare-cutting in New York City, however, resulted in substantial increases in the number of mass-transit passengers.

15–6. A model of the open-ended materials use system is shown in Figure E15–6a. In the open-ended system, each phase, including supply, use, recovery and disposal, contributes to environmental degradation. In the closed materials system, shown in Figure E15–6b, land reclamation is emphasized; air and water pollution are minimized and recycling is promoted in every phase to conserve materials and reduce discarding solid wastes.

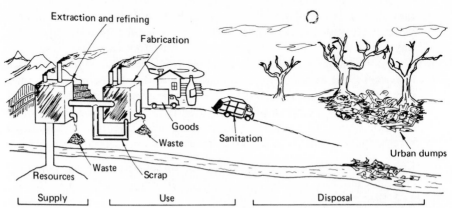

FIGURE E15–6a   Schematic drawing of model for open-ended materials use system generally employed at the present time in the United States.

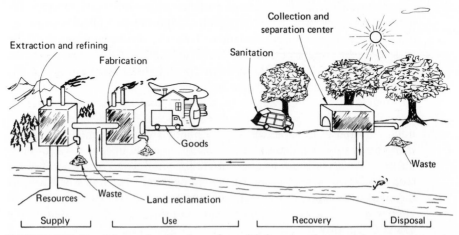

FIGURE E15–6b   Schematic drawing of model for closed materials use system, in which recycling, land reclamation and minimal pollution are emphasized.

The report of the United States National Commission on Materials Policy recommends the establishment of a closed materials system by economic means.[7] The recommendation calls for the nation to encourage everybody to prepare waste for recycling rather than dumping, and to create markets for materials recovered by recycling technology. The use of taxes and incentives should be considered. Prepare a proposal incorporating several economic steps in order to achieve the objectives listed above.

15-7. The recycling of used metals would reduce the need for additional raw materials and energy use. The recycling of steel offers some great advantages. At present only one-fourth of the scrap which the steel industry recycles comes from junkyards, dumps and recycling centers. The rest of the scrap comes from the steel mill itself or from machine shops. The primary deterrents to recycling scrap steel are economic, social and political.[8] Transportation costs, depletion allowances for raw materials, and the inclinations of steel makers are some of the difficulties.

Unfortunately, scrap receives poor railroad service and transportation charges are higher for scrap than for raw iron ore. Nevertheless, the production of steel from recycled material consumes only one-fifth as much energy per pound as production from ore. Examine the discriminatory railroad costs for scrap and the depletion allowances for raw iron ore. What would the reversal of these two economic factors have on the percentage use of recycled steel?

15-8. During the 1973-1974 gasoline fuel shortage, a discussion of how to ration or to allocate available gasoline supplies was pursued. The first of many proposals was utilizing rationing through nontransferable coupons assigned to motor vehicle owners on the basis of their occupations with larger allotments for salesmen and others dependent upon the auto for their livelihoods. This system would require a massive bureaucracy.

A second plan would distribute coupons to every licensed driver without attempting to assess need. These coupons would be transferable and a "white market" would result in the sale of coupons.

A third plan would place substantially higher taxes on fuel to decrease its use. Another possibility is to use a combination of tax plan and rationing.

Discuss the economic and social values of each plan and work out a plan for cutting automobile fuel consumption by 10%.

15-9. One solution to the shortage of fossil fuels in the United States and Europe is to remove all price ceilings and government controls. It is estimated that if there is 10% excess of demand over supply, that prices would rise 7%. Higher prices stimulate suppliers to explore for fuels and new alternative sources. Examine the effect of removing all price controls on fossil fuels in the United States and the resulting consequences to the purchaser and on the environment.

15–10. Another possibility of a partial solution to the energy shortage is to inform the society of the total cost of environmental control programs and their use of fuels, either through direct use of indirect use by means of increased fuel consumption of automobiles and industries for the control devices. Then, one alternative would be the selective removal of pollution control devices. Consider this possibility.

16

TECHNOLOGY AND SOCIAL PROBLEMS

THE USE OF TECHNOLOGY to improve steel productivity or increase the efficiency of electrical generating plants is fairly easy to understand. Through superior engineering design and development, technology can improve the technical qualities of machines, processes or production facilities. In recent years, new technologies have been sought for solving problems with social and political dimensions. (All engineering projects have social and political dimensions, although they may not be obvious at first glance, as has been demonstrated above, but such problems as those of population, housing, drug use and the like are primarily social, not technical.) These problems and the wise use of technology to solve them are the subjects of this chapter.

SHIFT FROM PRODUCTION TO LIFE STYLE

Providing people with adequate housing in Harlem, New York City, for example, requires an approach from a viewpoint incorporating social, political and technological analysis. It requires a marriage of social requirements and technical capabilities.

As we reorder our priorities during this decade for the investment of our

241

technology, our financial capital and our human resources, we must reconsider the investment in research and development of new technologies with the goal of meeting our social needs.

The focus of our attention is shifting toward a host of domestic social concerns such as housing, transportation, leisure and the environment. The shift appears to warrant a change from a standard-of-living measure such as Gross National Product to a quality-of-life measure which includes such factors as environmental quality, personal safety, educational opportunity, health care and others.[1] A good number of the newly-emphasized goals are not materialistic, yet technology can still assist in achieving them. The engineer is now challenged to incorporate within his knowledge the political and social sciences as well as knowledge of management and organization arrangements in order to achieve the broader goals of the new focus.

PHASES OF DECISION-MAKING

Engineering progresses through three phases in the pursuit of an objective:

(1) The objective phase—deciding what should be done
(2) The approach phase—deciding how it should be done
(3) The implementation phase—taking the necessary action to get it done.

In the area of social problems, public action is often impeded by the difficulty in achieving a consensus in Phases 1 and 3—that is, in setting objectives and taking the necessary action to do it. Many engineers have emphasized Phase 2, especially in the aerospace and electronics industries. But it is in the setting of the objectives that society needs assistance, as well as in the effective implementation of projects. Many social projects of the 1960s failed becasue they had poor implementation.

Engineers have encountered three major kinds of problems when attempting to solve societal issues. The first comes from their role in the decision-making process, as we have noted. The second kind of problem is that of qualitative and quantitative variables. It is difficult to quantify relevant, operational and reasonable variables for the design process. The final difficulty comes from developing and utilizing long-range plans in the social area. The shorter term view of elected officials, elected with two- or four-year terms, leads to an emphasis on a "fix-it-for-now" solution rather than long term solutions.

Some have said that the rate at which our society is changing has increased to the point where we have difficulty adapting to it.[3] It is also true that many approaches to social problem-solving are irreversible and could result in serious consequences if they are in error. Finally, a problem with technological approaches to social problems is that many of our organizations and social mechanisms are outdated and unwieldy, but difficult to change. In some sense, such outdated, inefficient organizations becomes overhead items in new designs to social problem-solving.[1]

A SYSTEMATIC APPROACH

In order to overcome our social problems by itilizing engineering approaches, we should:[1]

(1) Articulate the nature of the problems qualitatively and quantatively

(2) Provide variables based on social, political, institutional and economic costs

(3) Trade off qualitative *versus* quantitative, deferred *versus* immediate, and indirect *versus* direct benefits and consequences

(4) Set priorities among the social, economic and political objectives

Social programs in which engineers and social scientists devise and implement new programs must be allowed to go through the same time-consuming research and development sequences to which aerospace systems were subjected. In going to the moon, engineers did not jump from the objective-setting phase (Phase 1) to the implementation phase (Phase 3) while bypassing the research and development phase (Phase 2). Yet there is a tendency with social programs to move immediately from Phase 1 to Phase 3.

Also, there is a very significant difference between designing and implementating solutions to technical problems in a steel mill, for example, in contrast with designing and implementing solutions for social problems. For we must recognize, obvious as it is, that social problems intimately involve human beings who are not easy to change.[4] For example, in the area of cigarette smoking, the Surgeon General and the Federal government initiated a campaign in 1964 reporting the dangers of smoking. Over $27 million has been spent in the campaign and the effect on smoking has been slight. A small drop occurred from 1967 (11.73 cigarettes per day per person over 18) to 1969 (10.94 cigarettes per day). The average has again risen, however, to an average of over 11 per day per person over 18.[4] Another example is the campaign advocating drug abstention to drug addicts at a cost of $16 million per year. The modification of ingrown habits, personality traits and social values is a very difficult matter indeed. Yet the solutions for many social problems are greatly dependent upon people's attitudes.

EFFECTIVENESS

Cost-effective ways of solving social problems involve simple technological changes rather than social changes. For example, 59,220 Americans lost their lives on highways in 1970. A study revealed that it cost $88,000 per life saved to support driver education, an attempt to change driver attitudes and habits. However, seat belts were a thousand times more cost effective, because they cost only $88 per life saved. Similarly, we exhort people not to drive while under the influence of alcohol—with little result. Yet a breath-measuring device keyed to the ignition system could eliminate the drunken driver.

The use of technological devices to solve social problems is not a total solution, of course. Over the long run it is best to change people's attitudes the problem. An unchanged drunken driver will probably attempt to disconnect the breath measurement device from his auto or get another person to start the engine.

MAKING BEST USE OF NEW TECHNOLOGY

The New Technology Opportunities Program of the United States government was instituted in the early 1970s with the goal of initiating new technologies to solve social problems. Ideas such as cable-TV hookup so that patients in remote areas can consult with specialists in medical centers, electronic delivery of mail, use of tranquilizers instead of guns to control riots, and recycling of solid wastes have been proposed. The research and development budget of the United States is $16.3 billion. But almost half of this amount is for defense, $3 million is for the Atomic Energy Commission, and $3.2 billion is for space research. The remaining $2 billion is spent for research on health, transportation, agriculture, law enforcement, safety and other fields. The balance between defense, space and social issues must be reexamined and the priorities reordered.

In the following article, Dr. Alvin Weinberg proposes his concept of the "technological fix." As we encounter social problems, we should, in the view of the author, use modern technology to solve the social problem and not be confronted with the vastly more difficult task of altering people's attitudes and habits in order to solve the problems.[5] *

Can Technology Replace Social Engineering?

Alvin M. Weinberg

During the war, and immediately afterward, the federal government mobilized its scientific and technical resources, such as the Oak Ridge National Laboratory, around great technological problems. Nuclear reactors, nuclear weapons, radar, and space are some of the miraculous new technologies that have been created by this mobilization of federal effort. In the past few years there has been a major change in focus of much of our federal research. Instead of being preoccupied with technology, our government is now mobilizing around problems that are largely social. We are beginning to ask what we can do about world population, the deterioration of our environment, our educational system, our decaying cities, race relations, poverty. President Johnson has dedicated the power of a scientifically oriented federal apparatus to finding solutions for these complex social problems.

Social problems are much more complex than are technological problems, and much harder to identify: How do we know when our cities need renewing, or when our population is too big, or when our modes of transportation have broken down? The problems are, in a way, harder to identify just because their solutions are never clear-cut: How do we know when our cities are renewed, or our air clean enough, or our transportation convenient enough? By contrast the availability of a crisp and beautiful technological solution often helps focus on the problem to which the new technology is the solution. I doubt that we would have been nearly as concerned with an eventual shortage of energy as we now are if we had not had a neat solution—nuclear energy—available to eliminate the shortage.

There is a more basic sense in which social problems are harder than are technological problems. A social problem exists because many people behave, individually, in a socially unacceptable way. To solve a social problem one must induce social change—one must persuade many people to behave differently than they have behaved in the past. One must persuade many people to have fewer babies, or to drive more carefully, or to refrain from disliking Negroes. By contrast, resolution of a technological problem involves many fewer individual decisions. Once President Roosevelt decided to go after atomic energy, it was by comparison a relatively simple task to mobilize the Manhattan Project.

The resolution of social problems by the traditional methods—by motivating or forcing people to behave more rationally—is a frustrating business. People don't behave rationally; it is a long, hard business to persuade individuals to forego immediate personal gain or pleasure, as seen by the individual, in favor of longer-term social gain. And indeed, the aim of social engineering is to invent the social devices—usually legal, but also moral and educational and organizational—that will change each person's motivation and redirect his activities to ways that are more acceptable to the society.

The technologist is appalled by the difficulties faced by the social engineer; to engineer even a small social change by inducing individuals to behave differently is always hard even when the change is rather neutral or even beneficial. For example, some rice eaters in India are reported to prefer starvation to eating the wheat we send them. How much harder it is to change motivations where the individual is insecure and feels threatened if he acts differently, as illustrated by the poor white man's reluctance to accept the Negro as an equal. By contrast, technological engineering is simple; the rocket, the reactor, and the desalination plants are devices that are expensive to develop, to be sure, but their feasibility is relatively easy to assess, and their success relatively easy to achieve once one understands the scientific principles that underlie them.

It is therefore tempting to raise the following question: In view of the simplicity of technological engineering, and the complexity of social engineering, to what extent can social problems be circumvented by reducing them to technological problems? Can we identify Quick Technological Fixes for profound and almost infinitely complicated social problems, "fixes" that are within the grasp of modern technology, and which would either eliminate the original social problem without requiring a change in the individual's social attitudes, or would so alter the problem as to make its resolution more feasible? To paraphrase Ralph Nader, to what extent can technological remedies be found for social problems without first having to remove the causes of the problem? It is in this sense that I ask: "Can technology replace social engineering?"

The Major Technological Fixes of the past

To explain better what I have in mind I shall describe how two of our most profound social problems—poverty and war—have in some limited degree been solved by the Technological Fix, rather than by the methods of social engineering.

The traditional Marxian view of poverty regarded our economic ills as being primarily a question of maldistribution of goods. The Marxist recipe for elimination of poverty, therefore, was to eliminate profit, in the erroneous belief that it was the loss of this relatively small increment from the worker's paycheck that kept him poverty-stricken. The Marxist dogma is typical of the approach of the social engineer: One tries to convince or coerce many people to forego their short-term profits in what is presumed to be the long-term interest of the society as a whole.

The Marxian view seems archaic in this age of mass production and automation, not only to us, but apparently to many East European economists. For the brilliant advances in the technology of energy, of mass production, and of automation have created the affluent society. Technology has expanded our productive capacity so greatly that even though our distribution is still inefficient and unfair by Marxian precepts, there is more than enough to go around. Technology has provided a "fix"—greatly expanded production of goods—which enables our capitalist society to achieve many of the aims of the Marxist social engineer without going through the social revolution Marx viewed as inevitable. Technology has converted the seemingly intractable social problem of widespread poverty into a relatively tractable one.

My second example is war. The traditional Christian position views war as primarily a moral issue: If men become good, and model themselves after the Prince of Peace, they will live in peace. This doctrine is so deeply ingrained in the spirit of all civilized men that I suppose it is blasphemy to point out that it has never worked very well—that men have not been good, and that they are not paragons of virtue or even of reasonableness.

Although I realize it is a terribly presumptuous claim, I believe that Edward Teller may have supplied the nearest thing to a Quick Technological Fix to the problem of war. The hydrogen bomb greatly increases the provocation that would lead to large-scale war, not because men's motivations have been changed, nor because men have become more tolerant and understanding, but rather because the appeal to the primitive instinct of self-preservation has been intensified far beyond anything we could have imagined before the H-bomb was invented. To point out these things today, with the United States involved in a shooting war, must sound hollow and unconvincing; yet the desperate and partial peace we have now is far better than a fullfledged exchange of thermonuclear weapons. One can't deny that the Soviet leaders now recognize the force of H-bombs, and that this has surely contributed to the less militant attitude of the USSR. And one can only hope that the Chinese leadership, as it acquires familiarity with H-bombs, will also become less militant. If I were to be asked who has given the world a more effective means of achieving peace—our great religious leaders who urge men to love their neighbors and thus avoid fights, or our weapons technologists who simply present men with no rational alternative to peace—I would vote for the weapons technologist. That the peace we get is at best terribly fragile I cannot deny; yet, as I shall explain, I think technology can help stabilize our imperfect and precarious peace.

The Technological Fixes of the Future

Are there other Technological Fixes on the horizon, other technologies that can reduce immensely complicated social questions to a matter of "engineering"? Are there new technologies that offer society ways of circumventing social problems and at the same time do not require individuals to renounce short-term advantage for long-term gain?

Probably the most important new Technological Fix is the intra-uterine device for birth control. Before the IUD was invented, birth control demanded the very strong motivation of countless individuals. Even with the pill, the individual's motivation had to be sustained day in and day out; should it flag even temporarily, the strong motivation of the previous month might go for naught. But the IUD, being a one-shot method, greatly reduces the individual motivation required to induce a social change. To be sure, the mother must be sufficiently motivated to accept the IUD in the first place, but, as experience in India already seems to show, it is much easier to persuade the Indian mother to accept the IUD once than it is to persuade her to take a pill every day. The IUD does not completely replace social engineering by technology. Indeed, in some Spanish-American cultures where the husband's manliness is measured by the number of children he has, the IUD attacks only part of the problem. Yet in many other situations, as in India, the IUD so reduces the social component of the problem as to make an impossibly difficult social problem much less hopeless.

Let me turn now to problems which, from the beginning, have had both technical and social components—those concerned with conservation of our resources: our environment, our water, and our raw materials for production of the means of subsistence. The social issue here arises because many people by their individual acts cause shortages and thus create economic, and ultimately social, imbalance. For example, people use water wastefully, or they insist on moving to California because of its climate. And so we have water shortages; or too many people drive cars in Los Angeles with its curious meteorology, and Los Angeles suffocates from smog.

The water resources problem is a particularly good example of a complicated problem with strong social and technological connotations. Our management of water resources in the past has been based largely on the ancient Roman device, the aqueduct. Every water shortage was to be relieved by stealing water from someone else who at the moment didn't need the water or was too poor or too weak to prevent the theft. Southern California would steal from Northern California, New York City from upstate New York, the farmer who could afford a cloud-seeder from the farmer who could not afford a cloud-seeder. The social engineer insists that such expedients have gotten us into serious trouble; we have no water resources policy, we waste water disgracefully, and, perhaps, in denying the ethic of thriftiness in using water, we have generally undermined our moral fiber. The social engineer, therefore, views such technological shenanigans as being shortsighted, if not downright immoral. Instead, he says, we should persuade or force people to use less water, or to stay in the cold middlewest where water is plentiful instead of migrating to California where water is scarce.

The water technologist, on the other hand, views the social engineer's approach as rather impractical. To persuade people to use less water or to get along with expensive water is difficult, time-consuming, and uncertain in the extreme. Moreover, say the technologists, what right does the water resources expert have to insist that people use water less wastefully? Green lawns and clean cars and swimming pools are part of the good life, American style, 1966, and what right do we have to deny this luxury if there is some alternative to cutting down the water we use?

Here we have a sharp confrontation of the two ways of dealing with a complex social issue: The social engineering way which asks people to behave more "reasonably," the technologist's way which tries to avoid changing people's habits or motivation. Even though I am a technologist, I have sympathy for the social engineer. I think we must use our water as efficiently as possible, that we ought to improve people's attitudes toward the use of water, and that everything that can be done to rationalize our water policy will be welcome. Yet, as a technologist, I believe I see ways of providing more water more cheaply than the social engineers may concede is possible.

I refer to the possibility of nuclear desalination. The social engineer dismisses the technologist's simpleminded idea of solving a water shortage by transporting more water, primarily because in so doing the water user steals water from someone else—perhaps foreclosing the possibility of ultimately utilizing land now only sparsely settled. But surely water drawn from the sea deprives no one of his share of water. The whole issue is then a technological one: Can fresh water be drawn from the sea cheaply enough to have a major impact on our chronically water-short areas like Southern California, Arizona, and the eastern seaboard?

I believe the answer is yes, although much hard technical work remains to be done. A large program to develop cheap methods of nuclear desalting has been undertaken by the United States, and I have little doubt that within the next ten to twenty years we shall see huge dual-purpose desalting plants springing up on many parched sea coasts of the world. At first these plants will produce water at municipal prices. But I believe, on the basis of research now in progress at Oak Ridge and elsewhere, water from the sea at a cost acceptable for agriculture—less than ten cents per one thousand gallons—is eventually in the cards. In short, for areas close to the sea coasts, technology can provide water without requiring a great and difficult effort to accomplish change in people's attitudes toward the utilization of water.

The Technological Fix for water is based on the availability of extremely cheap energy from very large nuclear reactors. What other social consequences can one foresee flowing from really cheap energy eventually available to every country, regardless of its endowment of conventional resources? While we now see only vaguely the outlines of the possibilities, it does seem likely that from very cheap nuclear energy we shall get hydrogen by electrolysis of water, and thence the all-important ammonia fertilizer necessary to help feed the hungry of the world; we shall reduce metals without requiring coking coal; we shall even power automobiles with electricity, via fuel cells or storage batteries, thus reducing our world's dependence on crude oil, as well as eliminating our air pollution insofar as it is caused by automobile exhaust or by the burning of fossil fuels. In short, the widespread availability of very cheap energy everywhere in the world ought to lead to an energy autarchy in every country of the world, and eventually to an autarchy in the many staples of life that should flow from really cheap energy.

Will Technology Replace Social Engineering?

I hope these examples suggest how social problems can be circumvented or at least reduced to less formidable proportions by the application of the Technological Fix. The examples I have given do not strike me as being fanciful, nor are they at all exhaustive. I have not touched, for example, upon the extent to which really cheap computers and improved technology of communication can help improve elementary teaching without having first to improve our elementary teachers. Nor have I mentioned Ralph Nader's brilliant observation that a safer car, and even its development and adoption by the automobile industry, is a quicker and probably surer way to reduce traffic deaths than is a campaign to teach people to drive more carefully. Nor have I invoked some really fanciful Technological Fixes: like providing air conditioners, and free electricity to operate them, for every Negro family in Watts on the assumption, suggested by Huntington, that race rioting is correlated with hot, humid weather—or the ultimate Technological Fix, Aldous Huxley's "soma pills" to eliminate human unhappiness without improving human relations in the usual sense.

My examples illustrate both the strength and the weakness of the Technological Fix for social problems. The Technological Fix accepts man's intrinsic shortcomings and circumvents them or capitalizes on them for socially useful ends. The Fix is therefore eminently practical and in the short term relatively effective. One doesn't wait around trying to change people's minds: If people want more water, one gets them more water rather than requiring them to reduce their use of water; if people insist on driving autos while they are drunk, one provides safer autos that prevent injuries even in a severe accident.

But the technological solutions to social problems tend to be incomplete and metastable, to replace one social problem with another. Perhaps the best example of this instability is the peace imposed upon us by the H-bomb. Evidently the *pax hydrogenium* is metastable in two senes: In the short term, because the aggressor still enjoys such an advantage; in the long term, because the discrepancy between have and have-not nations must eventually be resolved if we are to have permanent peace. Yet, for these particular shortcomings, technology has something to offer. To the imbalance between offense and defense, technology says let us devise passive defense which redresses the balance. A world with H-bombs and adequate civil defense is less likely to lapse into thermonuclear war than a world with H-bombs alone, at least if one concedes that the danger of thermonuclear war mainly lies in the acts of irresponsible leaders. Anything that deters the irresponsible leader is a force for peace: A technologically sound civil defense would therefore help stabilize the balance of terror.

To the discrepancy between haves and have-nots, technology offers the nuclear energy revolution, with its possibility of autarchy for haves and have-nots alike. How this might work to stabilize our metastable thermonuclear peace is suggested by the possible political effect of the recently proposed Israeli desalting plant: I should think that the Arab states would be much less set upon destroying the Jordan River Project if the Israelis had a desalination plant in reserve that would nullify the effect of such action. In this connection, I think countries like ours can contribute very much. Our country will soon have to decide whether to continue to spend 5.5×10^9 per year for space exploration after our lunar landing. Is it too outrageous to suggest that some of this money be devoted to building huge nuclear desalting complexes in the arid ocean rims of the troubled world? If the plants are powered with breeder reactors, the out-of-pocket costs, once the plants are built, should be low enough to make large-scale agriculture feasible in these areas. I estimate that for 4×10^9 per year we could build enough desalting capacity to feed more than ten million new mouths per year, provided we use agricultural methods that husband water, and we would thereby help stabilize the metastable, bomb-imposed balance of terror.

Yet I am afraid we technologists will not satisfy our social engineers, who tell us that our Technological Fixes do not get to the heart of the problem; they are at best temporary expedients; they create new problems as they solve old ones; to put a technological fix into effect requires a positive social action. Eventually, social engineering, like the Supreme Court decision on desegregation, must be invoked to solve social problems. And of course our social engineers are right: Technology will never replace social engineering. But technology has provided and will continue to provide to the social engineer broader options, making intractable social problems less intractable; perhaps most of all, technology will buy time, the precious commodity that converts violent social revolution into acceptable social evolution.

Our country now recognizes—and is mobilizing to meet—the great social problems that corrupt and disfigure our human existence. It is natural that in this mobilization we should look first to the social engineer. Unfortunately, however, the apparatus most readily available to the government, like the great federal laboratories, is technologically, not socially oriented. I believe we have a great opportunity here for, as I hope I have persuaded the reader, many of our social problems do admit of technological solutions. Our already deployed technological apparatus can contribute to the resolution of social questions. I plead, therefore, first for our government to deploy its laboratories, its hardware contractors, its engineering universities, on social problems. And I plead secondly for understanding and cooperation between technologist and social engineer. Even with all the help he can get from the technologist, the social engineer's problems are never really solved. It is only by cooperation between technologist and social engineer that we can hope to achieve what is the aim of all technologists and social engineers—a better society, and thereby a better life, for all of us who are part of society.

The idea that technology may be utilized to overcome social problems is appealing. Without changing attitudes, technology might provide effective solutions. In the following article, Professor Etzioni discusses further the uses of technology to solve social problems; he documents this use for six technologies.[6] It appears that the use of technologies does work or achieve some goals for some significant portion of populations without additional inputs. When significant human labor is involved with the technological system, then often the economics originally realized by the technology are diminished.*

Technological "Shortcuts" to Social Change

Amitai Etzioni and Richard Remp

The idea that technological developments might be used to reduce the costs and pains entailed in dealing with social problems is appealing. A broad rationale for this approach is suggested by an analogy between the development of modern techniques of producing consumer goods and the search for new techniques of providing social services. Mass production and considerable reductions in cost per unit of consumer goods were achieved by an increased reliance on machines (broadly conceived to include communications satellites and computers) and a decreased reliance on muscle and brainpower, on persons. However, up to now in social services, in which performance is frequently criticized for falling far below desirable levels, most work has been unmechanized. Since the need for services in these areas is great, available resources low, and trained manpower in short supply, it seems useful to consider replacing the "human touch," at least in part, by new technologies.

A Methodological Note

To explore this question, we reviewed existing studies that evaluate the effectiveness of technological shortcuts in dealing with six distinct social problems. The term "technology" is construed here to apply to biological and physiological processes, as well as physical processes. This is in accord with R. S. Merrill's use of the term: "the concept of technology centers on processes that are primarily biological and physical rather than psychological or social processes." "Hard," or physical, technologies are emphasized

*From *Science*, Vol. 175, January 7, 1972, pp. 31–38. Copyright by the American Association for the Advancement of Science.

rather than "soft," or social-psychological, technologies because the shortcuts in question derive their efficiency not from the reorganization, but from the replacement of human services in the handling of social problems.

The technologies and problem areas selected were methadone in controlling heroin addiction; instructional television (ITV) in teaching; Antabuse (disulfiram) in treating alcoholics; gun control in reducing crime; the breath analyzer in highway safety; and the intrauterine device (IUD) in birth control. These technological innovations may be viewed as shortcuts because they either serve as a replacement for manpower (for example, the use of ITV instead of teachers) or they reduce the need for manpower (for example, methadone reduces the need for therapists, social workers, and guards in the treatment of heroin addiction).

The findings reported here are, of course, affected by the developmental status of the technologies studied. If we had selected technologies that were already in routine use, our findings might have been more optimistic. However, few of the technologies routinely used in the human services area aim at the core of the problem (although there are various auxiliary instruments—for example, teaching aids). We focused on procedures that would fundamentally affect the service in question. As a consequence, technologies still in various experimental stages were studied. Technologies other than the six reported were surveyed, although less intensively (such as the use of computers for instruction and cable television for conducting town hall-like dialogues); they do not differ significantly from those selected, from the viewpoint of the issues at hand.

The Main Findings

To the degree that the data permit us to conclude, each of the six technologies "works," in that it allows the handling of a significant part of the social problem faced at a considerable reduction in cost and in pains of adjustment. These conclusions are tentative, since the technologies are still experimental and limitations are inherent in evaluation research.

Methadone

Methadone is a narcotic drug, a synthetic, addictive opiate, which is being administered experimentally to heroin addicts to prevent the use of heroin. In one program, conducted in New York City and currently involving several thousand heroin addicts, patients have been maintained on methadone for periods of up to 6 years. Follow-up studies indicate that most of the individuals being maintained on methadone on an outpatient basis have not become heroin addicts again. Their involvement with the police and courts is over, and they hold jobs or study. That is, they have become socially rehabilitated. Among a group of 990 men examined in a follow-up study, the percentage of those employed or in school rose from 29 percent at admission to the methadone program, to 65 percent after 1 year on the program, to 74 percent after 2 years, and reached 92 percent by the end of their third year on the program.

At this time, the therapeutic use of methadone has not been federally authorized, and all the reports are based on what are legally defined as research and experimentation programs. Accordingly, there is little evidence on the usefulness of methadone under routine conditions of mass use. On the other hand, the number of addicts under treatment in these programs has risen rapidly in the last 2 years [to 3485 by 31 October 1970], and the research element has played a smaller role in several recent programs; thus, the technique is, in practice, no longer strictly experimental.

Antabuse

In contrast to methadone, which is a blocking drug reported to prevent individuals from getting a high from he use of heroin, Antabuse is a counterdrug that produces an aversion to alcohol. It makes those who consume alcohol while on the drug feel nauseated and otherwise uncomfortable. It has not been subjected to the extensive research or given the same attention by the press that methadone has, and it is much less "accepted" by public authorities or members of the medical profession. One reason is that, during the early 1950's, relatively high dosages (1.5 grams) of Antabuse were used, and the resulting troublesome and occasionally dangerous side effects suggested to observers that the drug's utility might be quite limited. However, subsequent research indicated that lower dosages (0.25 gram) might be used effectively, avoiding or rendering manageable most of the side effects while permitting the necessary discomfort-producing alcohol-Antabuse reaction to occur. In one follow-up study, 71 patients out of an original testing group of 118 patients were contacted 2 years after the start of treatment. Fifty-one percent of those contacted reported no relapses (17 of these 36 individuals were still taking Antabuse), 32 percent reported one or more relapses (21 of these 23 individuals were taking Antabuse), and 17 percent reported failure (none of these individuals was taking Antabuse). However, since many evaluation studies of Antabuse combine patients who are continuing to use the drug and patients who have stopped using it, it is unclear from "abstinence" figures the extent to which abstinence is maintained once administration of the drug has stopped. In any event, greater experimental use of the drug seems advisable.

Instructional Television

There are several hundred studies of the effectiveness of television for instructional purposes. There are even several summary reports of the ITV studies. The reviews report that "the vast majority of these studies has revealed 'no significant differences' in measured performance between students who were taught via television and those who were taught directly". For example, in 1956 ten lessons in physics and ten lessons in algebra were broadcast to 2405 students in 34 Chicago high schools. At the end of the lessons, examinations on the material were given both to students who had received televised instruction and to students who had received face-to-face instruction based upon course outlines prepared by the television instructors; no significant differences were found between the scores of the two groups. Although no direct inference can be drawn from the statistical finding of "no significant differences," the large number of studies in which this occurred implies great success, because it suggests that the carefully rehearsed, repeatedly usable, and economical videotaped instruction might be used where teachers are not available, or to release teachers for other tasks. A single lesson can be presented to many more students across space through closed circuit or broadcast television, and across time through videotape or film than through face-to-face instruction. This multiplicative capacity of ITV also implies that the average quality of contemporary instruction might be raised if the technology were used selectively, to extend the influence of the more competent instructors.

Unlike Antabuse, in whose case effectiveness is less well-established and where the extent to which it is established is not widely known, the effectiveness of ITV is fairly widely known—yet ITV is not widely used. A report prepared for the President and Congress in 1969 estimated that the use of televised instruction in the nation, as well as films, filmstrips, records, programmed texts, and computer programs, does not fill more than 5 percent of instructional time. It appears that ITV has a considerable capacity, not presently being exercised, for increasing the quantity and quality of instruction.

Breath Analyzer

The breath analyzer is a relatively simple device for detecting in a preliminary fashion whether or not a person is intoxicated. The subject blows into a tube or balloon, and a chemical reactant in the device changes color, thereby giving a rough indication of the proportion of alcohol in the subject's breath.

Since a high proportion of automobile drivers in fatal crashes (to 60 percent) have been found to have enough alcohol in their blood to significantly impair their driving ability [concentrations of more than 0.10 percent alcohol in the blood], it seems that if a significant portion of motorists could be prevented from driving while intoxicated the rate of fatal auto accidents might be reduced considerably. In Great Britain, when in 1967 a law was passed making it illegal to drive a car while intoxicated (having a concentration of 0.08 percent or more alcohol in the blood) and an enforcement program utilizing the breath analyzer was highly publicized, fatalities in auto accidents during the following year decreased by 15 percent, serious injuries decreased by 11 percent, and total auto accidents decreased by 10 percent. Although during the second year of the program the reductions in deaths (10 percent compared to the year prior to the program), injuries (9 percent), and accidents (10 percent) were somewhat smaller than they were during the first year, the figures continued to suggest that the approach did reduce auto fatalities and accidents, particularly during the hazardous nighttime drinking hours. There are fewer data on this technology than on the others, in part because a simple, portable, screening procedure, which is central to a workable auto safety program, was only recently developed.

Intrauterine Device

The intrauterine device is an object, made of various materials (often plastic or nylon) in various shapes (such as rings, spirals, or loops), which is inserted in the uterus for the purpose of preventing conception. Although effectiveness varies with the size and type of device and with the age and pregnancy history of the woman, "pregnancy rates for [women using] the IUD are quite low, typically with the best devices about 2 to 3 per 100 cases in the first year". The IUD seems to be more effective in controlling birth rates than are most other birth control measures, such as diaphragms or pills, since the latter require more continuous motivation on the part of the user and are apt to be more expensive.

The IUD, however, involves a number of difficulties: it is sometimes involuntarily expelled, and instances of medical complications are reported (for example, perforation of the uterus, excessive bleeding, the promotion of infection, and, possibly, cancer). One major follow-up study reported rates of 14.9 involuntary expulsions per 100 first insertions, and 22.0 removals per 100 first insertions for medical reasons 2 years after insertion.

National family planning programs, relying heavily upon IUD's, have been instituted in a number of developing countries during the past decade, with Taiwan and South Korea having particularly high per capita rates of IUD insertion. In 1966, 17 percent of the women from 20 to 44 years of age in South Korea were using IUD's, as were 13 percent of the women from age 20 to age 44 in Taiwan. There is disagreement on the implications of the characteristics and the IUD retention rates of participants in the programs for the long-term utility of the IUD. In addition, it is difficult to estimate the exact impact that large-scale IUD programs have upon national birth rates because of the difficulty of gauging the effectiveness of alternative birth control measures that might have been used by participants in the program. However, assuming that IUD acceptors (i) were as fertile as, or more fertile than, other married women in their age group and (ii) that they would not have used other contraceptive methods more effectively than the average woman in each age group, it has been estimated that, if the 547,000 IUD's being used in South Korea at the beginning of 1967 remained in situ for 1 year, there would be between 110,000 and 160,000 fewer births per year, or a drop in the yearly birth rate of 3.7 to 5.3 [South Korea had a crude birth rate of 44.7 in 1959]. Similar calculations indicate that Tawian's program would lead to a drop of 3.0 to 5.0 in the annual crude birth rate [Taiwan's crude birth rate in 1964 was 37.1] and that Pakistan's much less extensive family planning program would lead to a reduction of 1.0 in the nation's crude birth rate [which was 43.0 to 46.0 in 1964].

Gun Control

Since firearms are unusually efficient and deadly weapons [one study found firearms attacks to be five times more deadly than knife attacks] and since most homicides appear to be impulsive acts [Federal Bureau of Investigation statistics indicate that 82 percent of all murders in 1962 were committed within the family unit or among acquaintances], a significant reduction in the availability of domestic firearms should force many potential killers to use less efficient weapons or to express their aggression in other, less harmful ways. Gun control, in contrast to the other innovations discussed here, is a "negative" alteration in the technology of a social problem: the restriction or removal of a preexisting technology. Such procedures might be useful when a developing technology (or the changing context of an existing technology) leads to a widely recognized social problem *and* when that technology is susceptible to control or elimination through legislation.

It has been suggested that the gun control example is not comparable with the others because the action involved here is social legislation, not a technological intervention. This is clearly true. Our reasons for including the example are several. First, the net effect is achieved not by the change in the law, but by a change in the technology of violence available to citizens. If a law is passed and this technology is not changed (a rather common occurrence), gun control will not be affected. If the technology is changed in some other way without a change in law (if more powerful guns are bought or if there is a shortage of bullets), crime will be affected. Changes in the technology of violence are hence the salient and relevant variables. Perhaps "changing the level of armaments" is a better characterization of the variable, but "gun control" is the term commonly used. Second, as the discussion indicates, from every other viewpoint, the *effects* of this intervention and the *problems* it raises are very similar to those of the other interventions (for example, it assumes no change in the personalities of the criminals; it works better for some subgroups of the population afflicted with the problem and some goals than others; it is relatively inexpensive, compared to nontechnological approaches). Third, as the study of pollution control and food additives suggests, there are many other areas in which the main effect would be brought about by removing rather than adding a physical element. Hence the special interest in this intervention.

The relationship between the restrictiveness of gun control laws, including domestic disarmament, and rates of homicide and armed robbery has been investigated by comparing the crime statistics of regions with varying degrees of gun control. Overall, these comparisons indicate that restrictive gun control measures, such as prohibiting the private ownership of handguns, are associated with significantly lower rates of homicide, armed robbery, accidental death, and injury due to firearms.

This observation is based mainly upon international comparisons, since gun control laws within the United States are very limited in formal restrictions and the effect of them is vitiated by the largely unrestricted movement of firearms between jurisdictions with differing degrees of formal restrictions. International comparisons suggest that the availability or absence of firearms, particularly handguns, significantly influences rates of homicide and armed robbery.

In summary, although in most instances the technological shortcuts reviewed are not widely utilized, they all seem capable of contributing to the reduction of the relevant social problems.

Limits of the Evaluation Data

A detailed evaluation of the utility of the technologies reviewed here is inhibited because much of the needed information is missing. Evaluative studies frequently report a technology's achievement of a broadly defined objective, but bypass questions of how the result was produced and what other effects occurred. The favorable evaluations of methadone programs, for example, rest primarily on observations that patients usually end their involvement with the police and begin to hold steady jobs or study. There are, however, several competing theories as to how methadone wrks. One researcher argues that the primary effect of methadone is physiological (for example, methadone fills a metabolic deficiency that was created by heroin use); others see its major consequences as psychological (for example, it tranquilizes the user and suppresses psychic problems); still others believe the methadone program changes the social context (heroin addicts are in touch with criminals, methadone users with doctors and social workers). Thus, although the desirable results of the program are fairly evident, the processes involved are not clear.

The long-term consequences of the treatment are unclear also. Methadone is addictive and is used most often as a *maintenance* drug, which has to continue to be administered if it is to be effective. Programs in which low dosages of methadone are used or in which the drug is withdrawn are reported to be much less effective in curbing heroin use. It is not known what the long-term physical and psychological effects are on the person being maintained on methadone. Thus, when the methadone program is said to "work," the evaluation reflects chiefly the more visible and immediate effects, and not an understanding of how the program "works" or of what the long-term effects are.

Similarly, evaluation studies of ITV do not allow us to answer several key questions. For example, it is difficult to tell whether ITV is effective by itself or whether it requires human supplements, such as discussion groups; many studies have obscured this question by comparing "live" teachers to ITV *plus* various forms of supplementary live teaching. The remaining studies do not allow a sharp evaluative picture because of serious methodological limitations. A review of 250 comparisons of live and televised instruction contained in 31 reports showed that most of the comparisons were ill designed, used inadequate samples, misinterpreted the data, or suffered from other serious flaws. Nor do the studies indicate clearly which topics can be taught effectively over television and which cannot. Can one really communicate normative and esthetic values (a central objective of many courses in civics, history, art, and literature) as effectively as substantive information over the cold, impersonal, nonreactive medium?

The information about the other technologies is similarly fragmentary. The use of the breath analyzer in a British highway safety program is associated with a reduction of fatalities on the road, but it is not clear whether this was due primarily to the novelty of the program or to a lasting deterrent power of the program. Additionally, there are serious questions about the reliability of the breath analyzer as an indicator of alcohol levels in the blood. There are also some serious obscurities in the data on gun control. Some regions within the United States (for example, the Midwestern states) that have relatively few limitations on guns also have significantly lower rates of dangerous crimes. While the influence of factors other than the availability of firearms upon crime rates has never been doubted, if the presence of guns is a very significant influence these observations are difficult to account for. Finally, the precise manner in which the IUD works is not known, nor has the claim that it might encourage cancer in some women been fully rejected. Thus, beyond the indications of the technologies' overt effectiveness, the limitations of the data are lack of knowledge of the process involved and of long-term effects, simple incomplete evaluation (for example, it has not been established whether methadone really blocks the effects of heroin), contamination of variables, and the lack of sufficient "controls" or comparative data.

The causes of limitations of the knowledge available are numerous. Some are external to the research process, others are intrinsic. Jointly, they make the information not just incomplete and tentative, but fragmented and frequently difficult to rely upon. Moreover, it seems that highly reliable evaluations may be very difficult to produce, and we may need to act without them.

Extrinsic Factors

The research process is influenced from other areas of social activity, through economic pressures, political considerations, moral inhibitions, and legal constraints. While basic research is to a considerable degree protected from these forces, much of the research relevant to the societal application of new technologies is vulnerable. Such research lacks the ideological dedenses and, usually, the institutional protection available to basic research. The implications of applied research are usually apparent enough to arouse opposition among some of the people who would be affected by the adoption of the experimental procedure.

Political forces, most clearly illustrated by the gun lobbies, make it very difficult to obtain research funds for the consideration of, to say nothing of experimentation with, domestic disarmament—the type of gun control likely to be most effective for reducing violent crime. Experimentation with the extensive use of ITV tends to affront some faculty members, who are concerned about being replaced by machines.

Morally, the use of drugs to treat addiction is criticized for permitting patients to avoid developing the strong "willpower" that many people—including spokesmen of the medical profession—believe a person should have. Legal problems arise when the breath analyzer is administered involuntarily, and the subject is, in effect, made to incriminate himself.

The limited amount of support for such research and the manner in which it is administered also contribute to our fragmentary understanding of social technologies. A belief in the generally equal competence of all potential investigators and in their ability to determine the most useful direction research can take prevails among those who manage research in this country. Of course equality in the distribution of resources among researchers and institutions is not even approximated, but pressure for such a distribution exists. Hence, research by many people who have no business doing research is supported. The resulting poor quality of research is illustrated well by many of the ITV studies: they eat up funds, clutter up the libraries and computer memories, confuse policy-makers, and retard the developments in question. An arrangement whereby all ITV studies (or studies on any other subject) would be carried out by perhaps three competent research centers would be, to put it mildly, in sharp contrast to the procedure traditionally followed in this country. It would offend the community of researchers as well as Congress.

Equally problematic is the prevalent belief that the researchers alone should decide what to study. Unlike work ordered by the National Aeronautics and Space Administration and the Department of Defense, much of which is contracted quite specifically, especially when technologies are involved, the government and foundations have tended to shy away, at least until recently, from contracting for specific technologies in social areas. Most federal agencies concerned with domestic affairs do not have the research budget or the intellectual manpower needed to develop a clear research program that would allow them to approach a university or research corporation and order the studies needed to answer specific questions in their area of concern. Thus, instead of providing direct support to troubled areas, we frequently seed oceans of studies in the hope that bread will wash ashore where needed.

The limited knowledge of technologies also reflects the relatively small investment of resources and professional manpower in this area. From 1965 to .1970, the amount of money spent on research conducted in each of the areas examined can be safely estimated as below the million-dollar mark for the 5-year period, with the IUD being a possible exception (however, much of the expenditure on the IUD is not for research, but for demonstration projects). Since there is a correlation between investment in research and development and scientific output, this is a significant limitation on development in these areas. The relatively limited development of the behavioral sciences in general also limits potential understanding of technologies.

Intrinsic Factors

Even if there were more support and guidance of research concerned with technologies, intrinsic obstacles to understanding would still lie in the very process of producing knowledge. Scientific disciplines are primarily analytic in operation; they fragment the world into their respective sectors and then study these sectors as if the rest did not exist, or could be held constant. In doing applied research, one should be able to piece together different disciplines' findings into an interpretation of the actual system in which the interventions being studied take place. However, the integration of analytic observations from various fields is a difficult and hazardous task. What tends to happen is that the study of integrated, practical problems occurs in nonanalytic disciplines, such as engineering, medicine, social work, and education, which draw on their own traditions as well as on the analytic sciences. The process by which an applied discipline draws on the observations of analytic disciplines is very poorly understood, but it appears to be much less systematic and sustained than is often assumed. Research conducted in these applied disciplines seems to be oriented by two ranked criteria: effectiveness in advancing practical ends, and, in a secondary, supportive role, the canons of methodological adequacy and analytic precision characteristic of the basic sciences. The considerable reliance on nonanalytic disciplines in the development and evaluation of technologies is a major reason we know more about the extent to which a technology "works" (that is, is apparently effective) than about how it achieves its effects. The "how" requires theorizing that tends to be analytic. Hence, the experimental applications are often described as "approximations," evaluated on the basis of their most obvious consequences as "usable" or not so, but are full of the kinds of gaps in understanding discussed earlier.

In short, the combination of extrinsic pressures and intrinsic hindrances seriously inhibits the development of sophisticated knowledge in these areas. This situation suggests that evaluations of the effectiveness of technologies will continue to be based on relatively crude information and that policy-makers will have to continue to draw on such information in their decisions, although, of course, some improvement is possible.

Accordingly, one should not oversell technology assessment as promising, but instead recognize that technological shortcuts, like shortcuts across fields, are often crude, although useful, pathways.

Specification of the Effectiveness Proposition

In the six cases studied, and in others surveyed more informally, it appears that the shortcut works much better for one or more of the important subgroups of the population afflicted with the problem and for the remedial goals than it does for the others. Even when we cannot specify how the technology influences the people it does serve, we can often distinguish these people and effects from other subgroups of the population involved in the problem and from other remedial goals. Gun control appears more likely to save lives and avoid crimes in the cases of impulsive killers than it does in the case of determined criminals. Fortunately, impulsive men constitute the majority of convicted killers. The breath analyzer seems more effective in deterring the social drinker than in deterring the habitual drinker, who is the core of the drunken-driver problem. Thus, even crude ideas of the technological shortcut's operation and effects can lead to major discriminations about what proportion of the population afflicted with the problem is likely to be affected.

Similarly, in terms of goals, methadone seems to "work" *if* our purpose is *social* rehabilitation. However, it seems to have little value as a source of psychological rehabilitation, other than removing the effects of the taboos on heroin. That is, if a person uses heroin to deal with psychic stresses—because he wishes to escape reality, or because he is overdependent on his mother—methadone does not seem likely to make him mature enough to be willing to face reality, or cut off his undue attachment to his mother.

Thus, even under conditions of crude information, which may well be endemic to this type of evaluation, asking "Works for whom (which subgroup of the population afflicted with a given problem) and for what purposes?" clarifies the probable effectiveness of the shortcuts. The six shortcuts examined here seem, with the possible exception of the breath analyzer, to reach a significant subgroup of the population afflicted with a given problem and several of the goals.

Technology as an Alternative

A common way of evaluating a remedial social procedure is to ask how far it goes toward achieving the final goal—does it eliminate addiction, or end crime? Ignoring the fact that all remedial social procedures fall markedly short of eliminating the problems altogether obscures significant differences among them. A more profitable form of evaluation is to inquire how effective the technique is, relative to other procedures. Accordingly, it seems useful to review the efficacy of some alternatives to the use of technologies in the reduction of social problems.

One of the main alternatives to the use of social technologies in social problems is the use of punitive procedures. Heroin addicts are incarcerated; penalties for traffic violations have increased, and it has been suggested that people who have "too many" children be penalized by removing tax deductions for dependents or charging them a special tax. However, as criminologists M. Wolfgang and F. Ferracuti have observed, "punishment per se is certainly a confirmed failure as a treatment technique". The limited deterence provided by imprisonment has been well documented. Studies have shown that one-half to two-thirds of those confined in prisons, reformatories, and jails in the United States have served previous sentences in the same or other institutions. Penalties are certainly partly effeetive, probably particularly so with regard to first offenders. Still, the limited capacity of punitive procedures, operating alone, to deter motivated behavior and the difficulty of sustaining a high level of public support for such approaches have been repeatedly demonstrated. Prohibition is the best known case in point. Recently, laws prohibiting marijuana have been compared to those prohibiting alcohol; they are being eroded. While a minor reduction in the population afflicted with a problem might be achieved through the imposition of increased penalties, the procedure's direct operational expenses and the indirect costs, such as increased alienation and the growth of patterns of evasion, are likely to be high.

The information approach—educating people on the undesirability of smoking, drinking to excess, using narcotics, or being overweight—seems effective mainly when coupled with other techniques. By itself, the providing of information, particularly through the mass media, tends to be ineffective, since the reasons that people engage in these forms of conduct are often deep-seated and not apt to be altered by an encounter with new information. Contrary to a once widely held belief, studies show the mass media to be an effective means of persuasion only in matters in which there is little or no resistance to the adoption of new views. It is possible to use advertising to get people to switch from one brand of liquor to another, but not to get them to stop drinking.

The therapeutic approach—procedures such as psychotherapy, psychoanalysis, and rehabilitation counseling—has been generally regarded as ineffective, particularly with regard to specific pathologies such as alcoholism, heroin addiction, or criminal behavior. Even in the case of more diffusely defined psychological disorders, follow-up studies have shown psychotherapeutic treatments to be no more effective than custodial care in state hospitals, medical care that is not deliberately psychotherapeutic, or, in some studies, the absence of formal treatment altogether. Jerome Frank has observed that

... statistical studies of psychotherapy consistently report that about two-thirds of neurotic patients and 40 percent of schizophrenic patients are improved immediately after treatment, regardless of the type of psychotherapy they have received, and the same improvement rate has been found for patients who have not received any treatment that was deliberately psychotherapeutic.

Somewhat more positively, Frank notes that "by and large ... the effect of successful psychotherapy seems to be to accelerate or facilitate healing processes that would have gone on more slowly in its absence". In addition, the availability of psychotherapeutic treatment for problems such as heroin or alcohol addiction is limited. Psychotherapists, psychoanalysts, and social workers often refuse to treat alcoholics, in part as a consequence of small hopes for success. A number of psychotherapeutically oriented programs have attempted to treat heroin addicts. However, when addicts continue to live in their original communities and receive psychotherapy only during visits to program centers, the results are often discouraging. In one such program, which was examined in a comparative study, the attrition rate was high [47 to 57 percent of the patients were seen three times or less], and only a small percentage of the patients completed associated vocational training or held steady jobs.

Thus, for problems such as heroin or alcohol addiction, psychotherapy's discernible impact is quite minor. A major reason seems to be the individual's exposure to only a few hours of psychotherapy a week. When a person is integrated into a "total" therapeutic community, a community in which he lives or is otherwise deeply involved (for example, Synanon for heroin addicts or Alcoholics Anonymous for alcoholics), this form of therapy is much more effective. However, these groups have served only very small fractions of the populations afflicted, and it seems that there is little which can be done to significantly increase their number. These therapeutic groups appear to grow in a gradual, unplanned fashion, and do not seem susceptible to large-scale, controlled promotion. Overall, increased utilization of the therapeutic approach in the problem areas being considered is likely to require considerable investment in trained manpower and to offer only slight reductions in numbers of people involved.

The review of the main alternative means of handling social problems suggests that technological interventions are more effective, in that they can aid a larger proportion of the population afflicted, and are more economical on a per capita basis. A study of the cost effectiveness of various means of preventing deaths in auto accidents, for example, indicated that promoting the use of seat belts may save a life at an individual cost of $87, while driver education procedures may cost as much as $88,000 per life saved. Similarly, even if detaining the heroin addict in a rehabilitation center and maintaining him on methadone in the community both led to the same rate of social rehabilitation (steady employment and no involvement with the police), the costs of rehabilitative detention, both direct and indirect, would be much higher than those of methadone maintenance. Methadone costs only a few cents a day, and, although administration procedures increase expenses somewhat (administration by pharmacists has been tried), the procedure is still far less costly than a custodial program, particularly one involving psychotherapy [which costs $10 to $70 a day per patient].

Technological intervention is not only financially cheaper, its psychic costs are smaller. In many instances the technological intervention seems to reduce significantly the level of motivation required for progress, either by increasing the difficulty of undesired conduct (for example, under domestic disarmament it is very difficult to get a gun, especially on the spur of the moment), or by lowering the degree of self-control required for a desired behavior (for example, the availability of methadone makes forgoing heroin easier, and the IUD allows individuals who are fearful of birth control in general, or who are unable to use contraceptive devices either routinely or in a moment of impulse, to make a single, long-term decision).

Thus, while technological shortcuts are not absolutely effective—they do not deal with the full range of problems—they do seem more likely to increase effectiveness in handling major parts of these problems at lower economic and psychic costs than do punitive, therapeutic ("total" therapy being effective, but of limited applicability), or informational efforts. Although an advertising campaign about the dangers of drug addiction, increased legal penalties for heroin possession, and a considerable increase in psychotherapeutic treatment for addicts might each diminish addiction by a few percentage points, it seems possible that the proportion of heroin addicts susceptible to treatment with methadone may be as great as one third of the total group.

Combining Approaches

Will the effectiveness of a technology be increased if it is combined with one or more of the other approaches? Unfortunately, the available studies offer little guidance in this area. The relative merits of such combinations are obscured because many of the studies that might provide all or part of such comparisons are, for this purpose, contaminated; they include various mixes of remedial approaches with little attention paid to analytically specifying these effects. For instance, reviews of ITV studies which report that ITV is as effective as live instruction include studies with varying kinds and degrees of live adjuncts to the televised instruction. Antabuse and methadone reports also fuse varying degrees of group formation and counseling in their results.

The most we can tell is (i) the technologies "work," or achieve some goals for some significant subpopulation, *without* additional inputs; (ii) to the extent that "human" supplements are required, they tend to remove the main value of the technologies—their economy. If each ITV classroom requires an instructor to supervise discussion, and the instructor has to be well trained, the per capita costs of a televised lecture course may be higher than a live lecture course. If, besides employing a technology, psychotherapy is needed before individuals can learn to forgo heroin, excessive drinking, homicidal and suicidal driving, or be persuaded to change their family size preference, the costs of a remedial program will quickly rise by a factor of 100 or more.

On the other hand, some subgoals will not be achieved by the technological approaches. Methadone provides social, but not psychic, rehabilitation. It may open the door for personal growth without professional help, or leave personal problems untreated. Domestic disarmament will render many aggressive acts less harmful, and might reduce aggressive behavior somewhat, but it will not affect the main sources of aggression. Even if the nontechnological approaches are less cost-effective, they will, given sufficient resources, contribute to the solution. By and large, though, the case for mixing approaches is much less powerful than is often assumed, and the case much stronger for seeking and using technologies exclusively.

CENTRALIZING EFFORT

Professor Etzioni has proposed an Agency for Technological Development for Domestic Programs.[7] This agency is a domestic parallel to the National Aeronautics and Space Administration. The agency would be devoted to technological work specializing in solving domestic social problems. It is proposed that some of the excessive "defense" budget (over $85 billion) be devoted to domestic problems. Technologies for the solution of social problems are, in Etizoni's view, cost-effective. Many of our social problems arise from the fact that services previously sought by relatively small groups, such as high quality education or medical care, are now actively demanded by most citizens. The role of technology in bringing high quality services to the masses is illustrated by the potential of television and computer-aided instruction.

The technological needs of our domestic programs are now serviced by research and development efforts widely among federal and state agencies. Perhaps a concentrated effort in an agency for technology for social problems is required. It is essential to note that in the fields of space and defense the Federal Government is both the main *source* of research and development funds *and the consumer*. In the domestic sphere, on the other hand, often the consumer is a state, city or hospital or other corporate body. Nevertheless, an agency for technological development for domestic programs could present significant prototype solutions which the appropriate city or local institution could adopt if it wished. These prototypes of technological solutions offered to the many states, cities and local institutions would help these institutions avoid investing new technologies as a need arose. Such an agency might assist in the process of offering technological solutions to social problems.

CHAPTER 16 REFERENCES

1. G. Strasser, "Impediments to Societal Problem Solving," *IEEE Spectrum*, July 1971, pp.43–48.
2. H. Brown, "After the Population Explosion," *Saturday Review*, June 26, 1971, pp. 11–13.
3. A. Toffler, *Future Shock*, Random House, New York, 1969.
4. A. Etzioni, "Human Beings Are Not Very Easy To Change After All," *Saturday Review*, June 3, 1972, pp. 45–47.
5. A.M. Weinberg, "Can Technology Replace Social Engineering?" *Bulletin of the Atomic Scientists*, Dec. 1966, pp. 4–8.
6. A. Etzioni and R. Remp, "Technological Shortcuts to Social Change," *Science*, Jan. 7, 1972, pp. 31–37.
7. A. Etzioni, "Agency for Technological Development for Domestic Programs," *Science*, April 4, 1969, pp. 43–49.
8. P.W. Quigg, "Man's Unquenchable Thirst," *Saturday Review/World*, Oct. 9, 1973, pp. 39–40.

CHAPTER 16 EXERCISES

16–1. Dr. Weinberg believes it is easier to build a new technological device to overcome a social problem than it is to change people's attitudes. Can you give an example to counter this thesis?

16–2. Weinberg believes it is very difficult to get people to forego their short-term benefits in a manner that will be in the long-term interest of society as a whole. Can you give examples of cases where the citizens of the United States have foregone their short-term benefits in order to obtain a better quality of life over the long term?

16–3. Weinberg states that Dr. Teller may have supplied a technological fix to the problem of war. Examine this case closely and determine if it is that simple a matter.

16–4. Contrast the article by Weinberg with the views of Dr. Hardin given in the article "The Tragedy of the Commons," which appears in Chapter 10.

16–5. Some people believe that there are many social problems with no technological solutions. Indicate whether you agree or disagree and give an example to support your position.

16–6. List the three most important social problems of today, according to your belief, and describe whether you believe they are tractable to a technological solution.

16–7. Professor Etzioni has proposed the establishment of an Agency for Technological Development for Domestic Programs. Prepare a plan for such an agency and decide what the budget should be and how it should be allocated among the top problems.

16–8. In one of the most technologically-advanced countries of the world, the United States, over half of the water systems are judged to be inadequate. Approximately 23 million Americans are drinking water believed to be substandard and 8 million people are consuming water the Federal Government calls potentially dangerous. Over the world, perhaps a billion people do not have pure drinking water available. It is estimated that 500 million people suffer disabling diseases caused by unsafe water. Yet emphasis on making medical care more widely available often obscures the fact that this toll cannot be significantly reduced by providing more doctors and nurses, but rather by improving the water supply. If a dollar were diverted from health care to preventitive approaches, numerous increases in lives saved could be obtained.[8] Discuss the expenditure of the fiscal resources of a less-developed nation on obtaining safe drinking water *versus* the education of new physicians and nurses for the country.

16–9. One approach to improving the environment is to use new technologies to attack the problems. Another approach is to reverse the trends which placed us in the problem. The use of technology to overcome the problem is probably faster and easier than trying to reverse the original trend. For example, it is easier to add air-pollution control devices to power plants than to reduce the need for the power plants. Explore the use of technology to solve:

(1) The air pollution problem of power plants
(2) The air pollution problem of automobiles
(3) The problem of drunken drivers in autos
(4) The control of crime in the streets of cities.

16–10. Many people believe a new industrial revolution is called for. The United States and Europe, while heavily industrialized, are excessively dependent on outmoded technologies. The United States auto industry relies too heavily on the Otto-cycle engine. The last major innovation introduced into the auto industry was the automatic transmission, which became commonplace a generation ago. Construction is still tied to piecework techniques and has not developed mass production techniques. Papermaking is a backward industry. Coal mining still relies on methods from the last century. Prepare an agenda for a new industrial revolution. List in priority order the new technologies that are called for.

16–11. Many citizens of the United States believe that if we can put a man on the moon, we must be able to solve the problems of our cities—crime, housing and transportation. Critically examine this view and describe the key differences between the two kinds of problems.

17

TRANSPORTATION TECHNOLOGY

TRANSPORTATION TECHNOLOGY, its evolution, its future and its use in society comprise a vastly important case study of engineering in the United States, Europe and other industrialized countries. The transportation of goods and people is integral to the needs of any society, and the interactions between transportation systems, the structure of our communities, and the impact of transportation on the environment and our natural resources is profound. New transportation technologies for centuries brought apparently unrestrained growth, paced by advances in propulsion and new modes of travel. In the future, it will be necessary to achieve greater safety, greater capacity and much improved efficiencies with the aid of automation and other new technologies.

STOPPED IN THE PRESENT

Technology in transportation has evolved over many centuries. See Figure 17–1.[1] While transportation systems served society adequately in the past, the present system is unsatisfactory in terms of safety, environmental impact, social impact and energy usage. It has been advocated recently that only through the use of automatic control technology can the system be moved to a new improved level of service.[1] Beyond this, new modes of transportation will be required, new social arrangements will need to be worked out, new safety devices developed and new technologies for propulsion with greatly reduced emissions developed.

Digital computers used in automatic control systems may permit significant improvements in traffic movement and safety. Use of automated urban traffic-control systems is an example of a new technological approach. Sensors are placed in the streets to measure the state of all traffic flow from all directions. That information is used to compute continually what sequences of traffic light changes will produce the optimum flow of traffic. This sequence is applied to the traffic lights accordingly. Such systems are in operation in Tokyo, San Jose,

	Vehicles and Guideways				System Elements		
10⁶ BC	Water	Roads	Tracks	Air	Systems	Safety	Environment
	Logs	People, Walking					
	Rafts	Oxen, Horses					
		Camels					
	Sails	Wheels					
1000 BC		Carts					
500 BC	Galleys	Roads					
	Canals	Highways					
0 AD		System					
		Spoked			Magnetic		
		Wheels			Compass		
	Canal						
1700	Systems						
1800	Steam Engine						
	Fulton's				Telegraph		
	Steamboat	Electric Motor					
			Trains				
	Screw Prop						
	Iron Hulls						
	Internal Combustion Engine						
		Automobile	Refrigerator		Telephone		
			Cars				
1900							
		Trucks	Trolleys	Airplane	Traffic lights		
		Buses		Dirigible		Grade Crossings	
				DC-3			
1940				Helicopter		Divided Highway	
			Jet Engine		Computer		
					Control		
	St. Lawrence	Interstate			Air Traffic	Seat Belts	
	Seaway	System			Control		
		Containerized Freight		Commercial			
				Jets			
1968			High Speed	Boeing 707			
	Super Tankers		Trains			Seat Restraining	Emission
						Systems	Regulations
			Air Cushion	747		Anti-hijacks	
			Experimental	Jumbo Jets			Noise
			Vehicles			Experimental	Regulations
1972						Auto Air Bags	

FIGURE 17-1   The evolution of transportation techniques.

California and Charleston, South Carolina. The Charleston system resulted in improved flow, accompanied by a reduction of traffic collisions of approximately 15%.[1]

GETTING TO THE AIRPORT

The effects of transportation systems on communities are significant. New metropolitan airports provide good examples.* When a city decides it needs a new airport, it usually goes out beyond the line where the value of land is $500 per acre, and purchases sufficient relatively cheap land. By the time all the land has been purchased, the land value has risen $3,000 to $4,000 per acre in the neighborhood. This has been the pattern in the case of the new Dallas-Fort Worth airport in Texas. As the area matures, the airport becomes a center of commerce with highly increased land values and new uses for the land. The Dallas-Fort Worth airport and commercial area is planned to be the size of Manhattan Island.

EFFECTS FROM AUTOMOBILES

The automobile, like the airplane, has had a profound effect on society by establishing a strong utility for the people of the United States and Europe. Available generally for only fifty years, the automobile is now very widely used. In 1973 there were approximately 12 million vehicles produced by the auto industry, compared with 10.9 million in 1972. Before the Arab oil embargo, it was estimated that sales for 1974 would be approximately 11.4 million.[2] A summary of the effect of the United States ground transportation system is given in Table 17-1. The cumulative production of autos is shown for the period 1930 to 1970 in Figure 17-2. By 1930 approximately 40 million autos had been produced and the auto was in general use in the United States. Growth in dependency upon the auto in the United States is even more dramatically shown by plotting the passenger miles traveled each year, as shown in Figure 17-3. The motor vehicle passenger-miles (number of passengers times miles travelled) has grown from 0.45×10^9 in 1940 to 1×10^9 in 1970, or by a factor of 2.2 (a 120% increase).

The motor car has undoubtedly changed the shape of the cities. Within the past century, the population of the largest cities has grown, and also has become more spread out. In Boston, for example, the population density at the center of the city was 220 per acre in 1900 and fell to 69 per acre in 1950. The central density in London is now less than a fifth of what it was in 1801. The city worker has moved to the suburbs; he uses his car as a commuter conveyence. Americans have some $95 billion invested in 100 million passenger cars which travel 1,000

*A magazine article in March 1974 reported on the several ways to get to J.F. Kennedy Airport in New York. But, it added, the best way to *travel* is not to go to J.F.K. at all.

million miles annually. The average car uses 700 gallons of gasoline each year and until late 1973 it cost 14 cents per mile to operate. In addition, the United States spends about $20 billion on additions and repairs to our 3.7 million miles of roads.[3] Of the Americans who have steady jobs and private cars, 82%

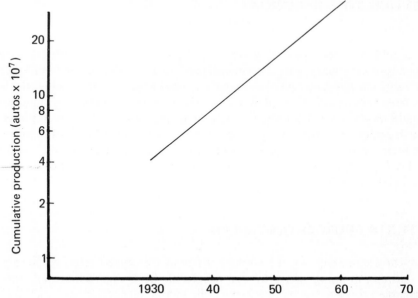

FIGURE 17-2 The cumulative production of automotive vehicles in the United States.

Table 17-1

The United States ground transportation system in 1970.

The United States ground transportation system consumed:
   75% of the rubber produced domestically
   56% of the petroleum
   29% of the steel
   53% of the lead
   27% of the cement
   25% of all energy

The system consisted of:
   3.73 million miles of rural and municipal highways
   108 million registered vehicles
   365,000 miles of railroad track
   176,000 miles of pipeline

The system produced:
   $196 million in expenditures
   10.3 million jobs
   1.1 trillion passenger-miles of intercity passenger traffic

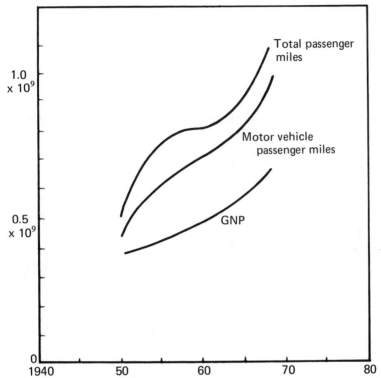

FIGURE 17-3   Domestic intercity passenger traffic (passenger miles) and Gross
National Product (1958 dollars) during the period 1950–1970 in the
United States. Source: *Problems of Our Physical Environment* by
J. Priest, Addison-Wesley Publishing Company, 1973. By permission
of the publisher.

commute by car, 56% of them alone. The automobile business accounts for 13%
of the GNP and one out of six jobs. Often as much as two-thirds of the land in
urban areas is devoted to freeways, parking, gas stations and auto-related uses.
Indirectly our autos have facilitated the move toward recreational second homes
and recreational vehicles.

The automobile changes our cities, causes air pollution and may cost more
than 50,000 lives each year in road accidents.* Yet what are the advantages of the
auto? It is an origin-to-destination method of transportation and it is normally
available at all times. It goes in any direction at the whim of the driver and it
doesn't have to stop to pick up other passengers. A great factor is that it provides
privacy and it safeguards against the annoyance of others.

Nevertheless, the auto is very inefficient in its use of energy and it pollutes
the air. It wastes space and is frequently run with only one occupant. Cars must
be stored at the end of a trip, requiring garages, parking lots and street parking.

*A happy consequence of the great gasoline shortage of 1973–1974 was an immediate reduction in the
loss of lives in highway accidents.

Autos are expensive to buy and operate and unsafe in some weather conditions. However, in the view of the auto user, the advantages appear to outweigh the disadvantages. Perhaps the individual use of automobiles is a distinct example of the Tragedy of the Commons. (See Chapter 10.)

Until 1974, the average weight and horsepower of automobiles in general continually increased. These increases are ostensibly to provide for easy and comfortable freeway driving at high speeds. The average horsepower of automobiles during the period 1900–1970 is shown in Figure 17–4. The average horsepower during that period increased from 18 to 180, or by a factor of ten. Yet most autos during commuting periods are driven at speeds averaging 30 miles per hour. At this speed the large engines are inefficient and produce significant emissions.

The costs of auto air pollution control devices are several. There is the cost of the device itself, the cost of maintaining it and the cost of added fuel used because the devices reduce the efficiency of the internal combustion engine. The estimated effect of the pollution control devices for autos is shown in Figure 17–5. The predicted actual results are less effective than the theoretical maximum effects because of troublesome maintainance of the devices as well as the predicted increase in autos in use in the later part of this century. Of course, this is a prediction which could be made obsolete by the move to an engine (other than the internal combustion engine) which would yield low emissions.

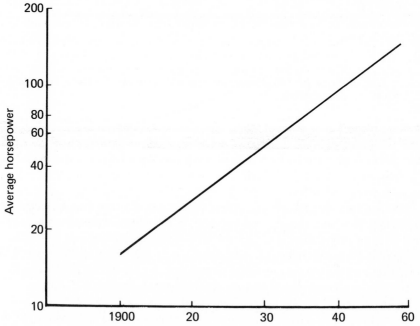

FIGURE 17-4 The average horsepower of automotive vehicles in the United States during the period 1900–1970.

Several legislators and engineers have suggested that the use-tax on the automobile be scaled to horsepower and weight. For example, a use-tax on an auto might call for a state tax of $0.50 per pound over 2,500 pounds and $1.00 per horsepower over 75 horsepower. Then a large popular car (such as the Buick Electra or the Cadillac Eldorado) would cost an added tax of approximately $1,000.

The Environmental Protection Agency (EPA) announced a sweeping series of antipollution regulations in 1973 and outlined a series of potential changes necessary in the United States brought about by our excessive dependency on the auto. They stated that new exhaust-control devices be required on old as well as new cars. Also, they recommended that gasoline rationing should occur in many cities in the next few years. By 1977 limits on gas sales could force a large proportion of the autos off the streets of Los Angeles, for example.[4] (This proposal was met with great disbelief and outcry in the Los Angeles area.) Among the EPA proposals are:

(1) In Boston, street parking would be banned in the central business district and a $5 per day surcharge instituted on new parking facilities.

(2) In New York, taxi cruising would be sharply reduced as would street parking. Truck deliveries would be allowed only at night and exclusive bus lanes would be created on parkways.

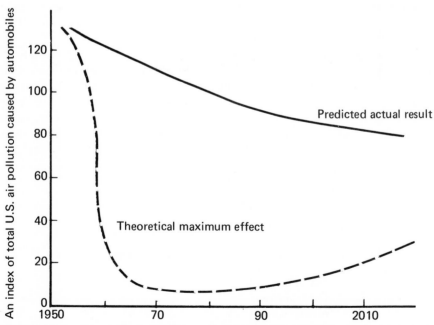

FIGURE 17-5 The estimated effect of pollution control devices for autos with internal combustion engines.

(3) In Pittsburgh and Philadelphia, special bus lanes on highways would be established.

POLLUTION AND ALTERNATE POWER SOURCES

The air pollution emitted by automobiles must be reduced. The Clean Air Act of 1970 sets out standards of clean air and calls for these standards by 1975, with extensions possible for qualifying regions until 1977. The sources of air pollution in 1966 are shown in Table 17–2. It is clear that the emissions of autos must be controlled to reduce general air pollution in the United States. The EPA has called for a 78% reduction in levels of carbon monoxide and an 87% reduction of photochemical oxidents in Los Angeles by 1977. (Smog results when Hydrocarbons and Oxides of Nitrogen are exposed to sunlight in stagnant air.) In northern New Jersey, for example, EPA calls for a reduction of 50% of smog emissions from autos. One method of doing this, as EPA proposes, is to ration the sales of gasoline in these areas.

The cost of pollution-control devices will add $700 to the price of a 1976 model car compared to a 1971 model, and these devices may not be sufficient for regions of high auto use such as Los Angeles. In 1975 and 1976 cars will have a catalytic converter for pollution removal from auto emissions. The Federal exhaust emission standards are given in Table 17–3. One control system uses a catalyst located in the exhaust manifold. The catalyst metals used are Platinum and Palladium. They increase the conversion of Carbon Monoxide to Carbon Dioxide, for example.

Alternate power sources for autos, buses and trucks may be developed in the future. They would be efficient, with low pollutant emission levels. Possibilities include diesel engines, steam cars, gas turbine engines and electric battery drives. Electric vehicles using battery power may be useful for urban traffic during the next decade.[11] While electric cars themselves are pollution free, they require electric power to charge their batteries. Most electric power plants pollute the air, but the pollutants would be significantly lower at a controlled power plant than for the equivalent gasoline-powered vehicles.[11] Of course, all the calculations

TABLE 17–2

Sources of air pollution in the United States in 1966.

Source	Percent
Industry	17
Power plants	14
Motor vehicles	60
Space heating	6
Refuse disposal	3

TABLE 17-3

United States Federal emission standards (grams per mile).

	Hydrocarbons	Carbon Monoxide	Oxides of Nitrogen
1960 (Precontrol)	15.0	90.0	5.0
1968	6.3	51.0	5.0
1972	3.0	28.0	5.0
1973	3.0	28.0	3.1
1975	0.41	3.4	3.1
1976	0.41	3.4	0.4

depend upon the efficiency of the batteries, the form of power plant and the weight of the urban vehicle. If an economical battery can be developed with an energy density approaching 100 watt-hours per pound of battery, then we could expect the construction of a significant number of electric autos. The present energy density of batteries of 10 to 15 watt hours per pound requires the auto to carry too many pounds of battery for reasonable cruising distances. Recharging of the auto batteries could be accomplished during the night when electric power plants have excess idle capacity. Electric autos would be very attractive if the problems of developing new sources of energy for electric power plants can be solved during the next 25 years. If safe, pollution-free nuclear power plants can be built, then the electric car is an important, low-pollution transportation alternative. The primary pollution from a nuclear power plant is radioactivity and the nuclear waste which must be disposed of. If these problems can be solved or an alternate source of electric power, such as solar energy, can be achieved, then electric cars, buses and trucks would be an answer to the need for low pollution transportation systems. Electric vehicles are low in noise and should be easy to maintain. Furthermore, electric vehicles are readily adaptable to guided roads and electronic control.

MASS TRANSIT SOLUTIONS

A solution to our urban transportation problems may lie in developing rapid mass transit systems. Mass transit systems use trains, trolleys, buses and other forms to move people rapidly, safely, inexpensively and without congestion. In the future, they should be able to move people with comfort and safety.[6]

At present about $5 billion per year is spent from the Highway Trust Fund for roads and the interstate highway system. It is only recently that the United States Federal Government has started to spend significant amounts for rapid mass transit. In 1973–74 the cities were permitted to use their share of the Fund to buy buses or build subways instead. This money may provide up to $500 million per year in addition to the current $1 billion a year Federal rapid transit budget under the direction of the Department of Urban Mass Transportation Administration. (UMTA).

Atlanta, for example, received $69 million per year in UMTA grants in 1973 and 1974 and expects to receive $140 million in fiscal 1975 (July 1974 to June 30, 1975).[7] Atlanta is presently constructing a two-county rail transit system. Currently, the Federal Government pays up to two-thirds of the cost of a project, with local funds paying the other third.

Rapid transit systems are important for large cities. Cities must have freeways to move traffic, but after a certain number of freeways are constructed, it is clear that freeways alone are not sufficient to move the commuters to the urban center. The number of vehicles using one freeway lane per hour is shown in Figure 17–6 as a function of auto speed. An urban freeway at rush hour has an average speed of approximately 38 miles per hour, which results in a capacity of 2,000 vehicles per hour per lane. If each auto averages 1.5 passengers, then we have 3,000 passengers per hour per lane as a maximum figure. Actually freeway lanes at rush hour have lower figures of passengers per hour passing on the lane because of starting and stopping of traffic and lack of a constant distance between autos. A comparison between the passenger-carrying capacity of freeway lanes and rapid transit tracks is shown in Figure 17–7. Using these figures, it takes 17.5 freeway lanes to move the same number of people as one rapid transit track.[8] Nevertheless, total public transit use of subways, trains and buses has declined from a high of 23 billion rides per year in 1946 to 8 billion rides per year currently.

A compromise between bus, auto and rapid rail transit is shown in Figure 17–8. For the bus, it is assumed that the buses travel on a dedicated lane with 50 passengers per bus at a time interval of 18 seconds. Therefore, for cities, dedicated bus lanes and rapid rail transit systems have very significant appeal.

THE EXAMPLE OF BART

Systems of buses, trains and auto parking lots at outlying rapid transit stations are needed. This is the case for the new Bay Area Rapid Transit System (BART). The station spacing is about two miles and a station stopping time is 20 seconds. BART is planned to operate with a maximum speed of 80 miles per hour and an average speed of 50 miles per hour. The translation of San Francisco's Bay Area Rapid Transit System from its initial planning phase in 1947 to its first partial operation in 1972 has required the solution of many political, economic and technical problems. The system has cost approximately $1.4 billion, and it is not fully operational as yet. BART was formed in 1957 by the legislature, initially to serve five counties with 52 stations. In 1962, two counties, Marin and San Mateo, withdrew from the BART district. Then the planning proceeded for San Francisco, Alameda and Contra Costa counties. The vote on the bond issue to construct the system received 61.2% of the votes in the three counties, barely exceeding the required 60%.

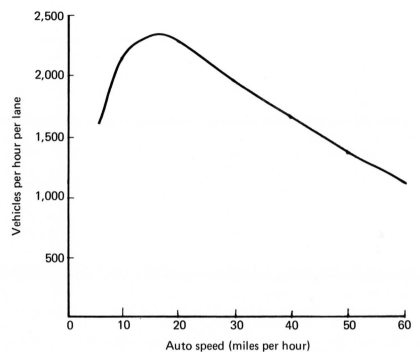

FIGURE 17-6 The vehicles that can use one freeway lane each hour as a function of vehicular speed. The curve is calculated for a braking acceleration of 20 miles/hour/second.

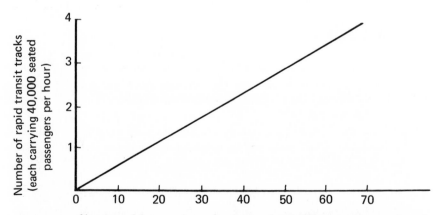

FIGURE 17-7 A comparison of the equivalent freeway lanes to rapid transit tracks. One rapid transit track is equivalent to 17.5 lanes of freeway. It is assumed that each train contains 10 cars, and each car seats 100 persons. The interval between trains is 90 seconds.

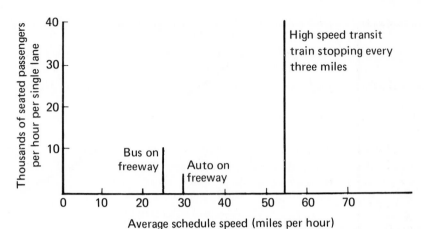

FIGURE 17-8 A comparison of passenger-carrying capacity of autos on a freeway, buses on a dedicated bus lane and a high speed transit train.

The BART system uses air-conditioned aluminum cars capable of speeds of 80 miles per hour. The trains are computer-controlled; they run (when fully operational) with 90 seconds between trains at rush hour. Inside the trains are carpeted floors, wide tinted windows and comfortably cushioned seats. The goal is to make the train ride as attractive as riding in one's automobile. A BART train is shown in Figure 17-9.

FIGURE 17-9 A BART train near San Francisco. Courtesy Bay Area Rapid Transit District.

The BART system has experienced considerable difficulty in its first years of operation arising from management and engineering conflicts. The 75 mile-long system is based on advanced technology, employing computer control and electronic sensing train location. Transit systems have heretofore relied on proven traffic systems and manual operation. The BART trains, when fully operational, will be automated, with a trainman riding on each train for emergency conditions only. The new control system, developed by Westinghouse Electric Company, has not been fully cleared of problems. This has resulted in a great wave of frustration and concern over the safety and operation of the system.[13] The most serious accident occurred in 1972, when, at the end of the line in Fremont, a train failed to stop as programmed and crashed into a sand barrier, injuring five people. As a result of the difficulties, the California Public Utilities Commission refused to allow BART trains to use the new tunnel beneath San Francisco Bay. Furthermore, the Commission has required the use of human dispatchers, who telephone from one station to the next to make sure the tracks are clear.

Regardless of its problems, BART has already made a significant impact on San Francisco and the surrounding counties. Property values in San Francisco, Oakland and along suburban BART routes have been rising dramatically for the past several years. Construction of new buildings and homes along the lines has been proceeding rapidly and the number of new high rise office buildings has grown rapidly in downtown San Francisco, apparantly in response to the access to be afforded by BART. In 1974 it is estimated that BART will have about 100,000 daily passengers from the three-county population of 2.4 million. BART projections show that for 1975, of a total of 2.5 million daily person-trips, BART will account for only 8%. BART's biggest impact will be on the crossbay traffic, where an estimated 44% of those who cross the bay will use BART.

An ultimate assessment of BART will need to consider the ease and cost of transportation in the region and the impact of transportation on the region's land use and the lives of its citizens. BART's main contribution may be toward the more efficient use of energy for transportation and its decreased effect on the air of the area. The system will be an interesting one to study throughout the decade of the 1970s.

Washington, D.C. and Atlanta are currently constructing new mass rapid transit systems. Los Angeles is now considering a plan for a mass transit system for the expansive Los Angeles County. However, in 1968 the voters of Los Angeles defeated a proposed $2.5 billion bond issue for a 90-mile rail transit system. Los Angeles is the freeway capital of the world and the nation's largest smog belt. It is very difficult to get Los Angeles citizens interested in rail rapid transit because, in part, the population is widely-dispersed and there is a lack of consensus on a wise route for such a rapid transit system. Interest remains in Los Angeles in depending upon subsidized bus lines to serve as a segment of a mass transit system.[9]

BUSING ALTERNATIVES

Many believe the bus is the only near-term alternative for rapid transit. In 1973, about 3,000 buses were purchased for mass transit systems, replacing those becoming obsolete or inoperable in the fleet of 50,000 buses in the United States. About 4,500 buses may be purchased in 1974. Clearly, the bus systems of the United States are remaining at essentially the same operating level for several years. Exclusive bus lanes are now in operations in many cities. Some dedicated lanes accommodate up to 500 buses per hour, yielding a mass transit index of approximately 20,000 passengers per hour per lane, which rivals the rail rapid transit index of 40,000 passengers per hour per track.

The long post-World War II decline in mass transit ridership appears to have reached its low point. Passenger use is increasing again. As a result of a program for subsidizing lower fares, increasing schedules, establishing bus lanes, and employing other innovations in the past few years, patronage apparently leveled off at over five billion passenger-trips during 1972.

EFFICIENCY OF TRANSPORTATION SYSTEMS

The efficiency of various means of locomotion is an important factor in examining the benefits of the different modes of transportation. The energy efficiency of traveling animals and machines is shown in Figure 17–10, where efficiency is defined as the energy required to move a unit weight a unit distance. For example, the light plane is three times as efficient as the helicopter in terms of energy expended to move a given weight over a given distance. A bicyclist ranks first in efficiency, because the efficiency of a man on a bicycle (.15 calories per gram per kilometer) is five times that of a man walking.[14] The bicycle was the first machine to be mass-produced for personal transportation; however, it was fully developed only about 100 years ago. The idea of changing the gear ratio of a bicycle to provide for pedaling uphill or against a head wind led to the *derailleur* gear of 1899. Bicycle manufacturing is a world-wide business, with a total production of about 40 million vehicles per year. Generally throughout the world, bicycles play a more important role than the automobile. For example, China, with its 800 million inhabitants, relies heavily on the bicycle for transportation of people and goods, as do the other countries of Asia and those of Africa. In Russia, which has about 1.5 million automobiles, the annual production of bicycles is 4.5 million. Currently in the United States there is a resurgence of interest in the bicycle for health and transportation needs. The construction of bicycle pathways is necessary in most towns and cities to provide safe travel.

The energy intensiveness of various means of transporting people is shown in Table 17–4. The most efficient mode of transport would have the lowest intensiveness. Note that automobiles provide, currently, about 89% of all passenger transportation in the United States.

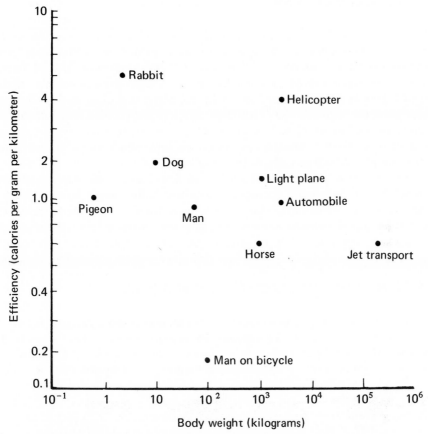

FIGURE 17-10   The energy efficiency of various means of transport.

TABLE 17-4

The energy intensiveness of various modes of transportation for people.

Mode	BTU/passenger-mile	Total Percent of Transportation Demand
Airplane	7150	9.3
Automobile	5400	88.8
Train	2620	0.7
Bus	1700	1.2
Walking (3 mph)	524	negligible
Bicycle (8 mph)	310	negligible

The transportation of freight and goods is an important part of a nation's transportation system. The efficiency of various modes of transportation of freight with respect to energy use is shown in Table 17–5. Pipelines are the most efficient for moving liquids or gases and waterways are the most efficient for moving goods and materials. The energy efficiency of transportation must become more important over the next decade as we experience energy shortages. Presently, 25% of the energy used in the United States is used for transportation, primarily by the auto and the truck. Note the forecast trends in transportation figures shown in Table 17–6.

Waterways are five times as efficient as road transportation, while railroads are four times as efficient as highway transport.

The total freight movement in the United States in 1973 was 2.1 trillion ton-miles.[12] Of this total, railroads carried 850 billion ton-miles, or approximately 40%. The rise in the tonnage moved by railroads, all modes of transport and the *per capita* increase in ton-miles is shown in Figure 17–11.[12] It is estimated, as shown in the figure, that the total freight ton-miles in the United States will increase by 145% during the period 1970–2000. Also, the *per capita*

TABLE 17–5

The efficiency of various modes of transportation of freight. with respect to energy use.

	Ton-miles/gallon of fuel	*BTU/ton-mile*
Pipelines	300	450
Waterways	250	540
Railroads	200	680
Roads	58	2340
Airplanes	3.7	37,000

TABLE 17–6

Trends in Transportation.

	1950	*1960*	*1970*	*1980 (estimated)*
Motor-Vehicle Registrations (millions)	49	65	105	150
Motor-Vehicle passenger-miles (billions)	450	710	1100	1500
Airplane passenger-miles (billions)	8	30	101	300

ton-miles will increase by 80% during the period 1970–2000. This increase will be a challenge to the efficient use of transportation technology. The freight ton-miles for four modes of freight transport as a per cent of the total are shown in Table 17–7 for 1943 and 1973.

Pipeline transportation already accounts for 25% of the freight moved in the United States.

The United States has 25,000 miles of navigable canals and is carrying 500 million tons of goods on the Federal and New York State canals alone.[10] However, the proportion of freight travelling by waterway has remained the same for many years. An advantage of waterway traffic is that larger vessels produce 373% less pollution, by volume, than diesel trucks based on equivalent tons of cargo carried. Canals in Great Britain and Europe are being used extensively and may be expanded. Russia has just completed a 3,500 mile waterway system in the European part of the country. Along with the improvement of canals has come a newly designed and built series of ships, called barge carriers. They carry barges which can be used on canals, and they can operate independently of ports. They can load and unload barges at the rate of four per hour. Canal barges are an old technology with newly-revived hopes for new new technologies and newly-realized economies of transportation.

RAILROADS

Travel between cities and towns in the United States is predominantly air and auto travel. However, in many parts of the world, train travel between cities is well patronized. In recent years high speed trains have been developed and operated at a profit. The Japanese opened their 515 kilometer (320 mile) Tokkaido-Shinkansen line in 1964 and operated it at a profit. The train averages 200 km/hour (124 miles/hour). It has showed that shorter travel time on a comfortable train will attract the passengers from airplanes and autos. Of course a high speed train is many times more efficient than autos or airplanes in terms of passengers transported per transit lane or in terms of energy consumed per passenger mile. The International Union of Railways has announced a master plan for Europe's railways involving construction of 6,000 kilometers of high

TABLE 17-7

Freight traffic in ton-miles for four modes of transport as a percent of the total traffic.

	1943	1973
Railroads	61%	39%
Trucking	10%	22%
Inland waterways	16%	5%
Pipelines	10%	23%

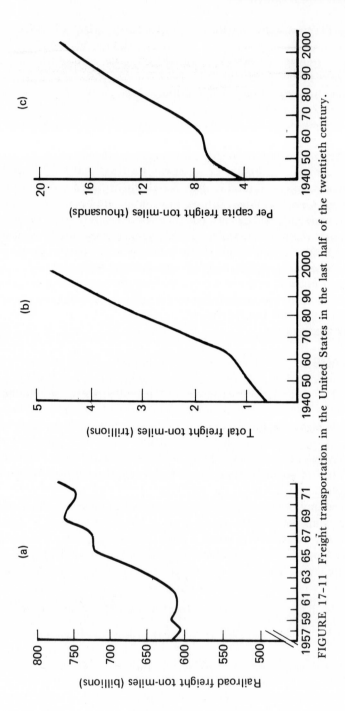

FIGURE 17-11 Freight transportation in the United States in the last half of the twentieth century.

speed lines.[15] The Italian Railways are busy building the first of these between Rome and Florence. In France a train for the Paris-to-Lyon route is being developed to travel at 300 km/hour. The United States Metroliner travels between Washington, D.C. and New York City, sometimes reaching a speed of 154 km/hour (96 miles/hour). It could achieve 200 km/hour if the track could be upgraded to accommodate that speed.

SUMMARY

The transportation system of the future must fit the needs of the urbanized, industrialized nations without radically affecting the social communities of the nation. The auto has been an important part of our technological society; it must be joined and, in part, replaced by well-developed and proven electric vehicles, mass rapid transit system, high speed intercity trains and expanded, efficient freight transportation systems such as railways and waterways.

CHAPTER 17 REFERENCES

1. R.H. Cannon, Jr. "Transportation, Automation and Societal Structure," *Proceedings of the IEEE*, Vol. 61. No. 5, May 1973, pp. 518–525.
2. L.G. O'Donnell, "Heard on the Street," *Wall Street Journal*, July 24, 1973, p. 32.
3. J. Belsey, "America's Automania," *Sierra Club Bulletin*, May 1973, pp. 15–16.
4. "Life Without Cars?" *Time*, June 25, 1973, p. 54.
5. "The EPA's Contribution to Better Cities," *Business Week*, Aug. 4, 1973, p. 21.
6. J.A. Kieffer, "The Automobile's Success: A Lesson for Its Critics," *The Futurist*, Feb. 1973, p. 21.
7. A.R. Karr, "Fiscal '75 Transit Aid and Trail Requests, Which Lines Can Double Under U.S. Law?" *Wall Street Journal*, April 8, 1973, p. 8.
8. E.L. Michaels, "Today's Need for Balanced Urban Transit Systems," *IEEE Spectrum*, Dec. 1967, pp. 87–91.
9. "Can Rapid Transit Compete with the Freeways?" *Business Week*, Feb. 10, 1973, pp. 65–66.
10. I. Breach, "Freightways of Tomorrow?" *New Scientist*, July 12, 1973, pp. 68–70.
11. J.T. Salihi, "The Electric Car—Fast and Fancy," *IEEE Spectrum*, June 1972, pp. 44–48.
12. "The Railroad Paradox: A Profitless Boom," *Business Week*, Sept. 8, 1973, pp. 54–63.

13. P. Barnes, "So-So Rapid Transit," *The New Republic*, Sept. 1, 1973. pp. 18–20.

14. S.S. Wilson, "Bicycle Technology," *Scientific American*, March 1973, pp. 81–91.

15. I. Yearsley, "Are High Speed Trains on the Right Track?" *New Scientist*, Sept. 6, 1973, pp. 546–548.

16. "API Seeks Delay of Auto-Emission Regulations," *The Oil and Gas Journal*, Sept. 24, 1973, p. 77.

17. K.E.F. Watt, *The Titanic Effect*, Sinaver Association, Inc., Publishers, Stamford, Connecticut, 1974.

CHAPTER 17 EXERCISES

17–1. It has been proposed that a sales tax be imposed on automobiles according to their gasoline consumption. Any car getting at least 20 miles to the gallon would not be penalized. The annual tax would be graduated to $300 at 13 miles per gallon and $900 at under 7 miles per gallon. Determine a suitable graduated tax and conduct a survey among your friends to determine if they would be induced to purchase an automobile with a gas consumption of better than 20 miles per gallon. What should the resulting tax income be used for?

17–2. The American Petroleum Institute estimated that fuel consumption was markedly increased by the addition of emission devices to automobiles. They estimate that a fuel economy penalty for the 1968 to 1970 models was 7% increasing to 12% for the 1971 to 1973 device additions. The penalty increases to 16% for 4,500 pound vehicles. The Institute estimates that proposed 1976 EPA controls could boost the penalty to 30%.[16] Discuss the possible compromises and alternatives to emission control devices and the resulting gasoline consumption penalties.

  Might the American Petroleum Institute have a vested interest in the information given? Explore this possibility.

17–3. A personal rapid-transit system has been constructed in Morgantown, West Virginia as a demonstration project by the United States Department of Transportation. The system incorporates a 3.2 mile automated guideway running small computer-controlled, rubber-tired self-service electric cars on a continuous round. Each car accommodates 15 passengers. What percentage of the daily commuting passengers could such a system attract in your city? Design a loop system for your town.

17–4. Air-cushion trains without wheels are prime contenders for high-speed ground vehicles in the range of 300 miles per hour. In France an Aerotrain runs on a 16 mile demonstration track outside Paris. A British train, called a Hovertrain, is being tested near Cambridge. Both trains are powered by almost noiseless, pollution-free electric linear motors. The trains hover over a guideway on a cushion of air which provides an almost frictionless path. Investigate air cushion trains for your state and estimate what percentage of inter-city traffic would use a 300 miles-per-hour train.

17–5. The current United States highway transportation system and the associated automobiles and trucks consume a significant portion of our wealth in the United States. Annual costs for vehicles are about $30 billion, and $20 billion are spent on the highways. A total of $90 billion are spent for vehicles, roads, fuel, operating, and maintenance. Also, approximately 56,000 people are killed each year and over 2,000,000 injured seriously enough to require hospitalization. Explore several transportation systems that would provide the comfort, privacy and door-to-door convenience of the automobile. Examine the costs and safety aspects of the proposed systems.

17–6. The great British engineer I.K.Brunel lived from 1806 to 1859. He designed and build the wooden ship *Great Western* of 2,300 tons in 1838, the iron *Great Britain* of 3,000 tons in 1845 and the iron *Great Eastern* of 18,900 tons in 1858. The latter ship was five times bigger than any ship then operating. It was used to put down the first transatlantic telegraph cable in 1865. Trace the increase in the size of ships from 1800 to the present. Has this means of transport reached its limit in size? What relationship does size have to efficiency?

17–7. The use of bicycles for transportation is important in many countries of the world. In the United States about 100 million Americans ride bikes, an increase of 20 million in two years. In 1973, over 15.4 million bikes were sold in the United States. About 25,000 miles of bikeways are already in use in the United States. Oregon has enacted a law allocating 1% of gasoline tax receipts for bikeways and California is spending $3.2 million a year for bike facilities. The Federal Highway Act of 1973 provides $120 million over the years 1973 to 1976. What are the beneficial effects of the increased use of bikes for transportation? If 10% of the passenger-mile traffic were to shift to bikes from autos, what would be the savings in petroleum use?

17–8. It has been stated, "A society which can send a man to the moon and back should be able to solve its urban transport problems." Do you believe that these problems are similar in difficulty and in their nature?

17–9. United States transportation is almost entirely dependent on oil for fuel. In 1972 imports accounted for about 25% of the domestic oil budget. The energy crisis of 1973–1974, caused in part by the mid-East conflict, demonstrated the difficulties of this dependence on oil. What forms of transportation would you propose that would alleviate this dependence on imported oil?

17–10. The railroad is about four times as efficient as trucks for shipping freight with respect to energy usage per ton-mile. The amount of freight moved by trucks increased rapidly during the period 1950–1970 while that on trains remained constant. Also, train passenger traffic fell dramatically during this 20 year period. What actions would you propose that could cause freight and passenger traffic to move back to trains with an attendant reduction in auto and truck transportation?

17–11. Electric vehicles on an automatic guideway are part of a possible transportation system for the future. The vehicles would carry from two to four people and would be routed and scheduled by computer. Investigate the feasibility of such a system and its advantages and disadvantages.

17–12. Video telephones arranged in a conference-call manner may reduce the need for many business and personal trips. Estimate what percentage of passenger trips could be eliminated by such a method. Do you think this approach will be common by 1985?

17–13. Since the transportation of freight is more efficient by train than by truck, why has truck freight grown as a percentage of the total? (See Table 17–7.) What suggestions do you have for the movement of an increased proportion of freight by train rather than by trucks?

FIGURE E17–15 The road system and parking lots of an urban area. Courtesy the United States Federal Highway Administration.

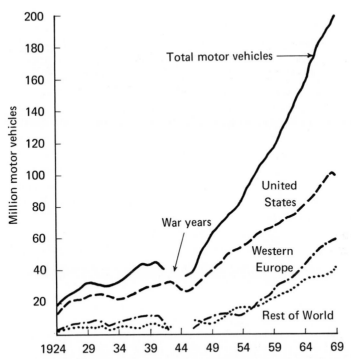

FIGURE E17-16   The total number of registrations of motor vehicles throughout the world during the period 1924-1969.

17-14. By 1980, two-wheeled vehicles may outnumber automobiles in the United States for the first time in some 60 years. It is estimated that more bicycles and motorcycles may be purchased than cars were purchased in 1973. By the end of this decade there may be 131 million bicycles and motorcycles in the United States, against 127 million cars. Over 20 million bicycles were sold in the United States in 1973, outpacing the sales of autos. Consider the social consequences of an increased use of two-wheeled vehicles. If half of all local trips of three miles or less are taken by bicycle or motorcycle, what effect will this have on the use of energy in the United States? Consider the potential for a combination of bicycles and mass transit for commuters.

17-15. The extensive use of automobiles and trucks in the United States has caused a great percentage of urban land to be used for roads, parking lots and related purposes, as is shown in Figure E17-15. Determine the percentage of land of a representative portion of your town or city that is dedicated to the automobile and truck. Estimate what this percentage could be reduced to if a dependence on bicycles and trains was developed over the next 10 years.

17–16. The total number of registrations of motor vehicles is rising rapidly throughout the world, as is shown in Figure E17–16. The dependence upon personal vehicles is growing throughout the less-developed portion of the world. What would be the projected effects of continuing growth of numbers, and dependence upon, motor vehicles worldwide? Discuss the effects upon natural resources, the need for roads, the effects of mobility, the impact upon economics generally, and also upon environment and other factors that come to mind.

18

ASSESSMENT OF TECHNOLOGY

IF IT IS WIDELY UTILIZED, new technology may have profound unforeseen consequences in any society. An example is the wide use of pesticides such as DDT. Another example is the introduction of automobiles throughout a society, with attendant development of road systems and the resulting growth of suburbs.

We have found that devices such as power plants, communication systems, airplanes and oil refineries are the source of profound secondary impacts on our society. In recent years there has been an attempt to develop mechanisms to forecast the potential effects of new technology and thus to be able to choose the alternative technological approach not only on the basis of the immediate benefits, but also on the basis of potential longterm deleterious side effects.

THE CONCEPT OF TECHNOLOGICAL ASSESSMENT

The systematic study of the effects on society that may occur when a technology is introduced, extended or modified with special emphasis on the impacts that are unintended, indirect or delayed is called *technology assessment*. Ideally, the assessment of technology should anticipate and evaluate the impacts of any new technology on all sectors of society. Technology assessment emphasizes the secondary or tertiary effects of a new technology, because in the long run they may be the most significant and undesirable consequences. Furthermore, undesirable consequences may often be prevented by proper planning. In building a bridge, dredging a canal or choosing a power plant site, the first-order effects are planned, since these are the goals of the project. Technology assessment focuses on the question of what else may happen when the technology is introduced.

A technology assessment should provide a balanced look at all the alternatives, options and possible outcomes. It should consider all the aspects of the technology and identify certain areas for intensive examination. An assessment usually requires an interdisciplinary team effort to obtain a comprehensive and balanced report. It is very important that the team carrying out the assessment maintain sufficient independence from the sponsoring firm or agency. Technology assessment is seen as "an early warning system to protect man against his own inventions," in the words of Jerome Wiesner, President of the Massachusetts Institute of Technology.[2]

The growing complexity of society and the interconnectedness of the various elements in our society require that undesirable side effects are anticipated and avoided if possible. Our ability to anticipate the future consequences of our actions reliably is aided by the growing availability of scientific knowledge. In addition, new research and development is stimulated as potential side effects are discerned. The large engineering enterprises of today require huge capital investments and long planning periods; they can engender large and often irreversible consequences.

As Toeffler noted, unless we probe ahead in time with all our intelligence and imagination, we may be compelled to endure an undesirable future.[1]

As a result of a technology assessment, a project may be modified or cancelled. Alternatively, it may proceed, but a monitoring system may be instituted to determine the consequences of the project and to forestall any significant deleterious effects. Often an assessment will stimulate new research and development to determine means to overcome the negative consequences of a project or to find alternative methods or meeting the original need without the negative effects. Regulations, taxes, prohibitions and other controls may be instituted when an assessment indicates that a particular technology would have undesirable side effects.

The selected impacts of the automobile on the life of Americans and Europeans in the twentieth century are shown in Table 18–1. Note the range of impacts of this technology, which was widely introduced in the United States after 1910.

ASSESSMENT AGENCIES

Technology assessment is carried out by several organizations within the government and the professions. Numerous Federal and state agencies have been

TABLE 18-1

Selected impacts of the automobile.

Values
  Geographic mobility
  Expansion of personal freedom and privacy

Environment
  Noise pollution
  Air pollution

Society
  Changes in patterns of courtship, use of leisure time
  Substitution of auto for mass transit

Demography
  Population movement to suburbs

Institutions
  Land use for highways
  Changes in insurance, education, travel conventions

Economy
  Mainstay and prime mover of American economy in the twentieth century
  Depletion of oil reserves

established to regulate the activities of technology-bases industries or clusters of industries. Examples are the Federal Communication Commission, which regulates the communications industry, and the Federal Aviation Agency, which regulates the airlines. Other regulatory agencies are the Interstate Commerce Commission, the Federal Power Commission and the Environmental Protection Agency. The decision-making apparatus of government has come to play a large role in the assessment of new technology and the regulation of its use.

To some extent the professional societies are influential in technology assessment. Voluntary industry-wide cooperation on specifications, or in the case of the American Society for Testing and Materials, represents another assessment prototype. Nevertheless, with few exceptions, the central question asked of a new technology is often what it would do to the economic or institutional interests of those who are deciding whether or how to exploit it.

In a less affluent and technologically sophisticated society than that of the United States or Europe, it is occasionally necessary to discount the future in favor of survival. However, in the United States and Europe, we can afford to pay more attention to thinking about the implications of our actions for future generations. Thus the question arises: How do we weigh the potential dangers? To prohibit use of every proposed pesticide until its benefits are proved to outweigh its suspected dangers would be extremely costly, and would cause development to stagnate. Technology assessment must not become technology arrestment, although much of the present regulatory system fails to give all affected interests effective representation in the crucial processes of decision. The introduction of technology assessment into our regulatory and legislative processes must result in an open and effective process with minimum of delays added to the decision process. Competition in the industrial world rarely rewards and often penalizes behavior that is socially desirable in a larger and longer-term context. New mechanisms to alter this factor would be of great benefit.

FEDERAL LAW FOR ASSESSMENT

The Environmental Policy Act of 1969 (Public Law 91–190) is a law requiring technology assessment. The law requires every United States government agency that is planning a project to file with the Council on Environmental Quality an assessment of the impact that the project may have on the environment. The law has the effect of requiring private firms to prepare environmental impact statements through judicial interpretation of the Act. In addition, many states such as New York and California require public agencies and individual firms to file environmental impact reports with appropriate bodies for review.

The EPA of 1969 requires that each report shall set forth "current and foreseeable trends in the quality, management and utilization of such environments and the effects of those trends on the social, economic and other requirements of the Nation." In the words of Section 102 of the Act, each agency must, "...include in every recommendation or report on proposals for legislation and other major Federal actions significantly affecting the quality of the human environment, a detailed statement by the responsible official on:

(i) the environmental impact of the proposed action; (ii) any adverse environmental effects which cannot be avoided should the proposal be implemented; (iii) alternatives to the proposed action; (iv) the relationship between local short-term use of man's environment and the maintenance and enhancement of long-term productivity, and (v) any irreversible and irretrievable commitments of resources which would be involved if the proposed action should be implemented."

The courts have interpreted Section 102 quite stringently, and many environmental impact reports have been closely examined by hearing bodies such as zoning commissions and city councils. For example, in the case of Calvert Cliffs Coordinating Committee vs. The Atomic Energy Commission, the United States Circuit Court of Appeals in 1971 found the AEC standards for licensing of nuclear power plants invalid. Similarly, the courts have been involved, as has Congress, in the review of the environmental impact report for the proposed Trans-Alaska Pipeline. The pipeline report cost $7 to $8 million to prepare and was finally accepted by the United States Congress, which approved its acceptability in 1973, as is discussed further in Chapter 20. The Trans-Alaskan Pipeline companies were required to prepare a statement in cooperation with the United States Department of the Interior, since the pipeline was to pass through Federal lands.

A summary of an environmental impact report may be shown in matrix form as shown in Figure 18–1. The person preparing the report enters a brief (several words) summary at each intersection in the matrix. For example, in the case of a proposed electric power plant the effect of the fuel extraction on land use would be entered in Column 3 and Row 2a.

An environmental impact report should answer questions such as, could the project affect:

(1) The use of a recreational area or area of important aesthetic value;

(2) The potential use, extraction or conservation of a scarce natural resource;

(3) The existing features of the region;

(4) The noise level in the area.

		Proposed Activities Which May Cause Environmental Impact									
		1. Modification of Regime	2. Land transformation and Construction	3. Resource Extraction	4. Processing	5. Land Alteration	6. Resource Renewal	7. Changes in Traffic	8. Waste Emplacement and Treatment	9. Chemical Treatment	10. Accidents
Existing Factors and Conditions of the Environment	1a. Earth Resources										
	1b. Water and Air Resources										
	1c. Biological Resources										
	2a. Land Use Activities										
	2b. Social Factors										
	3. Ecological Relationships										
	4. Other										

FIGURE 18-1 A matrix form for a summary of an environmental impact report.

ASSESSMENT BY THE FEDERAL GOVERNMENT

In recent years Congress has increasingly been required to make decisions about unfamiliar new technologies. In the 92nd Congress, decisions were made concerning the SST, nuclear power development, pesticide regulation and military weapons. To deal with such considerations, the United States Congress established a Congressional Office of Technology Assessment (O.T.A.) which was signed into law on October 13, 1972. Federal technological projects such as the S.S.T. and the breeder reactor are planned in the Executive branch and sent to Congress for authorization and appropriation. In the past Congress was rarely presented with an explanation of alternative programs or a technology assessment of the proposed project.

The O.T.A. bill as passed in 1972 called for a board with 13 members: six Senators and six Representatives and the O.T.A. Director. The bill authorized a budget of $5 million for fiscal years 1973 and 1974. Special panels of experts advise the O.T.A. and assessments on projects will be prepared by private consulting firms, research institutes and universities. An annual report will be submitted by the O.T.A. to Congress. It is expected that the O.T.A. will carry out assessments of new energy technologies, the electric car, nonlethal weapons, invasion of privacy by computers, ocean resources, pesticide alternatives, drug additives to food, urban mass transit and toy design—among other subjects.

In preparing an assessment, the O.T.A. is to:

(1) Identify existing or probable impacts;

(2) Where possible, establish cause-and-effect relationships;

(3) Determine alternative technological methods of implementing specific programs;

(4) Identify alternative programs for achieving requisite goals;

(5) Estimate and compare the impacts of alternative methods and programs;

(6) Present findings of completed analysis to the appropriate legislative authorities;

(7) Identify areas where additional research of data collection is required.

It is planned that O.T.A. will use the resources of the Library of Congress, the National Science Foundation and the General Accounting Office.

HOW ASSESSMENT WORKS

The methodology os technology assessment currently contains seven steps.[3] The first step is to define the assessment task, establish the scope of the assessment and discuss the relevant issues. Then the major technology and other supporting technologies are described. The third step is to identify and describe major technological factors influencing the application of the relevant technologies. Then the areas of impact are identified. For example, major areas of impact might be the environment, demography or social values. The fifth step is to make a preliminary impact analysis. Next, the possible action alternatives are identified. This step calls for the development and analysis of various programs for obtaining maximum public advantage from the assessed technologies. The final step is to complete the impact analysis by analyzing the degree to which each action alternative would change the specific societal impacts of the technology being studied.

Of course each step in making a technology assessment is closely linked to every other step. Also, an iterative procedure may be the most complete. The multidisciplinary systems approach is well-suited to preparing technology assessments. In order to be accepted by the public, technology assessment must be reliable and credible. This requires systematic methods for developing assessments—methods which can be repeated.

In preparing assessments, questions of value arise. It becomes difficult to quantify or to make objective all the questions in an assessment. If this is true, then it may be equally difficult to prepare credible assessments which can be repeated. (There is some fear that anyone advocating a particular project can justify it through a technology assessment.)

Because technology assessment, in its broadest definition, is the anticipation and evaluation of the impacts of a technological development on all sectors of society, including the economic, environmental, social political, legal and institutional components, it will be difficult to satisfy this definition in an objective or bias-free manner.

Also, technology assessment is limited in its scientific basis. In extrapolating from the past to the future, there is no firm basis to assure us that the future will resemble the present or the past. The political, social and institutional changes in our society as a result of technological changes are only partially understood. We did not foresee the effect of the automobile on our social institutions or the long–term effects of pesticides, let alone the effects of electronic surveillance devices on our politics, ethics and morality.

THE DELPHI ASSESSMENT

One method of assessing the potential effects of a new technology is the Delphi method, originally developed by Olaf Helmer and Norman Dalkey at the RAND Corporation. This is a method for soliciting and collating informed judgments on a particular topic systematically. Under this procedure, participants respond to a series of questionnaires interspersed with summaries of the responses to earlier questionnaires by group members. Communication and consensus with the group is facilitated with the aid of computers. Each participant completes his questionnaire and returns it. The summary of all group member responses is returned to him and he then completes a second copy of the questionnaire. An example of a Delphi questionnaire for a second or later round is given in Figure 18–2.

1. Development	2. Consensus to date	3. Reasons for minority opinions	4. Your previous 50% date	5. Indicate when you think the development will take place A = Earliest possible (10% probability) B = Most likely (10% prob.) C = Almost certainly before (90% prob.)								6. Implications Assume the development occurs. What social, environmental and technological changes might follow?
				1975-80	1980-85	1985-90	1990-2000	2000-25	Later	Never		
Laboratory operation of automated language translators capable of coping with idiomatic syntactical complexities	Most likely in 1990 Earliest possible by 1980	Difficulty with syntactical complexity	1980–1985		A		B	C				1. Increased scientific and technical communications 2. Decrease in the number of extant languages

Please complete these items

FIGURE 18–2 An example of a Delphi questionnaire for a second or later round of the survey. Note that items 2, 3 and 4 report the results of previous issues of the questionnaire. The respondent completes items 5 and 6 as shown.

Automation of the Delphi process by substituting computers for the usual paper, pencils and mailing of questionnaires results in a reasonably efficient process. A committee of experts can examine a given technology and its possible consequences with reasonable dispatch and efficiency with the aid of a computer.[4]

A recent Delphi study has estimated the probability of several events occurring by 1985. They are listed in Table 18–2.[5] This Delphi study, carried out by the Institute for the Future, used a computer to automate the processing of the replies to the questionnaire. The spread of opinion narrows from the first to the last questionnaire, and face-to-face discussion actually produces less accurate results than a Delphi study.[1]

The consensus of a Delphi panel that convened in the mid-1960s is shown in Figure 18–3.[17] This panel estimated the times when new scientific break-throughs would occur. The median estimate is indicated by the top of the box and the right and left edges indicate the quartile estimates. It is interesting to note that item 2 has already occurred, while item 1 is still not achieved.

INTUITIVE FORECASTING

Another form of forecasting the social consequences of a technological development is based on intuitive methods which an individual uses to assess some aspect of the future. This method depends solely on the inspiration and imagination of the forecaster. Thus, an individual with sharpened insight draws on his experience and imagination to state what he thinks may occur. This form of forecasting was carried out by Jules Verne, Aldous Huxley, H.G. Wells, George Orwell and A.C. Clarke, among others.

Forecasts of potential effects of new developments may also be made by assuming that trends established in recent history will continue. This method assumes that the forces at work to shape the trend in the past will continue to work in the future. Thus, this method will be useful for surprise-free projections,

TABLE 18–2
Events likely to occur by 1985.

Event	Probability of Occurrence
Many chemical pesticides phased out	95%
Spending on environmental quality exceeds 6% of GNP	90%
Insect hormones widely used as pesticides	80%
Wide use of computers in elementary schools	25%
Autos banned in central areas of at least seven cities	20%

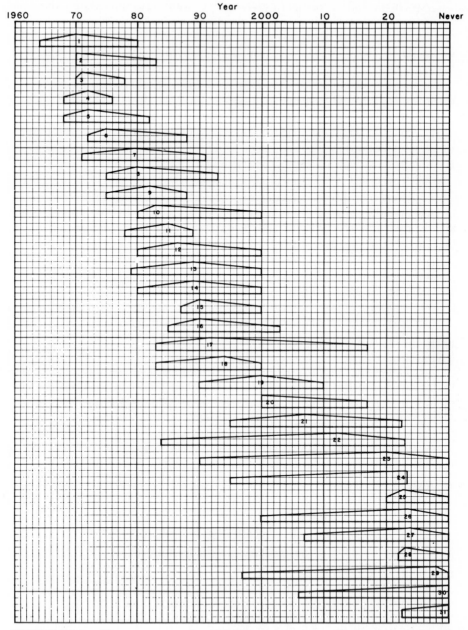

FIGURE 18–3a   The consensus of a Delphi panel on scientific breakthroughs. The medians and quartiles are indicated.

1. Economically useful desalination of sea water
2. Effective fertility control by oral contraceptives or other simple and inexpensive means
3. Development of new synthetic materials for ultra-light construction
4. Automated language translators
5. New organs through transplanting or prosthesis
6. Reliable weather forecasts
7. Operation of a central data storage facility with wide access for general or specialized information retrieval
8. Reformation of physical theory, eliminating confusion in quantum-relativity and simplifying particle theory
9. Implanted artificial organs made of plastic and electronic components
10. Widespread and socially widely accepted use of nonnarcotic drugs (other than alcohol) for the purpose of producing specific changes in personality characteristics.
11. Stimulated emission ("lasers") in X and Gamma ray region of the spectrum
12. Controlled thermo-nuclear power
13. Creation of a primitive form of artificial life (at least in the form of self-replicating molecules)
14. Economically useful exploitation of the ocean bottom through mining (other than off-shore oil drilling)
15. Feasibility of limited weather control, in the sense of substantially affecting regional weather at acceptable cost
16. Economic feasibility of commercial generation of synthetic protein for food
17. Increase by an order of magnitude in the relative number of psychotic cases amenable to physical or chemical therapy
18. Biochemical general immunization against bacterial and viral diseases
19. Feasibility (not necessarily acceptance) of chemical control over some hereditary defects by modification of genes through molecular engineering
20. Economically useful exploitation of the ocean through farming, with the effect of producing at least 20% of the world's food
21. Biochemicals to stimulate growth of new organs and limbs
22. Feasibility of using drugs to raise the level of intelligence (other than as dietary supplements and not in the sense of just temporarily raising the level of apperception)
23. Man-machine symbiosis, enabling man to extend his intelligence by direct electro-mechanical interaction between his brain and a computing machine
24. Chemical control of the aging process, permitting extension of life span by 50 years
25. Breeding of intelligent animals (apes, cetaceans, etc) for low-grade labor
26. Two-way communication with extra-terrestrials
27. Economic feasibility of commercial manufacture of many chemical elements from subatomic building blocks
28. Control of gravity through some form of modification of the gravitational field
29. Feasibility of education by direct information recording on the brain
30. Long-duration coma to permit a form of time travel
31. Use of telepathy and ESP in communications

FIGURE 18–3b The list of expected scientific breakthroughs for which data are reported in Figure 18–3a.

but will entirely miss discontinuties in technological developments such as those caused by the introduction of the transistor.

A scenario or "future history" is a method which uses a narrative description of a potential course of development which might lead to a state of affairs. In their book *The Year 2000*, Herman Kahn and Anthony Wiener provide a significant set of scenarios.

Another method of forecasting the future consequences of a new development is to prepare a decision tree. A decision tree is a graphic device which displays the potential results of alternative approaches. Of course no path is actually as clear cut as a tree or a graph may imply, and actual consequences may occur over a very complex series of stages. Nevertheless, if a series of events can be outlined, the resultant consequence may be discerned.

In order to measure or predict the effects of a project, it is necessary to use a commonly agreed-upon standard of quality of life or of the environment. In order to determine the effect of a new technology, we must agree on what is undesirable. Smoke belching from a power plant means jobs to some and pollution to others. The Gross National Product (GNP) is a summing up of transactions and is too broad a measure; it includes a great many factors. For instance, the GNP reflects how much the citizens of a country produce for the marketplace but not whether they are healthier or happier than others. Also, GNP does not account for air pollution or wastes in the rivers. The GNP of several countries is given in Table 18–3.[6]

SOCIAL INDICATORS

Indicators or measures of social and environmental quality are difficult to develop. A social indicator is a quantitative measure of socially-important conditions of society. A panel of experts under the sponsorship of the United

TABLE 18–3

The Gross National Product of several countries in 1973.

U.S.A.	$1,280 billion
The Common Market	$1,114 billion
U.S.S.R.	465
Japan	438
West Germany	389
France	282
Britain	158
Italy	140
Canada	117
Netherlands	65
India	75
Brazil	62

States Department of Health, Education and Welfare issued *Toward a Social Report* in early 1969.[7] This was a first attempt to develop social measures for use in the assessment of new projects. Unfortunately, the development of adequate and reliable social measures of our society are still not available for reliable use in technology assessment. The impacts and amenities are difficult to measure or quantify, but are nevertheless real and should be as integral to decision-making as quantifiable technical and economic considerations.[11]

As we have observed, technology influences the future through new developments and unforeseen consequences. The use of technology assessments and environmental impact reports will aid in our society's anticipation and control of technology. In the following article, Dr. Mesthene analyzes several ways that technology will shape the future.[8]* We must incorporate these effects in our planning of new technological developments so that the maximum benefit will be obtained over the long term.

How Technology Will Shape the Future

Emmanuel G. Mesthene

There are two ways, at least, to approach an understanding of how technology will affect the future. One, which I do not adopt here, is to try to predict the most likely technological developments of the future along with their most likely social effects. The other way is to identify some respects in which technology entails change and to suggest the kinds or patterns of change that, by its nature, it brings about in society. It is along the latter lines that I speculate in what follows, restricting myself largely to the contemporary American scene.

New Technology Means Change

It is widely and ritually repeated these days that a technological world is a world of change. To the extent that this statement is meaningful at all, it would seem to be true only of a world characterized by a more or less continuous development of new technologies. There is no inherent impetus toward change in tools as such, no matter how many or how sophisticated they may be. When new tools emerge and displace older ones, however, there is a strong presumption that there will be changes in nature and in society.

I see no such necessity in the technology-culture or technology-society relationship as we associate with the Marxist tradition, according to which changes in the technology of production are inevitably and univocally determinative of culture and social structure. But I do see, in David Hume's words, a rather "constant conjunction" between technological change and social change as well as a number of good reasons why there should be one, after we discount for the differential time lags that characterize particular cases of social change consequent on the introduction of new technologies.

The traditional Marxist position has been thought of as asserting a strict or hard determinism. By contrast, I would defend a position that William James once called a "soft" determinism, although he used the phrase in a different context. (One may also call it a *probabilistic* determinism and thus avoid the trap of strict causation.) I would hold that

*From *Science*, Vol. 161, July 12, 1968, pp. 135–143. Copyright 1968 by The American Association for the advancement of Science. Note: The last sentence of this excerpt has been altered slightly at the suggestion of the author.—Ed.

the development and adoption of new technologies make for changes in social organization and values by virtue of creating new possibilities for human action and thus altering the mix of options available to men. They may not do so necessarily, but I suggest they do so frequently and with a very high probability.

Technology Creates New Possibilities

One of the most obvious characteristics of new technology is that it brings about or inhibits changes in *physical* nature, including changes in the patterns of physical objects or processes. By virtue of enhancing our ability to measure and predict; moreover, technology can, more specifically, lead to controlled or directed change. Thus, the plow changes the texture of the soil in a specifiable way; the wheel speeds up the mobility (change in relative position) of people or objects; and the smokebox (or icebox) inhibits some processes of decay. It would be equally accurate to say that these technologies respectively *make possible* changes in soil texture, speed of transport, and so forth.

In these terms, we can define any new (nontrivial) technological change as one which (i) makes possible a new way of inducing a physical change; or (ii) creates a wholly new physical possibility that simply did not exist before. A better mousetrap or faster airplane are examples of new ways and the Salk vaccine or the moon-rocket are instances of new possibilities. Either kind of technological change will extend the range of what man *can* do, which is what technology is all about.

There is nothing in the nature or fact of a new tool, of course, that requires its use. As Lynn White has observed, "a new device merely opens a door; it does not compel one to enter". I would add, however, that a newly opened door does *invite* one to enter. A house in which a number of new doors have been installed is different from what it was before and the behavior of its inhabitants is very likely to change as a result. Possibility as such does not imply actuality (as a strict determinism would have to hold), but there is a high probability of realization of new possibilities that have been deliberately created by technological development, and therefore of change consequent on that realization.

Technology Alters the Mix of Choices

A correlative way in which new technology makes for change is by removing some options previously available. This consequence of technology is derivative, indirect, and more difficult to anticipate than the generation of new options. It is derivative in that old options are removed only after technology has created new ones. It is indirect, analogously, because the removal of options is not the result of the new technology, but of the act of choosing the new options that the technology has created. It is more difficult to anticipate, finally, to the degree that the positive consequences *for* which a technology is developed and applied are seen as part of the process of decision to develop and apply, whereas other (often negative) consequences of the development are usually seen only later if at all.

Examples abound. Widespread introduction of modern plumbing can contribute to convenience and to public health, but it also destroys the kind of society that we associate with the village pump. Exploitation of industrial technology removes many of the options and values peculiar to an agricultural society. The automobile and airplane provide mobility, but often at the expense of stabilities and constancies that mobility can disturb.

Opportunity costs are involved in exploiting any opportunity, in other words, and therefore also the opportunities newly created by technology. Insofar as the new options are chosen and the new possibilities are exploited, older possibilities are displaced and older options are precluded or prior choices are reversed. The presumption, albeit not the necessity, that most of the new options will be chosen is therefore at the same time a presumption that the choice will be made to pay the new costs. Thus, whereas technology begins by simply adding to the options available to man, it ends by altering the spectrum of his options and the mix or hierarchy of his social choices.

Social Change

The first-order effect of technology is thus to multiply and diversify material possibilities and thereby offer new and altered opportunities to man. Different societies committed to different values can react differently (positively or negatively, or simply differently) to the same new possibilities, of course. This is part of the explanation, I believe, for the phenomenon currently being referred to as the "technological gap" between Western Europe and the United States. Moreover, as with all opportunities when badly handled, the ones created by technology can turn into new opportunities to make mistakes. None of this alters the fact that technology creates opportunity.

Since new possibilities and new opportunities generally require new organizations of human effort to realize and exploit them, technology generally has second-order effects that take the form of social change. There have been instances in which changes in technology and in the material culture of a society have not been accompanied by social change, but such cases are rare and exceptional. More generally:

... over the millennia cultures in different environments have changed tremendously, and these changes are basically traceable to new adaptations required by changing technology and productive arrangements. Despite occasional cultural barriers, the useful arts have spread extremely widely, and the instances in which they have not been accepted because of pre-existing cultural patterns are insignificant.

While social change does not necessarily follow upon technological change, it almost always does in fact, thus encouraging the presumption that it generally will. The role of the heavy plow in the organization of rural society and that of the stirrup in the rise of feudalism provide fascinating medieval examples of a nearly direct technology-society relationship. The classic case in our era, of course—which it was Karl Marx's contribution to see so clearly, however badly he clouded his perception and blinded many of his disciples by tying it at once to a rigid determinism and to a form of Hegelian absolutism—is the Industrial Revolution, whose social effects continue to proliferate.

When social change does result from the introduction of a new technology, it must, at least in some of its aspects, be of a sort conducive to exploitation of the new opportunities or possibilities created by that technology. Otherwise it makes no sense even to speak of the social effects of technological change. Social consequences need not be and surely are not uniquely and univocally determined by the character of innovation, but they cannot be entirely independent of that character and still be accounted consequences. (Therein lies the distinction, ultimately, between a hard and soft determinism.) What the advent of nuclear weapons altered was the military organization of the country, not the structure of its communications industry, and the launching of satellites affects international relations much more directly than it does the institutions of organized sport. (A change in international relations may affect international competition in sports, of course, but while everything may be connected with everything else in the last analysis, it is not so in the first.)

There is a congruence between technology and its social effects that serves as

intellectual ground for all inquiry into the technology-society relationship. This congruence has two aspects. First, the subset of social changes that can result from a given
technological innovation is smaller than the set of all possible social changes and the
changes that do in fact result are a still smaller subset of those that can result—that is, they
are a sub-subset. In relation to any given innovation, the spectrum of all possible social
changes can be divided into those that cannot follow as consequences, those that can (are
made possible by the new technology), and those that do (the actual consequences).

It is the congruence of technology and its social consequences in this sense that
provides the theoretical warrant for the currently fashionable art of "futurology." The
more responsible practitioners of this art insist that they do not predict unique future
events but rather identify and assess the likelihood of possible future events or situations.
The effort is warranted by the twin facts that technology constantly alters the mix of
possibilities and that any given technological change may have several consequences.

The second aspect of the congruence between technology and its social consequences
is a certain "one-wayness" about the relationship—that is, the determinative element in it,
however "soft." It is after all only technology that creates new *physical* possibilities
(though it is not technology *alone* that does so, since science, knowledge, social
organization, and other factors are also necessary to the process). To be sure, what
technologies will be developed at any particular time is dependent on the social
institutions and values that prevail at that particular time. I do not depreciate the
interaction between technology and society, especially in our society which is learning to
create scientific knowledge to order and develop new technologies for already established
purposes. Nevertheless, once a new technology is created, it is the impetus for the social
and institutional changes that follow it. This is especially so since a social decision to
develop a particular technology is made in the principal expectation of its predicted
first-order effects, whereas evaluation of the technology after it is developed and in
operation usually takes account also of its less-foreseeable second- and even third-order
effects.

The "one-wayness" of the technology-society relationship that I am seeking to
identify may be evoked by allusion to the game of dice. The initiative for throwing the dice
lies with the player, but the "social" consequences that follow the throw are initiated *by the
dice* and depend on how the dice fall. Similarly, the initiative for development of
technology in any given instance lies with people, acting individually or as a public,
deliberately or in response to such pressures as wars or revolutions. But the material
initiative remains with the technology and the social adaptation to it remains its
consequence. Where the analogy is weak, the point is strengthened. For the rules of the
game remain the same no matter how the dice fall, but technology has the effect of adding
new faces to the dice, thus inducing changes in society's rules so that it can take advantage
of the new combinations that are created thereby. That is why new technology generally
means change in society as well as in nature.

Technology and Values

New technology also means a high probability of change in individual and social
values, because it alters the conditions of choice. It is often customary to distinguish rather
sharply between individual and social values and, in another dimension, between tastes or
preferences, which are usually taken to be relatively short-term, trivial, and localized, and
values, which are seen as higher-level, relatively long-term, and extensive in scope.
However useful for some purposes, these distinctions have no standing in logic, as
Kenneth Arrow points out:

One might want to reserve the term "values" for a specially elevated or noble set of choices. Perhaps choices in general might be referred to as "tastes." We do not ordinarily think of the preference for additional bread over additional beer as being a value worthy of philosophic inquiry. I believe, though, that the distinction cannot be made logically ... .

The *logical* equivalence of preferences and values, whether individual or social, derives from the fact that all of them are rooted in choice behavior. If values be taken in the contemporary American sociologist's sense of broad dominant commitments that account for the cohesion of a society and the maintenance of its identity through time, their relation to choice can be seen both in their genesis (historically in the society, not psychologically in the individual) and in their exemplification (where they function as criteria for choice).

Since values in this sense are rather high-level abstractions, it is unlikely that technological change can be seen to influence them directly. We need, rather, to explore what difference technology makes for the choice behaviors that the values are abstractions from.

What we choose, whether individually or as a society (in whatever sense a society may be said to choose, by public action or by resultant of private actions) is limited, at any given time, by the options available. (Preferences and values are in this respect different from aspirations or ideals in that the latter can attach to imaginative constructs. To confuse the two is to confuse morality and fantasy.) When we say that technology makes possible what was not possible before, we say that we now have more options to choose from than we did before. Our old value clusters, whose hierarchical ordering was determined in the sense of being delimited by antecedent conditions of material possibility, are thus now subject to change because technology has altered the material conditions.

By making available new options, new technology can, and generally will, lead to a restructuring of the hierarchy of values, either by providing the means for bringing previously unattainable ideals within the realm of choice and therefore of realizable values, or by altering the relative ease with which different values can be implemented—that is, by changing the costs associated with realizing them. Thus, the economic affluence that technological advance can bring may enhance the values we associate with leisure at the relative expense of the value of work and achievement, and the development of pain-killing and pleasure-producing drugs can make the value of material comfort relatively easier of achievement than the values we associate with maintaining a stiff upper lip during pain or adversity.

One may argue further that technological change leads to value change of a particular sort in exact analogy to the subset of possible social changes that a new technology may augur (as distinct from both the wider set that includes the impossible and the narrower, actual subset). There are two reasons for this. First, certain attitudes and values are more conducive than others to most effective exploitation of the potentialities of new tools or technologies. Choice behavior must be somehow attuned to the new options that technology creates, so that they will in fact be chosen. Thus, to transfer or adapt industrial technologies to underdeveloped nations is only part of the problem of economic development; the more important part consists in altering value predispositions and attitudes so that the technologies can flourish. In more advanced societies, such as ours, people who hold values well adapted to exploitation of major new technologies will tend to grow rich and occupy elite positions in society, thus serving to reinforce those same values in the society at large.

Second, whereas technological choices will be made according to the values prevailing in society at any given time, those choices will, as previously noted, be based on the foreseeable consequences of the new technology. The essence of technology as creative of new possibility, however, means that there is an irreducible element of uncertainty—of unforeseeable consequence—in any innovation. Techniques of the class of systems analysis are designed to anticipate as much of this uncertainty as possible, but it is in the nature of the case that they can never be more than partially successful, partly because a new technology will enter into interaction with a growing number and variety of ongoing processes as societies become more complex, and partly—at least in democratic societies—because the unforeseeable consequences of technological innovation may take the form of negative *political* reaction by certain groups in the society.

Since there is an irreducible element of uncertainty that attends every case of technological innovation, therefore, there is need for two evaluations: one before and one after the innovation. The first is an evaluation of prospects (of ends-in-view, as John Dewey called them). The second is an evaluation of results (of outcomes actually attained). The uncertainty inherent in technological innovation means there will usually be a difference between the results of these two evaluations. To that extent, new technology will lead to value change.

Contemporary Patterns of Change

Our own age is characterized by a deliberate fostering of technological change and, in general, by the growing social role of knowledge. "Every society now lives by innovation and growth; and it is theoretical knowledge that has become the matrix of innovation."

In a modern industralized society, particularly, there are a number of pressures that conspire toward this result. First, economic pressures argue for the greater efficiency implicit in a new technology. The principal example of this is the continuing process of capital modernization in industry. Second, there are political pressures that seek the greater absolute effectiveness of a new technology, as in our latest weapons, for example. Third, we turn more and more to the promise of new technology for help in dealing with our social problems. Fourth, there is the spur to action inherent in the mere availability of a technology: space vehicles spawn moon programs. Finally, political and industrial interests engaged in developing a new technology have the vested interest and powerful means needed to urge its adoption and widespread use irrespective of social utility.

If this social drive to develop ever more new technology is taken in conjunction with the very high probability that new technology will result in physical, social, and value changes, we have the conditions for a world whose defining characteristic is change, the kind of world I once described as Heraclitean, after the pre-Socratic philosopher Heraclitus, who saw change as the essence of being.

When change becomes that pervasive in the world, it must color the ways in which we understand, organize, and evaluate the world. The sheer fact of change will have an impact on our sensibilities and ideas, our institutions and practices, our politics and values. Most of these have to date developed on the assumption that stability is more characteristic of the world than change—that is, that change is but a temporary perturbation of stability or a transition to a new (and presumed better or higher) stable state. What happens to them when that fundamental metaphysical assumption is undermined? The answer must be sought in a number of intellectual, social, and political trends in present-day American society.

CHAPTER 18 REFERENCES

1. A. Toffler, *The Futurists*, Random House, New York, 1972.
2. A.H. Cahn and J. Primack, "Technological Foresight for Congress," *Technology Review*, March/April 1973, pp. 39-48.
3. M.V. Jones, "The Methodology of Technology Assessment," *The Futurist*, February 1972, pp. 19–22.
4. M. Turoff, "The Delphi Conference," *The Futurist*, April 1971, pp. 55–57.
5. "A Think-Tank That Helps Companies Plan," *Business Week*, August 25, 1973, pp. 70–71.
6. "World Economies: Strong Growth Rates Tempered By Inflation," *Business Week*, July 28, 1973, p. 34.
7. United States Department of Health, Education and Welfare, *Toward a Social Report*, United States Government Printing Office, Washington, D.C., 1969.
8. E.G. Mesthene, "How Technology Will Shape the Future," *Science*, July 12, 1968, pp. 135–143.
9. D. Turtle, "LSI—Plug Into the Electronic Future," *New Scientist*, May 24, 1973, pp. 476–479.
10. M. Kenward, "Assessing the Impact of the Video Telephone," *New Scientist*, September 20, 1973, pp. 693–694.
11. M.S. Baram, "Technology Assessment and Social Control," *Science*, May 4, 1973, pp. 465–473.
12. D.E. Kash *et al*, *Energy Under the Oceans*, University of Oklahoma Press, Norman, Oklahoma, 1973.
13. R.A. Bauer, *Second Order Consequences*, The M.I.T. Press, Cambridge, 1969.
14. P. Baran, "30 Services That Two-Way Television Can Provide," *The Futurist*, October 1973, pp. 202–210.
15. D. Shapeley, "Auto Pollution: EPA Worrying That the Catalytic May Backfire," *Science*, October 26, 1973, pp. 368–371.
16. L.J. Carter, "Environment: Academic Review for Impact Statements," *Science*, November 2, 1973, p. 462.
17. T.J. Gordon and O. Helmer, *Social Technology*, Basic Books, New York, 1966.

CHAPTER 18 EXERCISES

18–1. The transistor was invented in 1948. Since then the technology has advanced so that many thousands of transistors, which are solid-state devices, may be put onto a single very small, cheap component. In 1965 it was possible to place 10 transistors on one component, while it is now possible to place 10,000 transistors on one component. The technology, often called Large-Scale Integration (LSI), enables small computers, small calculators and digital wrist watches to be built. As LSI components become cheaper and easier to design, more consumer applications become possible.[9] Prepare a technology assessment for the wide implementation of LSI.

18–2. The social impact of the video telephone, often called the Picturephone, may be significant. The Picturephone, under development by Bell Telephone Laboratories, is shown in Figure E18–2. The widespread use of video telephones awaits the introduction of a reliable, low cost device.[10] One impact of wide-spread use will be on business organizations. The video phone could speed the trend toward decentralization of office activities. The secondary effect of moving work away from the center of towns could be significant on existing transportation systems. A study of the air passenger traffic in the New York region found that 57% of all trips by air for were for business purposes. However, several visits by video phone may substitute for an in-person visit. Determine the favorable impacts in terms of energy and resources and the potentially undesirable consequences such as loss of privacy and the high cost to the society in capital investment.

18–3. At the present time, oil and gas account for almost 70% of the total energy consumption in the United States, which imported 14% of its energy resources in 1973. This percentage is expected to grow. The Outer Continental Shelf of the United States holds vast potential oil and gas resources. By 1985, approximately 25% of the domestic oil and gas could come from the OCS. The location is favorably close to the population centers of the United States. Nevertheless, environmental concerns about oil spills and blowouts are equally important.[12] Prepare an outline of an environmental impact statement for the leasing of the United States Federal rights to oil and gas off the American shores.

18–4. Fluoridation of public water supplies is a deliberate adjustment of the fluoride content to a level of approximately one part per million to reduce tooth decay. Commonly a chemical such as Sodium Fluoride is used. This process is favored by medical and dental authorities and the United States Public Health Service. Nevertheless, the fluoridation of water remains a controversial subject. Although experience seems to indicate the efficacy of fluoridation, voters reject it in referendums across the country. Prepare a technology assessment for the fluoridation of your community water supply suitable for submission to the voters at a referendum to decide whether to introduce fluoridation.

FIGURE E18-2   The picturephone may transmit graphic information, as shown, or a
view of the person speaking on the phone. The control unit on the
desk below the television screen gives the user fingertip control of
speaker volume, screen brightness and the operation of the transmitting camera, which can be seen mounted directly above the
viewing screen. Courtesy of Bell Telephone Laboraties.

18-5. Cable television (CATV) is growing rapidly in use throughnout the United
States. CATV uses a coaxial cable to bring television signals and
locally-originated programs to a person's home or office. Two-way
interactive transmission is possible by means of CATV systems. A recent
Delphi study considered the effects of CATV and the potential demand
for new services.[14] This survey determined that the middle estimate for
when shopping transactions might take place routinely via CATV would
be 1985. Person to person communication with the person paid to work at
home is estimated to occur in 1985. Prepare a Delphi survey for these
events and others of interest to you and carry out a survey with a suitable
group of "experts." Also prepare a list of questions to assess the social
cousequences of such new technologies.

18-6. United States automakers are planning to install catalytic converters on automobiles in order to meet emission standards. While reducing the emission of hydrocarbons and carbon monoxide, the converter may emit sulfuric acid and small particles of platinum.[15] Prepare an assessment of the catalytic converter.

18-7. The digital computer may be used in a wide variety of tasks. Recently the computer has been applied to tasks at retail stores, banks and super-markets. For example, retail stores by the thousands are converting mechanical cash registers into electronic point-of-sale machines that are linked to powerful central computers. Prepare a technology assessment of digital computers in retail stores.

18-8. Electric power transmission lines require relatively small amount of land area for the towers. Nevertheless, the visual impact of power lines may be significant, as is shown in Figure 18-8. Prepare an environmental impact statement for the power transmission lines that serve your town by visiting the site and discussing their effects with local residents and the power company

18-9. The Environmental Impact Statements prepared by United States Federal agencies each year now number over 1,500. Each statement is circulated for comment among other interested Federal agencies. The Environ-mental Protection Agency is routinely one of the reviewing agencies. Also, the Federal courts appear to be very willing to entertain lawsuits brought under the provisions of the N.E.P.A. It has also been proposed that an independent review of draft statements be undertaken by faculty members of universities.[16] All of these reviews serve to improve the quality of the statements, but also may serve to delay projects important to the health or safety of the nation. Discuss the Environmental Impact Statement process and indicate whether you find the process equitable and valuable.

18-10. Highways, and particularly highway interchanges, have a profound impact on a city. See Figure E17-15. Examine the costs and benefits of a highway interchange in or near your town. Talk with local residents and the highway department. Prepare an outline for an environmental impact statement for the selected interchange.

18-11. The consensus of a Delphi panel on automation is shown in Figure E18-11.[17] Review this result in light of recent developments. How successful was the Delphi panel? Estimate the year of success for items 7 through 13 based on the increased knowledge we now have. Item 11 calls for automated transit. Review the discussion of BART in Chapter 17 and note the accuracy of the median estimate in light of the expected date of successful automatic control of BART trains.

18-12. Technology assessment could become a stifling influence on progress if emphasis is placed only on the dangers of proposed new technologies rather than on their potential for good. Innovators often have quite a lot of difficulty in introducing their ideas without added inhibitors. Do you think technology assessment will inhibit progress? What is the balance

FIGURE E18-8   Electric power transmission lines on California's rolling hills. Courtesy of Rondal Partridge.

between change and conservatism? Describe a balance, in the initiation and control of new technologies, which you would propose.

18–13. There is a history of technology assessment in the United States. As early as 1830 our Federal Government began to assess the causes and effects of boiler technology on steam boats as a result of explosions on several boats. The tragic sinkings of *U.S.S. Thrasher* and *U.S.S. Scorpion*, the disastrous fire in the Apollo spacecraft and the grounding of the *Torrey Canyon* oil tanker all resulted in investigations. Explore the past history of technology assessment.

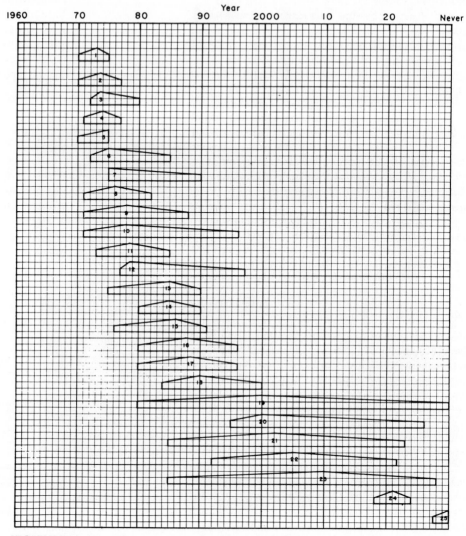

FIGURE E18-11a   The consensus of a Delphi panel on automation. The median year
is shown by the tip of the box and the edges of the box indicate
the quartiles.

1. Increase by a factor of 10 in capital investment in computers used for automated process control

2. Air traffic control—positive and predictive track on all aircraft
3. Direct link from stores to banks to check credit and to record transactions
4. Widespread use of simple teaching machines
5. Automation of office work and services, leading to displacement of 25% of current work force
6. Education becoming a respectable leisure pastime
7. Widespread use of sophisticated teaching machines
8. Automatic libaries, looking up and reproducing copy

9. Automated looking up of legal information
10. Automatic language translator—correct grammar
11. Automated rapid transit
12. Widespread use of automatic decision-making at management level for industrial and national planning
13. Electronic prosthesis (radar for the blind, servomechanical limbs, etc.)
14. Automated interpretation of medical symptoms
15. Construction on a production line of computers with motivation by "education"
16. Widespread use of robot services, for refuse collection, as household slaves, as sewer inspectors, etc.
17. Widespread use of computers in tax collection, with access to all business records—automatic single tax deductions
18. Availability of a machine which comprehends standard IQ tests and scores above 150 (where "comprehend" is to be interpreted behavioristically as the ability to respond to questions printed in English and possibly accompanied by diagrams)
19. Evolution of a universal language from automated communication
20. Automated voting, in the sense of legislating through automated plebiscite
21. Automated highways and adaptive automobile autopilots
22. Remote facsimile newspapers and magazines, printed in home
23. Man-machine symbiosis, enabling man to extend his intelligence by direct electromechanical interaction between his brain and a computing machine
24. International agreements which guarantee certain economic minima to the world's population as a result of high production from automation
25. Centralized (possibly random) wire tapping

FIGURE E18-11b The list of categories judged by the Delphi panel on automation reported in Figure E18-11a.

19

ENERGY RESOURCES, USE AND POLICY

IF THE MACHINE HAS FREED man from slavery to toil, then the fuel of this change has been the abundant energy stored in the natural resources of the planet. In the nineteenth century man developed a capability not only to manufacture materials and goods, but also to manufacture and control power. More correctly, we should state that man learned to release the vast amounts of stored energy contained within the fossil fuels of the world and to control this energy and put it to use in the technological processes of industry, commerce and the home. A century after the invention of the first successful oil well, only 1% of America's physical work is done by man himself. The rest is done by machines fueled with the energy released from natural resources.

THE QUESTION OF CAPACITY

The problem of the vastly increasing use of energy resources may be the finite nature of the energy resources of the world—or at least our ability to tap only a limited amount of them. Energy is the capacity for doing work. The fuel resources of our world contain the capacity to accomplish work, such as moving an automobile. In the most general sense, work can be defined as a force, F, moving an object through a distance, d, so that

$$W = Fd \qquad (19\text{--}1)$$

The unit of force times distance is kilogram-meter²/second². It is called a *joule*. The units of work or energy are joules. However, there are several equivalent units of energy. They are shown in Table 19–1.

Energy is called *potential energy* until it is released. Water at the top of a dam is potentially available to fall down a pipe and drive a turbine. Similarly, energy stored in gasoline may be potentially released by a combustion process. When energy is put to use it is *kinetic energy*.

When energy is put to use it is consumed at some rate over a period of time. The rate of use of energy is defined as power. Thus, we have over a period of time, t, a power P as follows:

$$P = \frac{W}{t} \qquad (19\text{--}2)$$

with the units of joules per second. The unit of one joule per second is called one watt. Several equivalent units of power are given in Table 19–2.

SOURCES OF ENERGY

About 90% of the energy consumed in the United States is derived from fossil fuels—coal, oil and natural gas. All fossil fuels required millions of years to be formed. At the rate we consume them, they will not replace themselves. Thus fossil fuels are a finite or limited source. We may define Q_R as the proven reserves of a fossil fuel and Q_P as the total accumulated quantity of the fuel removed from the ground up to any time. The cumulative discoveries, Q_D, represents the

TABLE 19-1

Units of energy.

1,055 joules = 1 British thermal unit (BTU) = 0.252 kilocalories

3,412 BTU = 1 kilowatt-hour = 3,600,000 joules

TABLE 19-2

Units of power.

One joule/second = 1 watt

1 kilowatt = 1,000 watts

1 megawatt = 1,000,000 watts = 1,000 kilowatts

1 horsepower (HP) = 746 watts

quantities removed from the ground plus the recoverable quantities remaining in the ground. Therefore,

$$Q_D = Q_P + Q_R \qquad\qquad (19\text{--}3)$$

If Q_D is eventually limited or finite, we have an effect portrayed in Figure 19–1.[1] As time progresses, the cumulative discoveries Q_D, reach the finite limit. As production, Q_P, increases, eventually the proved reserves, Q_R, reach a maximum and begin to reduce. As an example, let us consider the use of the United States reserves of oil. America was producing 4.2 billion barrels (b.b.) per year in 1970 from its own oil resources. It is estimated that the finite amount of oil in the United States is 195 b.b. and 90 b.b. have been used up to 1970. Then, if these estimates are correct, the proved reserves, Q_R, are dropping; if we consumed United States domestic oil at the rate of 4.2 b.b. per year we would exhaust this resource in $105/4.2 = 25$ years. Over the past decade the new discoveries, especially in Alaska, have increased the cumulative discoveries and thus the ultimate finite amount of oil. Nevertheless, if the fuel is limited, we eventually reach a point of near-exhaustion of the fuel. In the United States this point may be estimated to be 30 years for oil, 50 years for natural gas and several hundred years for coal. Of course, the reserves are never fully exhausted, since as the reserves become very low and scarce, the price of this fuel will rise and consumption will decrease. As Figure 19–1 shows, the reserves are reduced slowly as time progresses toward the far right of the curve for Q_R. In other words, fossil fuels are limited and current rates of use do indicate that during the next decade

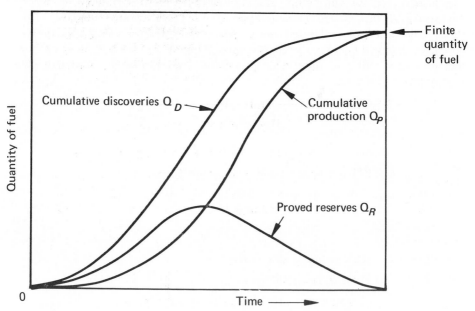

FIGURE 19-1   The consumption of a limited fuel over the lifetime of the fuel.

new policies for the use of fossil fuels for supplying energy to our industries and homes and for transportation will be required.

ENERGY CONSUMPTION

Consumption of all kinds of fuel is bound to increase for several decades at least. There are many people in the world in need of warmth, transportation and machinery. Economic development of countries implies that each person uses more energy each year. In the decade 1959–1969, the amount of energy used by each person in the world increased by 36%. With the growth of the world population during the same period, the total amount of energy consumed increased by 63%. However, the amount of energy consumed varies enormously from one country to another. In South America the energy used *per capita* is only 7% of that of North America.

The consumption of energy *per capita versus* the Gross National Product *per capita* is shown for several nations in Figure 19–2. Notice the trend of developing countries to use more energy. The correlation between increased GNP and increased energy *per capita* is very strong. For example, Yugoslavia has three times the GNP *per capita* of Syria with approximately three times the energy-use *per capita*. Similarly, the United States has a GNP *per capita* twenty times that of Syria and an energy use *per capita* of twenty times. Thus, if the wealth of a nation is to be increased measurably, it appears that the trend of increased use of energy per capita must be followed.

ELECTRIC POWER

Much of the total energy used by a nation is in the form of electricity. Electrical power is readily distributed to homes and factories and is a valuable source of energy to an industrialized society. Electrical energy is primarily generated by using fossil fuels to generate steam which then drives electric generators. The electrical energy consumption *per capita versus* the GNP *per capita* for several nations is shown in Figure 19–3.

The United States uses approximately 8,200 kilowatt-hours per person while the U.S.S.R. uses 3,200 kilowatt-hours. Less-developed countries like Brazil use 450 kilowatt-hours. Thus the United States uses over 18 times the electrical power *per capita* compared to Brazil, while achieving a GNP 16 times greater. The electrical energy consumption *per capita* for several nations during the period 1930 to 1970 is given in Table 19–3. Over this 40-year period, the consumption increased by a factor of 10.8 and 20.3 for the United States and Yugoslavia respectively. As nations become industrialized and increasingly affluent, the demand for electric power grows very significantly.

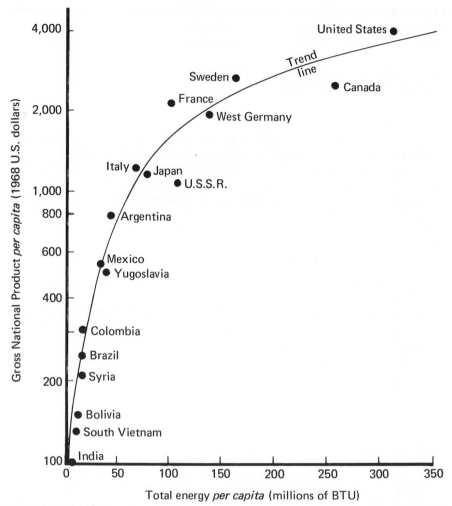

FIGURE 19-2   The total consumption of energy *per capita versus* the Gross National
Product *per capita* for several nations. Note that the vertical scale is
logarithmic.

As we found in Table 19–3, the electrical energy consumption *per capita* grows significantly. In addition, the population of many of the newly industrialized nations has also been growing, thereby increasing the demand for energy. In the United States, production of electrical energy has continued to grow faster than the population has increased. This growth of electric power production is shown in Figure 19–4 for the period 1900–1970 with projections to the year 2000.

The total energy use in the United States as well as the *per capita* consumption has grown over the past century. Total energy consumption includes energy for all uses, including transportation, the generation of electricity, residential heating, and industrial use among others. The *per capita* energy consumption in the United States during the decade 1962–1972 is shown in Figure 19–5. During that decade the energy consumption *per capita* increased by 35%.

The primary sources of energy for the United States have changed greatly

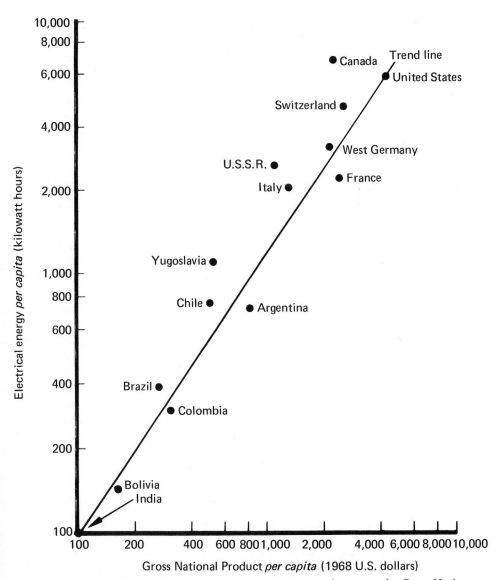

FIGURE 19-3 The electrical energy consumption *per capita versus* the Gross National Product *per capita* for several nations. Note that *both* scales are logarithmic.

TABLE 19-3

Electrical energy consumption *per capita**
(in kilowatt/hours *per annum per capita*).

	USA	UK	Japan	Hungary	Netherlands	Yugoslavia	Fed. Republic Germany
1930	740	280	230	100	210	60	—
1940	1300	620	470	200	420	50	—
1950	2500	1120	550	320	730	150	860
1960	4600	2630	1270	760	1440	490	2220
1970	8000	4440	3100	1460	3300	1220	4000
Increase during the period 1940–70	10.8	15.9	13.5	14.6	15.7	20.3	—

*Source: "Changes in the World induced by Technology" by H. Chestnut, General Electric Report, February, 1972.

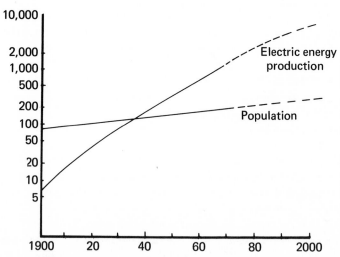

FIGURE 19-4 The growth of the United States population (in millions) and of electric energy production (in billions of kilowatt-hours) over the period 1900-1970 with projected growth for the period 1970-2000. Note that the vertical scale is logarithmic.

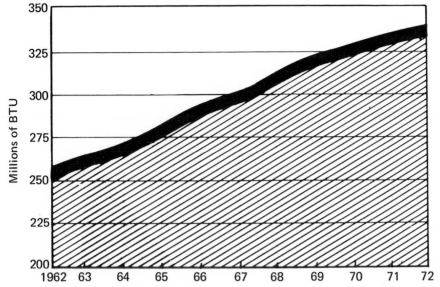

FIGURE 19-5 The *per capita* energy consumption in the United States during the decade 1962-1972.

over the past century. The various sources of energy used in this time are shown in Figure 19–6.[2] The use of wood as fuel has greatly decreased, but has not become insignificant. Coal has become an important fuel. More recently oil and gas have become important sources of energy in the United States. Note that water power as a source of energy has always been significant, but remains only 4% of the total energy sources. Nuclear power plants have become significant only in the last decade; they provide less than 1% of the total energy now. It has been estimated that in 1970 gas supplied 35.8% of the energy in the United States, as shown in Figure 19–7.[3] Also, coal and oil supplied 21.8% and 41% of the energy respectively in 1970. One estimate of the supply of energy in 1985 by various fuels is shown in Figure 19–7b.[3] This estimate projects that 17% of the energy in 1985 will be supplied by nuclear power plants. Transportation will account for approximately 25% of the use of fuel in 1970 and 1985, according to these estimates. Currently residential and commercial uses for energy account for 19%. Residential and commercial uses are primarily for lighting and hearting. The projections shown in Figure 19–7 assume a growth in energy of 73% from 1970 to 1985, with a growth in the percentage supplied by oil from 41% to 46%. Whether the United States can import sufficient oil in 1985 to meet this estimated demand is really doubtful. The estimate, originally published by the Joint Committee on Atomic Energy of the United States Congress in 1973, states that while the United States imported 1.28 billion barrels of oil in 1970, or 25% of its total oil supply, it will import 5.33 billion barrels in 1985 or 57% of its total oil supply.[4] Whether this percentage of imported oil can be obtained in 1985 remains in grave doubt considering the economic and political consequences of such a large importation. One forecast of the total use of energy in the United States is shown in Figure 19–8. This forecast must be taken with serious reservations, since the energy supply may not meet the demands.[23]

Transportation is a large user of oil, particularly in the form of refined gasoline. Approximately three-fourths of the energy used in transportation is consumed by automobiles. Furthermore, 57% of auto passenger service is of a

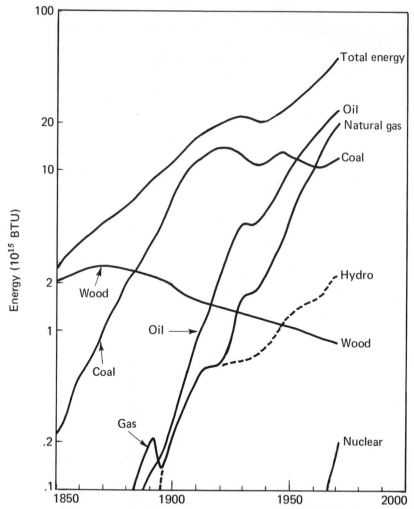

FIGURE 19-6   Energy use in the United States over the period 1850–1970. Note: The vertical scale is logarithmic. Reprinted from *New Energy Technology* by H.C. Hottel and J.B. Howard by permission of the M.I.T. Press, Cambridge, Mass. © by The Massachusetts Institute of Technology.

short, urban trip. The increased use of mass rapid transit and the design of automobiles with relatively lower gasoline consumption could reduce the oil consumption of the United States significantly. Transportation is 96% dependent on the availability of oil as a fuel because of the predominance of the internal combustion engine. Perhaps the electric automobile and the mass rapid transit

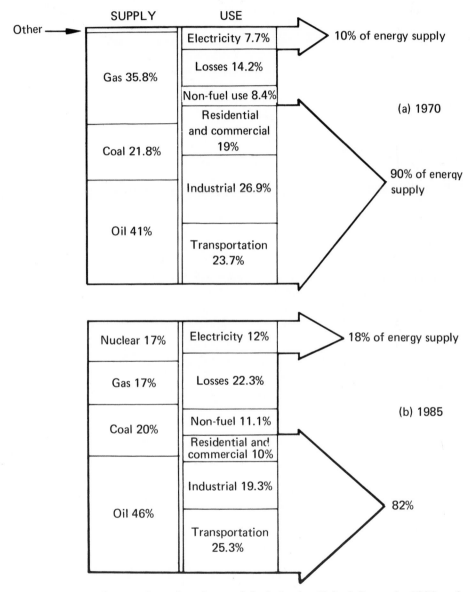

FIGURE 19-7 The supply and end use of fuels in the United States in 1970 and projected for 1985.

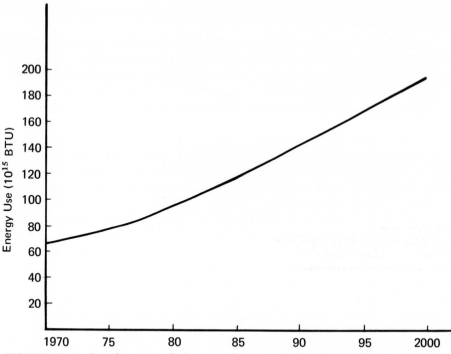

FIGURE 19-8   One forecast of the use of energy in the United States during the period 1970–2000.

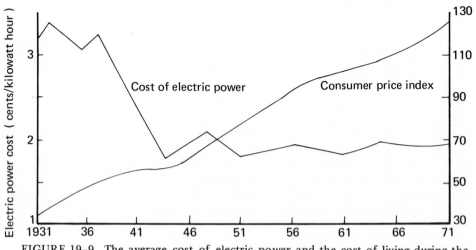

FIGURE 19-9   The average cost of electric power and the cost of living during the period 1931–1971.

systems will shift the transportation system of the future to dependence upon other fuels.

At the place of production, fossil fuels are relatively inexpensive. Before the 1974 Arab embargo, oil was $0.65/million BTU at the place of production. Of course, scarcity during the next decade will drive oil prices higher. Over the decade 1962–1972 the prices of coal, oil and gas remained essentially constant. To the price of production must be added the cost of transporting the fuel to the user. Pipelines are used for oil and gas, while coal is transported primarily by train. Electric power is transmitted by electrical transmission lines. Nevertheless, it is approximately three times more costly to transport electricity via wires than it is to transport an equivalent amount of energy through an oil or gas pipeline. Furthermore, pipelines can be buried and do not mar the visual environment as do large power transmission lines.

The cost of electric power, as shown in Figure 19–9, has decreased during the period 1931–1971 because of economies of scale and improved technology. With larger and improved power plants electric power was reduced in cost. Now, however, as the prices for the fuel to supply the power plants increases, we can expect the price of electricity to rise throughout the 1970s. Total generation of electric energy was 1.83 trillion kilowatt-hours in 1972.

AFFLUENCE, PRICES AND COSTS

One of the reasons for the affluence of the developed nations is the low cost of electrical energy. As the cost of electricity has declined, the use has increased. The United States uses more than one-third of the entire world's electric power generating capacity, and the continuing rate of growth doubled the use over the past two decades and promises to do so for the next decade.

In the United States, 33% of the total electricity consumed is used for residences and 42% of the total is used for industrial purposes. Approximately 23% of the electricity consumed in the United States is used for commercial purposes such as store lighting and heating and advertising signs. Electricity consumption for residential and commercial heating and air conditioning is growing at an accelerated rate. It is estimated that 17 million tons of new air conditioners were installed in 1973—an increase of 18% in that year.

The environmental impact of energy usage is vast and widely distributed. Automobiles introduce air pollution as a result of the combustion of gasoline. Coal mines introduce waste, and the burning of coal introduces sulfur dioxide into the air as well as soot and fly ash. Nuclear power plants yield radioactive waste which must be stored or disposed of. The stripping of coal from the surface, often called strip mining, results in degradation of the soil and the deforestation of the mining area. While industry and power plants produce 31% of the air pollutants, automobiles produce 51%. Therefore, environmental degradation is a result of a growing use of energy for transportation and industry.

In addition, all consumption of fuel for energy purposes results in the disposal of large amounts of waste heat. This waste heat is dumped into rivers or lakes or into the air.

The cost of controlling the environmental effects of energy use can be very significant. These costs will be passed on to the consumer in the increased costs of energy delivered. As the cost of energy increases, the economics of energy usage will change and the user may shift to less energy-intensive uses.

COAL

One of the greatest sources of energy in the United States is its coal reserves. American coal reserves are virtually untapped, but most are too deep for recovery by economic and conventional methods. The most economical process of obtaining coal is strip mining, but the long term effects of strip mining can be very costly to the environment. Reforestation and reclamation are subjects of current controversy. Furthermore, the most desirable coal has a low sulfur content and is located primarily in the west, far from its potential users. Shipment of western coal to the eastern United States would result in high transportation costs. Furthermore, coal-mine safety is an important consideration as new underground mines are developed. As a result of these problems with coal mining, the mining of coal has decreased over the past several years. Coal's share of the United States energy market has fallen from 35% in 1950 to 17.7% in 1971. Approximately 600 million tons of coal were mined in the United States in 1972.

Nevertheless, the use of coal would increase if accepted improvements in strip mining could be developed and sulfur could be removed from the coal. One possible process currently being developed is the gasification of coal for shipping via pipelines to the users. The process of making gas from coal is still, however, in the experimental stage. It may be 1990 before gas from this source would yield 10% of the total gas supply in the United States. Because the United States has approximately 50% of the world's known coal reserves, this source may yield one long-term supply for American energy consumers.

STRIP MINING

At the present time about 50% of American coal is strip mined, and this percentage is increasing. In the eastern states, strip mining disturbs about 0.6 acres per 1,000 tons of coal removed. The land area disturbed annually by coal strip mining in the United States is shown in Figure 19–10. Strip mining in the United States has dug up to two million acres, mostly in Appalachia, only half of which have been reclaimed

The United States has tens of billions of tons of low sulfur coal in the Upper Missouri Basin under small overburdens of earth. In eastern Montana and parts

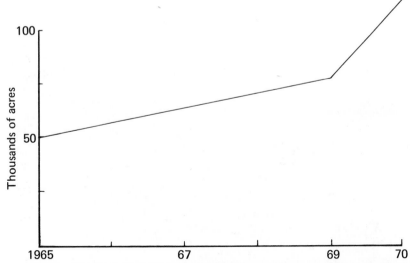

FIGURE 19-10   The land area disturbed annually by coal strip mining in the United States. Source: *Environmental Quality,* Third Annual Report of the Council on Environmental Quality, 1972.

of North Dakota, coal fields promise up to 80,000 ton yields per acre. With the increasing demand for this coal for city power plants, the strip mining of the western states is increasing rapidly.

The environmental effects of stripping are, among others: acid and mineral seepage into the water table; soil bank slippage and landslides; silting; soil sterility and the reduction of vegetation on the land, resulting in increased run-off and flooding.

The overburden of soil (above the coal) is shattered by explosive charges and then removed by a machine such as the one shown in Figure 19-11. This machine is used near Braidwood, Illinois by Peabody Coal Company. It is suitable for removing soil and soft types of rock in relatively flat land. This machine can move 5,200 tons of spoil per hour. Today, the biggest machine in operation, Big Muskie, has the capacity to move four million cubic yards of earth a month, removing the overburden of earth up to 60 feet deep.[10] An example of a strip-mined area in the State of Washington is shown in Figure 19-12. Strip-mined land can be reforested and reclaimed by planting suitable vegetation in topsoil replaced over the stripped area. Reclamation costs vary, but can run up to $1,000 an acre. The most difficult areas to reclaim are steep slopes where the soil has little stability on the slope and rain runoff can cause erosion and the loss of topsoil. Also, acid seepage from the mining can cause, after reclamation, later destruction of the trees, vegetation and streams.

As new reclamation techniques and mining technologies are developed, strip mining may become less harmful to the environment. The United States needs

FIGURE 19-11 A giant coal stripping machine. This machine was manufactured in Germany and is used in Illinois. This machine removes soil and rock from above the coal and moves 5,200 tons of spoil per hour. Courtesy of the Mid-Appalachian Environment Service.

the 45 billion tons of strippable coal, mostly of the low sulfur type, to meet its demands over the next 30 years. The question is: can strip-mined coal be obtained without great harm wrought on the environment and the persons who live near strip mining sites?[10]

The largest strip mine in the United States is located at Black Mesa in northeastern Arizona, where the Peabody Coal Company has leased 64,858 acres of land from the Navajo and Hopi tribes. The Black Mesa mine feeds into a six-plant complex (The Four Corners facility), which is operated by a combine of 23 utilities. The tribes will get about 25 cents a ton for the coal, some $75 million over the life of the project. The Four Corners project lies near the junction of Colorado, Utah, New Mexico and Arizona. In 1972 a total of 2.2 million kilowatts were generated each day; the plants were using 25,000 tons of stripped coal daily. In full production the project will supply, via electric transmission line, 14 million kilowatts to the metropolitan areas of Los Angeles, San Diego, Las Vegas, Phoenix and Tucson. The plants until recently have been heavy air polluters, spewing coal, soot and sulfur dioxide into the air from thirty-story smokestacks. This was a case of the metropolitian areas simply moving their waste products to another area, the Four Corners, while consuming the power themselves. The first plant began operation in 1963 and the full project may be operational by late in this decade.[11] The Four Corners plant produces more fly ash and soot than all sources in Los Angeles and New York City combined. The smoke from the Four Corners plant has been traced as far as 215 miles from the plant.[17] A photo of the Four Corners plant is shown in Figure 19–13.

EFFICIENCY OF COMBUSTION ENGINES

The conversion of stored energy in a fossil fuel or in the nucleus of an atom is accomplished by a combustion process or a nuclear reaction, in the case of an atom, with released energy in the form of heat. This heat is used to drive a piston

FIGURE 19–12   A strip mine at the Centralia, Washington power plant. Courtesy of David C. Flaherty and *Quest* magazine.

FIGURE 19-13   The Four Corners electric power project. The stripped mine is shown
               in the foreground and the power plant in the background. This plant
               is located near the conjunction of Colorado, Arizona, New Mexico
               and Utah. Courtesy of the Central Clearing House.

in an internal combusion engine or to heat fluid in a steam engine or turbine. In
the process of the conversion of energy from its stored form, only a fraction of it
is converted to energy of motion. For example, only 20% of the energy stored in
gasoline is converted to energy at the wheels of an automobile. Therefore,
increasing the efficiency of automobile engines and other energy-using devices
might reduce the need for increasing energy sources. The efficiency of a heat
engine is

$$\epsilon = \left(1 - \frac{Q_{cold}}{Q_{hot}}\right) \qquad (19\text{-}4)$$

where ϵ = efficiency and Q_{cold} and Q_{hot} are the heat released and the heat
absorbed, respectively. For the particular case of the Carnot engine, we have
$Q_{cold}/Q_{hot} = T_{cold}/T_{hot}$ where T = temperature in degress Kelvin and where
Kelvin temperature is 273° plus the temperature in degrees centigrade.

$$\epsilon = \left(1 - \frac{T_{cold}}{T_{hot}}\right) \qquad\qquad (19\text{--}5)$$

For example, in a steam turbine where T_{hot} = 825° K and the environment is 300° K or 23° Centigrade, the efficiency of this engine is then

$$\epsilon = \left(1 - \frac{300}{825}\right) = 0.64$$

or 64% efficient. A perfect heat engine would convert all the thermal energy that it takes in into mechanical work. The Second Law of Thermodynamics states that, for a system operating in a cycle between heat sources designated as T_{hot} and T_{cold} it is impossible to extract a net amount of heat from the hot sources and convert it entirely into work. Some heat must be rejected to the cold source. This fundamental law results in our desire to raise the temperature of the hot source as much as possible. We are of course limited by the melting point of the metal of the engine. Nevertheless, even if T_{hot} = 1,300° K, we have,

$$\epsilon = \left(1 - \frac{300}{1300}\right) = 0.769$$

assuming the environment is at a temperature of 300° K or 23° C. Even if an increase could be achieved to T_{hot} = 1,500° K, the efficiency only increases to ϵ = 0.80. Thus, 20% to 70% of the energy used in an engine or an electric generating plant is lost as waste heat. This waste heat is added to a river or lake or the air.

The actual steam cycle in an electric power plant using fossil fuels is not a Carnot cycle but rather a less efficient cycle known as Rankine cycle. Some heat is lost around the boiler also as well as at the cold temperature sink in the Rankine cycle. The best fossil-fuel generating plants now in use have an efficiency of about 40% and the upper limit obtainable has been estimated as 46%.

WASTE HEAT

Waste heat is wasted or useless only because it is not economically practical to use it. Actually a scheme could be devised for using waste heat to heat buildings or greenhouses for agriculture, for example. If water from a river or lake is used to cool the heat sink (Q_{cold}) then the water is usually returned to its source some 10° to 20° Centigrade warmer. If the body of water cannot cope with this added heat, the temperature of the river or lake will rise. The temperature of the entire Yellowstone River below Billings, Montana rose 1° to 1.5° Farenheit as a result of operation of the 180 megawatt electric power plant near Billings.

Energy in the fuel is changed to energy in steam in a steam boiler for a power plant. Waste heat is rejected through the cold sink such as cooling water. As energy is used it is reduced to a less useful state such as waste heat. Energy is reduced to less useful forms, but it is never lost. The natural process is from a state of order to one of disorder.

Entropy, a concept of thermodynamics, can be thought of as a measure of the order of a system. Entropy increases as the disorder increases. Entropy is a measure of the potential energy of a system or fuel. In natural thermodynamic processes, such as in the use of fossil fuels, entropy always increases.

The accumulation of wastes in an industrialized society is an example of increasing entropy. Municipal wastes such as paper and wood could be used as fuel, however, since they still have stored energy at a lower level of entropy than in their original state. Instead of burying wastes in landfill, they could be burned and the heat used to generate electricity and to heat city buildings. The city of Palo Alto, California is planning to burn its 400 tons of solid waste each day to produce 15 megawatts of electric power. It is estimated that if the 160 million tons of garbage in the United States were burned efficiently, it could produce 10 million kilowatts or approximately 2.5% of the United States consumption of electricity.

The world operates on energy degradation, or increased entropy, and the residue of energy and waste is rejected to the environment. Energy fuels are good or useful when the energy level with respect to some reference level (e.g. when the remaining energy) of no economic usefulness is high and the entropy of the fuel is correspondingly low. When the energy level is only marginally above the reference state, or equal to it, then we can extract no useful energy. Heat is contained in ocean waters and inside a ship's furnace, yet it is the latter that is infinitely more useful. Waste products are useful at their level of entropy. However, it will require additional energy to raise the level of energy of the waste, (or decrease its entropy) and thus return the waste to its earlier state. This brings us to the question of recycling and energy. Newsprint paper, after it has been used in a newspaper, is then used in various kinds of paperboard, since the gray color of the old printed paper is not harmful. To return the used newsprint to its whiter state for reuse as newsprint would require added energy to process the paper. In this case, what is often called "recycling" is actually using the paper at its current level of entropy. Similarly, old glass bottles are used in road material rather than to add energy to obtain new glass bottles from the old glass fragments.

Consider the use of pollution control devices with automobiles. It is estimated that the devices added each year have increased the fuel consumption per mile by 7% per year. Between 1970 and 1975 it is expected that the fuel consumption per mile will have been doubled in order to meet the emission standards. Since energy utilization to move an auto a given distance will have doubled, we shall have increased the waste of the given area and increased the entropy of the earth, but it will presumably be easier to live with.

The use of energy to reduce the effects of pollution resulting from energy conversion is now significant in the United States. A recent survey of 87 electric utility companies revealed that in 1971 the industrial customers of these utilities used 8.8 billion kilowatt-hours of electric power for environmental protection and pollution control. This energy was 7.3% of the total electrical power sold to the industrial customers. Electricity for environmental protection is expected to reach 10% of industry's annual requirements by 1977.

AVAILABILITY OF OIL

As the United States becomes increasingly dependent upon imported oil during the next decade, several significant effects on the nation will occur. The balance of trade payments will be difficult to maintain and the United States dollar may be held down in value as the Mid-Eastern nations hold significant amounts of dollars. In addition, as imports increase, the need will grow for new large tankers, ports, refineries and pipelines. In addition, the availability of oil will become increasingly a matter for diplomatic maneuvering. For example, King Faisal of Saudi Arabia in 1973 used his country's oil, representing 21% of the world's proven reserves, as a weapon in influencing the Middle Eastern policy of the United States.[5] By 1980, the United States may be spending $15 to $20 billion for imported oil, at which time half of the oil consumed in the United States will be imported.

HYDROELECTRIC POWER

One common and often desirable form of electric energy generation is by means of hydroelectric projects. Hydroelectric developments use dams and waterways to harness the energy of falling water to produce electric power. As of 1972 the total potential hydroelectric power capacity of the United States was estimated to be 178.6 million kilowatts, capable of generating 702 billion kilowatt-hours annually. Approximately 53.4 million kilowatts have been developed or about 30%. In addition, about 125 million kilowatts of plant capacity are under development, with new dams and waterways. There has been a five-fold increase in the installed hydroelectric generating capacity of the United States since the 1930s. Hydroelectric plants currently contribute 16% of the United States electric generating capacity.

Hydroelectric plants use a renewable resource, water, and do not contribute to air pollution. However, the dams do significantly disturb the environment by flooding canyons or other areas behind the dam, and they destroy or block the navigational qualities of the river or waterway.

Two photos of paintings of the Orville Dam and power plant are shown in Figure 19–14. These paintings express the dynamic quality of electric energy and water power.

FIGURE 19–14 Photographs of two paintings by Professor Roland Peterson of the University of California, Davis. The photograph on page 335 shows the dam at Oroville, California. The photograph above shows one of the large generators at the Oroville plant. Both paintings express the dynamic quality of water power and electric energy. Courtesy of Professor Roland Peterson.

NATURAL GAS

Estimated world production of natural gas in 1973 totaled 48 trillion cubic feet. Almost 50% of this total is produced and consumed in the United States. Gas supplied approximately one-third of the total energy of the United States in

1973. The increased demand for natural gas has pushed producers to the limits of their capacity and many distributors no longer accept new customers. Renewed interest, therefore, has been generated in the research and development of new processes for the gasification of coal. Also, interest has been kindled in the importation of gas from suppliers in the Middle East and from the U.S.S.R. It is estimated that 28% of the natural gas consumed in 1985 in the United States may be imported. If this large amount of gas is imported, the United States balance of payments and diplomatic relations may be significantly affected.

NUCLEAR POWER

Nuclear power plants for the generation of electricity may be the long-term solution to the energy needs of the United States and Europe. However, long delays in the construction and operation of nuclear power plants means that only in the last decades of this century will any significant power be supplied by nuclear plants. Nuclear fission energy is released in the form of heat and is transferred to a conventional steam cycle which generates electricity. It is estimated that nuclear power plants supplied approximately 1% of the American power in 1973 and will supply 17% and 22% in 1985 and 1990 respectively. From 1956 until recently, Britain led the world in the generation of power by means of nuclear reactors. However, there are currently more reactors in the United States and recent plans call for Japan, Germany and France to build a system of nuclear power plants equal to that of Britain and the United States. It may be that nuclear power plants will be the only way to supply the industrialized nations' demand for increased electric energy. (The use of electric energy in the United States has grown at a much more rapid rate than the growth of total energy consumption. The use of electric energy has grown approximately 7% per year during the past decade in the United States.)

Even nuclear fuel in the form of Uranium-235 is limited in availability and would be exhausted by extensive use of nuclear power plants throughout the world. Therefore, the successful development of a nuclear breeder reactor may be of critical importance to the future. The breeder reactor uses the more commonly available isotope Uranium-238 and further produces fuel in the form of the isotope Plutonium-239.

Nuclear fission reactors accumulate radioactive products which must be removed and stored. The British place these wastes at the bottom of the oceans. The United States Atomic Energy Commission plans to place them deep in salt mines. Nevertheless, safety factors of the transportation and storage of these wastes are still in question. Also, there is a remaining question of the safety from a catastrophic breakdown of a nuclear reactor (as discussed in Chapter 9). Even if the probability of a catastrophic event is very low, can we afford the risk of such a disaster as a spillage of radioactivity over an area? And what about theft of Plutonium? A reasonably intelligent person could make a nuclear bomb from reactor wastes stolen or diverted enroute to storage or use.

Present plans call for nuclear power to supply 22% of the total energy to the United States by 1990 and 30% by 2000. Yet the current concerns with the environmental and safety aspects of nuclear power plants place that prediction in real jeopardy. One projection of the growth of nuclear power production of electricity for the world is shown in Figure 19–15.[6] This projection estimates that 18% of the world's production of electricity will be provided by nuclear power plants by 1985. A current record of nuclear power plants operational, under construction, or on order and planned for operation by 1983 shows that the total nuclear power capacity for the world in 1983 will be 262,974 megawatts with a total United States nuclear capacity in 1983 of 159,917 megawatts.[12] The United States Atomic Energy Commission offers a more optimistic projection which states that the operating nuclear capacity in 1985 in the United States will be 280,000 megawatts.[22]

A nuclear power station under construction in Washington is shown in Figure 19–16. The capacity of the nuclear power plants operational throughout the world in 1974 are shown in Table 19–4.

The primary deterrents to the installation of new nuclear power plants are the environmental effects and the concern for safety. One solution to these problems may be to utilize nuclear plants offshore and to float the plants on artificial islands. The ocean water could then be used for cooling. One plan to

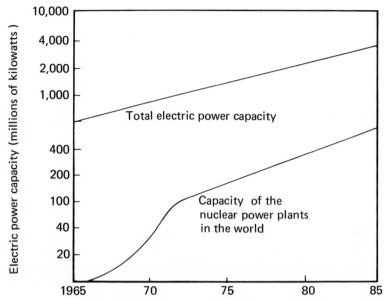

FIGURE 19-15   The trend of electrical power capacity in the world and that produced by nuclear power stations projected for the period 1965-1985.

TABLE 19-4

Nuclear power plants operational in 1974
Capacity in megawatts (10^6 watts).

Canada	France	Britain	United States	U.S.S.R.
2,504	2,964	9,931	41,576	3,179

float the nuclear power plants off the Atlantic coast would place 1.2 million-
kilowatt power stations on 140,000 ton floating stations three miles off the
coast.[7] The floating islands would be placed inside breakwaters in order to
maintain them safe from ships, hurricanes and tidal waves. Another approach to
solve the safety problem would be to bury the nuclear power plants several
hundred feet below ground in artificial caverns.

NUCLEAR FUSION

Nuclear power may be derived from one of two processes—fission or fusion.
The sun, the stars and the hydrogen bomb derive their energy from thermonuclear fusion. In this process, energy is released when the nuclei of light

FIGURE 19-16   A nuclear power station under construction near Longview, Washington on the Columbia River. The nuclear reactor containment spheres and cooling-water towers can be seen. Courtesy of David C. Flaherty and *Quest* magazine.

elements fuse together to form nuclei of heavier elements. The most suitable fusion fuels are available from hydrogen, which is very abundant.

The fusion process would occur at hundreds of millions of degrees centigrade and fusion power plants could be 90% efficient. The heat loss would be low and the environmental effect of heating would be lower than that from fission plants. Also, theoretically, the radioactivity leakage could be controlled to a low and acceptable level. Theft would be impossible. Unfortunately, research on fusion reactors is only in the early stages, and fusion may not be available for several decades.

The achievement of a stable, high temperature gas reactor operating by the fusion process has escaped scientists and engineers since the first fusion experiments in the 1950s. However, steady progress has been made, especially since 1958, when all security restrictions were removed by the United States, Russia and other nations in order to facilitate international efforts toward the development of a fusion reactor. It is significant that the experimental fusion reactor in operation at Princeton University in 1974 carries a Russian name, *Tokomak*. It is built to specifications developed by Russian fusion experts.

NEW TECHNOLOGY FOR ENERGY

The United States has a potential for over 1,000 billion barrels of oil from the oil shale deposits in Colorado, Utah and Wyoming. If new technology to extract the oil from the shale can be successfully designed and put into operation by 1985, we may expect to obtain approximately 500,000 barrels per day or 0.2 billion barrels per year. This amount, while helpful, would be only a small fraction of the amount of oil the United States is currently consuming. Furthermore, the environmental effects of oil-shale processing may be significant. Vegetation would be destroyed during operation and the landscape altered. Mine wastes would have to be deposited, perhaps in box canyons. Finally, air quality might decline as a result of the mining and processing operations.

Another large potential source of energy is the geothermal heat within the earth, as shown in Figure 19–17. Geothermal steam or superheated water is produced when the earth's heat energy is transformed to subsurface water from rocks in the earth's crust.

Geothermal energy could possibly supply 10% of the total energy used by the world each year. One electric power plant at the Geysers, California is the only geothermal generating facility in the United States in 1974. This plant currently generates 396 kilowatts and it is planned to generate 900,000 kilowatts by geothermal steam by 1976. A geothermal plant near Lardarello, Italy generates 380,000 kilowatts. At present, geothermal use is in its infancy. Substantial technical problems remain to be solved. Prospecting techniques are in their early stages. Also, controlling of corrosion from mineral-laden hot water creates difficult problems. Geothermal power plants may also cause air and water pollution and subsidence of land following removal of the hot water. Furthermore, the hot water and steam are at a relatively low temperature, 205° C at The Geysers, compared to the 550° C used to generate steam in many fossil-fueled plants.

Solar energy is an inexhaustable energy, as far as the timescale of the earth. Solar energy can be utilized with a minimum of environmental impact, if new conversion technologies are fully developed.

The power of solar radiation intercepted by the earth is almost 100,000 times the power capacity of all electric utilities in the United States. Nevertheless, its low density over the surface of the earth makes the direct use of solar power an

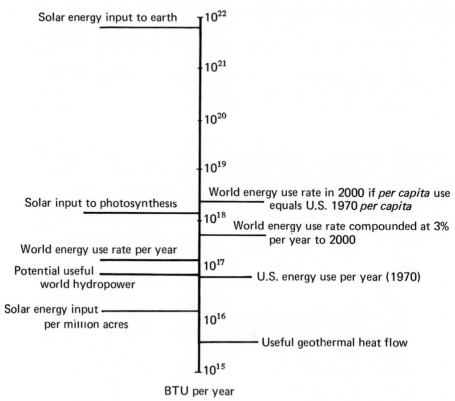

BTU per year

FIGURE 19-17   A comparison of some energy-use rates per year with the total solar
energy input to the earth.

economic and technological challenge. It is physically possible to cover the
necessary area with energy collecting devices and to store, transmit and
transform the energy, but problems of costs, special technologies and environmental impact have not yet been solved. If we could convert 0.01% of the
solar energy reaching the earth, the world's current energy needs could be met, as
is shown in Figure 19-17.

Over a period of a year, one square mile in the United States receives about
15×10^{12} BTU.[8] The United States consumed about 75×10^{15} BTU of energy
in 1973. Thus about $75 \times 10^{15}/15 \times 10^{12} = 5,000$ square miles of United States
land received the equivalent of all the United States energy needs. If a device for
converting solar energy to electricity with an efficiency of 10% were commercially available at a satisfactory cost, 50,000 square miles would have been
required to supply the United States needs for 1973. This area is about 1.5% of
the total land of the 48 contiguous states. Furthermore, the presently-obtainable
efficiency of solar cells is less than 10%, and they are relatively expensive. The
problems associated with the use of solar energy during the next several decades,

then, arise from the low efficiency of currently available conversion devices and
the large area required for these collecting cells. Furthermore, storage of collected
solar energy would be required, since solar energy is available only eight to 12
hours per day at most.

Nevertheless, it is important to consider the development of solar heating
and cooling systems for buildings in the near future. Mass-produced economical
heating and cooling systems for buildings would operate at a cost comparable to
those of other sources as shown in Figure 19–18. If 35% of the buildings could be
heated and cooled by solar energy systems by 2000, the savings could be $2
billion per year in fossil fuels.[8] While the current low efficiency solar cells
makes them too expensive for common use, new technologies for solar heating
and cooling systems have become economically feasible.

The collection of solar energy for the generation of electric power on the
earth presents several problems. The low density of energy over the area of the
earth is a significant factor. Also, weather conditions and day-night cycles place
important obstacles before the use of solar energy. One proposal to overcome
these problems calls for the placement of a solar-energy conversion satellite in
space.[9] The satellite would be located at 36,000 kilometers (22,300 miles) from
the earth in an orbit which places the satellite stationary relative to a point on the
earth. This orbital position is called synchronous orbit, since the satellite moves
at the same speed that a reference point on the earth rotates below it. The satellite

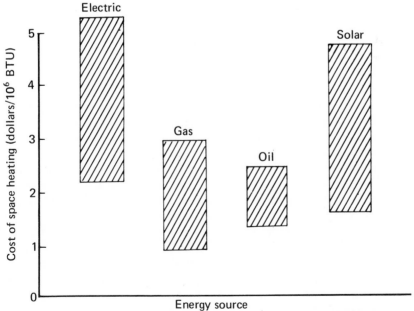

FIGURE 19–18   The costs of space heating for four sources in dollars per million
BTU's.

is then fully illuminated by the sun for 24 hours every day. Also, the incidence of solar energy in space is 130 thermal watts/ft², while it is only 17 thermal watts/ft² on the ground due to reductions from the atmosphere. Solar-cell collectors on the satellite convert solar energy to electricity. The electricity then supplies energy to a microwave (very high frequency) generator, and microwave energy is beamed to a ground station to convert it back to electricity, as shown in Figure 19–19.[13] One study shows that a satellite could feasibly deliver approximately 3,000 megawatts. However, much of the technological development remains to be done. If a satellite solar station could supply 3,000 megawatts (3 million kilowatts), then approximately 150 satellites and their associated receiving stations would be required to supply the electrical energy capacity of the United States in 1973. Of course, if it is economically feasible, 15 satellites could supply 10% of the American electrical generating capacity available in 1973—which makes solar energy still a very significant source of power.

Solar energy can provide a significant portion of the total heating and cooling load of buildings as well as electric power. Extensive research and development efforts are required, but it is estimated that extensive solar heating for buildings could be in use by 1978 and solar cooling systems could be in use by 1980. There are many applications of solar heating and cooling at this writing, both in the United States and abroad, but most must still be classified as experimental. Furthermore, electric power from converted solar energy could be available in 10 to 15 years.[13] In order to achieve these goals, however, the Federal Government must take a leading role in the research and development effort. Low cost, low maintenance solar conversion devices are needed for earth or satellite solar-electric systems.

FUTURE ENERGY CONSERVATION

As has been noted, the annual consumption of energy in the world has increased significantly over the past 50 years. From this fact one might observe that the future use of energy will be a projection of the past rate of increase. This projection for continual increase usually assumes a supply of cheap and abundant energy. Others see the exponential increase of energy use as basically wasteful, because the energy resources available to mankind are limited. Whatever one's view, it is generally recognized that a strategy of conservation of energy and wise use of our resources is desirable. There is growing agreement in the world community that efficiency and conservation in the use of fuels must be built into our way of life. There should be rewards for those who build and use more efficient devices, and penalties to those who do otherwise. For instance, it has been suggested that there be a tax on the amount of waste heat emitted from a power plant.

Transportation systems are presently inefficient, although they serve the desires of passengers. If 50% of urban auto passenger traffic could be shifted to

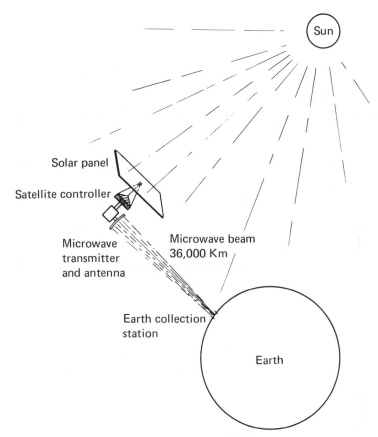

FIGURE 19-19   A space station in synchronous orbit. This device is designed to pro-
               duce 10,000 megawatts. The solar panels are oriented to face toward
               the sun at all times. Each solar panel would be four km square;
               normally panels would be mounted in pairs. The sending antenna for
               microwaves would be one km in diameter, and the receiving station
               seven km in diameter.

well-loaded urban buses, it is estimated that an energy savings of 12% of the total
used in the United States could be obtained.[14] If the present urban passenger-
load per vehicle could be increased from the current 1.4 to 2.0, then a savings of
9.5% of the total United States energy could be obtained.

The relation between miles per gallon and the auto weight in city driving is
shown in Figure 19–20. If the large United States auto could be reduced in weight
from 5,000 pounds to 2,500 pounds, an increase of 6.25 miles per gallon could be
obtained for an improvement of 53%.[15] It has been estimated that improving
the average efficiency of the American car from 12 to 18 miles per gallon would
save 25 billion gallons of gasoline annually. That is equivalent to saving 1.2
billion barrels of oil annually in the United States. This amount of oil is

approximately 20% of the oil consumption of the United States in 1973. An additional 450,000 barrels per day of aircraft turbine fuel could be saved by increasing the average load factor from 55% to 80%.

Within the residential and commercial market, the largest use of energy is for space heating and cooling. If recent insulation standards for homes are widely adopted (and if older houses are brought up to standard), homeowners could achieve sizable savings. In addition, new insulation standards need to be adopted for commercial buildings. The total savings in the residential and commercial market could amount to 0.3 billion barrels annually in the United States from use of improved insulation. In addition, we must burn about twice as much natural gas or oil as as an electric power plant to put a unit of heat into a home or office that uses electric space heating as would be required if those fuels were burned directly in a home or building furnace. For gas heating, for example, the efficiency of home gas heating is about 60% while the efficiency of generating and distributing electricity is about 30%. Therefore, as far as it is possible, gas and oil should be used for home and building heating rather than for generating electricity for space heating. In fact, some conservationists have proposed the abolition of electric space heating.

One report notes that if: (1) three of four new units constructed would have proper thermal insulation treatment; (2) one of four existing single-family homes would have added improved insulation; and (3) one of eight single-family dwellings previously constructed would add storm windows and doors, there

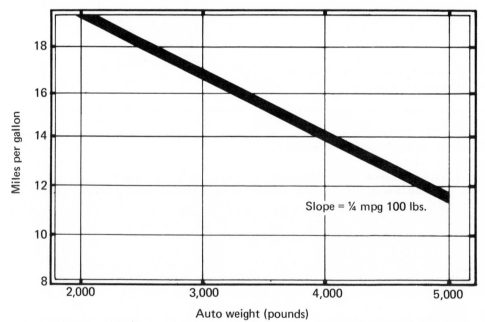

FIGURE 19-20 The effect of automobile weight on fuel economy during city driving.

would be 1,500 trillion BTU conserved in the year 1982 alone and 8,200 trillion BTU in the decade 1974–1983. In consumer savings this represents $3.1 billion in 1982 and $17.2 billion over the decade. The cost of such a program is estimated to be $6.37 billion, about one-third of the consumer savings.

Industry could adopt more energy-efficient processes, perhaps with the encouragement of a tax incentive. Using aluminum in cars instead of steel reduces the weight of the cars. The lighter cars can save up to 26 times the energy needed to make the aluminum. Furthermore, increased efficiency of the aluminum-smelting process would be a real gain. One new process may save up to 30% of the power consumption in the manufacture of aluminum.[16]

Other measures to achieve reduced use of electricity might be to black out billboards and other outdoor advertising at night. Also, the hours of shopping centers could be limited to 12 hours a day and the lights turned off after 9P.M. Night sporting events could be rescheduled for the day. People could be asked to allow the temperature of their homes and offices to rise to 78° F in the summer and to drop to 68° F in the winter. This reduced use of heating and cooling energy would have a significant impact on the use of fuels.

EFFECTS ON CULTURE

The use of energy in the United States is an underlying contributing factor to the high American standard of living. The United States has one of the highest energy-uses *per capita*, and this index is increasing because of affluence and growing population. In the following article, Mr. Boyd Keenan considers energy use in the United States and its meaning for America.[24]

**The Energy Crisis and Its Meaning
for American Culture***

Boyd Keenan

The so-called energy crisis cuts to the heart of American culture. But our preoccupation with it tends to blind us to the fact that it is but one aspect of a global problem facing all mankind—the problem, namely, of the universal scarcity of natural resources. Our own country finds itself in an unprecedented situation.

In the first comprehensive assessment of the nation's mineral resources since 1952, the U.S. Geological Survey declared (May 8, 1973) that of 64 vital minerals many already are or soon will be in short supply. This shortage, it warned, puts "not merely U.S. affluence, but world civilization . . . in jeopardy."

The energy dilemma, then, cannot be considered alone. But energy is the immediate question, and we must come to grips with it. Our politicians know as much. Thus every conceivable candidate for presidential nomination in 1976 has made dramatic speeches about the matter. Not atypical were grim predictions of one aspirant, Indianapolis Mayor

Richard Lugar, who told an audience in Peoria, Illinois, on May 13 that the energy crisis—not Watergate or today's other headlines—would be the crucial issue in 1976. "People think somehow we're going to muddle through," he said. "They expect to wake up someday now and read a headline that says the energy shortage is over and everything is okay. But it isn't okay. We're plunging on day by day, uncertain about what is to occur."

Categories of Concern

The issues relating to the energy dilemma are so complex as to boggle the mind. Can they be clustered into any rational grouping? Probably not. Let us try nevertheless to discern broad categories of concern. Three such categories suggest themselves: (1) national security, (2) domestic inflation and related economic questions, and (3) possible adjustments in life style.

National Security. Though some of our best minds decry it, the sovereign "nation state" is still the basic unit of power in world politics. Virtually everywhere, "national well-being" is measured by the availability of energy sources. In the U.S. those who govern believe that they just cannot allow the nation to be without energy for driving the engines of our technological system. Therefore they will have to see to it that, whatever the costs, energy is supplied. This requirement was not frightening so long as prospects appeared good that we would be able to furnish the fuel needed. But now the experts are telling us that we overestimated the potential of nuclear energy and that foreign oil must pick up much of the slack resulting from the miscalculation.

Worse, at the very time when the nuclear dream is being shattered, the dependable fossil fuels long abundant at home have become scarce for one reason or another. The high sulphur content of much of our coal prohibits its use in many areas where ecology-minded lawmakers have set up stringent standards. Geologists generally assert that natural-gas reserves are low. And—perhaps most important politically—the oil corporations insist that they cannot afford to locate and extract crude oil at today's prices. Thus almost all knowledgeable observers agree that over at least the next decade the U.S. will have no alternative to importing oil from foreign ports, notably the Middle East. In that case, we shall find our national security—broadly defined—dependent on a politically unstable region of the world.

Add to this the fact that most of the countries allied with us since World War II are themselves increasingly dependent on Middle Eastern oil. Hence our friends have become our competitors in the search for energy fuel, and our desire for national security puts us potentially in conflict with our allies. The frustrating stinger in the total national-security picture is our future posture toward Israel and the oil-producing Arab states in the Middle East. No wonder that the U.S. defense establishment is haunted by the question: Over the next decades will we possess the petroleum products necessary to satisfy the thirst of the huge military apparatus which, as the cliche goes, is our "first line of defense"?

Further, nations are teaming up around the international oil market as they have not done since World War II. Distinguished students of world affairs differ in their analyses, but they all agree that unprecedented power blocs are forming. The power of the producing lands is centered in the Organization of Petroleum Exporting Countries (OPEC), a consortium of 11 nations. Complex cross-relationships have discouraged the consuming countries from creating equally powerful consortia.

In short, it takes little imagination to devise international scenarios that have frightening implications for the U.S. That is why alert presidential aspirants have identified the energy crisis as the major issue for 1976.

Inflation and Related Economic Issues. The complexity of the energy dilemma becomes even more evident when one considers the relation of the international situation to domestic problems. World monetary experts tell us that the increasing U.S. payments to foreign countries for oil exacerbate our own inflationary spiral. It doesn't take a professional economist to understand that a scarcity of fuel would cause ripples in every sector of American life. For instance, if our steel manufacturers lacked fuel to fire their furnaces, hundreds of other manufacturers would have to cut down production at once and lay off thousands of workers. The result might be political chaos of a scope never before experienced in this country.

Life-Style Changes. As gasoline prices soar and the threat of rationing looms, a tired cliché is sounded: "The situation requires nothing less than a total refashioning of life styles." Even the oil companies that peddle petroleum products and the utilities firms that sell electric power have taken up this cry. On their part, our philosophers and theologians attempt to trace the energy dilemma to ideological roots. Some of them argue that Western religion is the culprit. They point to pronouncements in "holy" writings which—so they say—have encouraged Western man to build extravagant industrial systems that gulp up the world's resources. Lynn White and Alan Watts (to cite only two distinguished scholars) contend that a "conquer the earth" syndrome, deriving chiefly from Judeo-Christian Scriptures, is somehow a uniquely Western characteristic. Therefore, they suggest, one hope for world civilization lies in Eastern religions, which emphasize meditative as opposed to material rewards. This approach seems apt in view of the fact that the United States, with about 6 per cent of the world's population, is now consuming over 35 per cent of the planet's total energy production. The average American uses as much energy in a few days as does each of half of the world's people in a year.

The basic question here is whether U.S. consumption is a function of its philosophical and theological underpinnings or whether many other complex factors are critical variables. Put this way, the question may seem abstract, but in the effort to understand the world energy chain it may be the most practical question yet faced by world civilization. In less cosmic terms, it has significance for every American who pulls his automobile into a service station or flips an electric light switch. Yet to anyone who has observed businessmen from the Orient in hot pursuit of Western technological dollars, a theory that fastens total responsibility for the world energy problem on Judeo-Christian ideologies appears inadequate.

Where then can we find help in making sense of what is clearly a world problem, not just an issue affecting American culture? The psychological pressure for comfort tempts us to accept at face value the propagandistic pronouncements of various interest groups. And indeed it is far too early to assess the sincerity or wisdom of either the huge corporations or the militant spokesmen for ecological concerns.

But it is not too early—in fact, it is already past time—to ask if we are at the mercy of these groups which offer such bizarrely diverse interpretations of the problem. We shall betray the best of our political traditions if we blindly accept the explanations of either the newly emerging total-energy conglomerates or the recently aroused environmentalists. After all, the energy dilemma did not come upon the U.S. full-blown late in the 1960s. For years American writers had been predicting a time when our materialistic society would succeed in enlisting the nation's economic, technological and even moral forces in an almost worshipful search for energy sources. Many of these thinkers were as confused as we seem to be about solutions to the energy dilemma. But if we had heeded their warnings, we might not now be so paralyzed with shock that we cannot think clearly about our predicament.

Henry Adams as Major Prophet

Outstanding as a prophet of today's dilemma was Henry Brooks Adams (1838–1918), great-grandson of John Adams, revolutionary hero and second President of the United States, and grandson of John Quincy Adams, fifth President of the U.S.

It was late in his life, as he experienced a sense of failure in not achieving a status befitting a member of one of America's "first families," that Adams conceived a theory of history centering on the role of energy in civilization. Of his many books, his autobiographical *The Education of Henry Adams* (first printed only for his friends in 1907) was probably the most eerily prophetic. The following passage from this book should be read in the light of certain events of May 11, 1972: President Nixon's energy message and some of the most dramatic of the Watergate disclosures.

> The work of domestic progress is done by masses of mechanical power—steam, electric, furnace, or other—which have to be controlled by a score or two of individuals who have shown capacity to manage it. ... The work of internal government has become the task of controlling these men, who are socially as remote as heathen gods. ... Most of them ... are forces as dumb as their dynamos, absorbed in the development or economy of power. They are trustees for the public, and whenever society assumes the property, it must confer on them that title; but the power will remain as before, whoever manages it, and will then control society without appeal, as it controls its stokers and pitmen. Modern politics is, at bottom, a struggle not of men but of forces. The men become every year more and more creatures of force, massed about central power-houses. The conflict is no longer between the men, but between the motors that drive the men, and the men tend to succumb to their own motive forces.

"Dumb as their dynamos"—a striking phrase. To Adams, the dynamo was a symbol of an accelerating industrial society. His interest in this symbol dated from Chicago's Columbian Exposition in 1893, at which he spent a fortnight studying the exhibits of new and powerful engines. "Chicago was the first expression of American thought as a unity," he wrote later. "One must start there."

Exactly 80 years after Adams identified Chicago as the symbolic home of the energies America adored, that city remains the focal point for new dynamos. Commonwealth Edison, the city's electric public utility, has built the world's largest nuclear power plant at nearby Zion, and is now besieging the U.S. Atomic Energy Commission for a license to make the plant fully operative. Meanwhile, numerous environmental groups have filed suits to prevent the plant from operating at all until there is proof that it can do so safely. Thus the battle between Adams's "dumb dynamos" and people who fear them continues at the very spot where he envisioned energy as a common denominator for the various forces at work in the universe.

Oil corporation executives, environmentalists, political leaders and indeed all of us would do well to read about the agonies Adams experienced in developing his idea of energy as an almost universal factor that could explain the march of civilization. It took him seven years to realize that the dynamo—that ingenious channel for "conveying somewhere the heat latent in a few tons of poor coal"—was in truth a symbol of infinity and that, as such, it took on the characteristics of a moral force, much as the cross did for the early Christians. He even predicted that, ultimately, men would pray to the dynamo.

Energy and the Agonies of Eugene O'Neill

Like Henry Adams, Eugene O'Neill sensed the quality of infinity in energy, and in his play *Dynamo* he labored to give it spiritual and intellectual content. The plot revolves around a boy in his late teens, the son of an overbearing clergyman, who rebels against the religious teachings of his childhood and embraces the religion of electricity. Driven mad by the conflict between his lust for a girl and his desire to attain oneness with the god Electricity, he murders the girl and, flinging himself upon the dynamo, is electrocuted.

Dynamo was produced on Broadway in 1929, but closed after a few performances. Reviewers panned it mercilessly, to the playwright's deep disappointment. In an unsigned letter to the *Times Literary Supplement* (May 8, 1937), he said his intention had been to get at the roots of a sickness involving "the death of an old God and the failure of science and materialism to give any satisfying new one for the surviving religious instinct."

Many lines in *Dynamo* seem to draw heavily on Henry Adams. Today, 40 years after O'Neill failed to get his point across to the intelligentsia, ambitious politicians are drawing on ideas very similar to his to appeal to the general electorate. Tracing the circuitry of ideas from seminal thinker to artist to politician represents an intriguing intellectual exercise. In 1907 Adams could say that following the track of energy was the historian's main business. But both he and O'Neill, if they were alive today, would probably join those who say that tracking energy is the chief business of government. For technological experts are virtually unanimous in asserting that there is no real scarcity of energy resources; that there is, rather, a crisis in the management of energy, both within and among nation-states, that only governments can deal with.

Ironically, the heads of the big American oil companies have always protested against any government intervention in energy affairs. Actually, of course, these same companies have themselves functioned as governments, singly or collectively. Several of them are larger, in terms of wealth, than many nations; and their long-standing cooperative networks possess great political power. So they are indeed "governments" by any imaginable criterion.

Theories and Faith for an Energy Age

That aside, however; if the energy issue is worthy of a top spot on the agenda for debating the merits of 1976 presidential candidates (and most aspirants have declared it is), our society has an unprecedented task on its hands. The chore of educating thinking people to the enormity and complexity of the energy question will be the most demanding ever undertaken by a democratic system. Our failure to succeed in this assignment will no doubt be interpreted by world intellectuals as final evidence that a representative system of government cannot control technology. And unless our leaders can create the kind of passion felt by Adams and O'Neill, the educational effort will fail.

The problem we face is not a bureaucratic one requiring only a bureaucratic mentality. Instead, we shall be thrashing about in a thicket of ignorance in an attempt to understand what are perhaps the most mysterious characteristics of the universe. We need to devise theories that both explain the industrial and technological revolutions and project approaches for a supersensuous age of energy acceleration. In other words, we need to construct a dynamic formula of history.

Never before has world civilization been called upon to sustain itself while fashioning such a cosmic program. But hitherto the tests have never been programmatic or technological; they have been tests of faith. At times it may have been provincial or even perverse faith; yet the happenings of history were inspired by it. From the days when

Constantine set up the cross through World War II days when American scientists built the atomic bomb, a blending of political and religious ideologies provided much of the faith undergirding Western political achievements. Now, however, the very term "Western civilization" is an anachronism. As a result of technology, there is only one civilization: world civilization. If there is one distinguishing characteristic to the energy crisis, it is its worldwide scope. Hence faiths built only on Western foundations are, by definition, inadequate for the challenge posed by the energy dilemman.

Let me quote Adams once more, for he offers hints concerning the character of the faith required to attack energy problems. In 1907 he wrote that the forces of energy represent "a sort of God compared with any former creature of nature." A heady and perhaps heretical point of view. But it is what the energy crisis is all about and what American culture must ponder if it is to survive.

FORECASTING FUTURE ENERGY USE

The fossil fuels that supply most energy today are finite in supply but are not in danger of exhaustion for several hundred years.—at least, in the case of coal. Nevertheless, because of the unequal distribution of these fuels over the world and the environmental hazards of using them, the supply of energy may exceed the demand in several countries such as the United States during the last decades of this century. In general, a crisis of supply and demand can be met by increasing supply or by stabilizing demand. In the United States during the next 30 years, there will be attempts to follow both approaches. Much energy is now wasted, and conservation measures could close the supply-demand gap. During the next 10 years the United States will depend upon imported oil, strip-mined coal, and conservation methods to maintain a supply to meet the energy demands. Then in the last 15 years of the century, the increased availability of nuclear power plants will provide an increased supply of energy to meet the demands and replace our dependence on strip-mined coal; then imported oil can be reduced. By the year 2000, it is hoped that efficient solar energy conversion plants will be available as well as high-efficiency fusion reactors. The electricity produced by these devices will in part be converted to a hydrogen fuel which will be transmitted through pipelines and also used for transportation.

One forecast of the sources of energy that will constitute the total energy use in the United States during the period 1970–2000 is shown in Figure 19–21. This forecast shows a decreasing use of domestic oil and gas over the remaining years of the century. It also assumes that Alaskan oil will be available during the period 1977–1989. According to this projection, coal provides about 15×10^{15} BTU per year after 1975. Nuclear fission increases as an energy supply, providing about one-third of the total energy in the year 2000. Imported oil and natural gas are expected to provide 22×10^{15} BTU per year after 1975. Solar heating and cooling of homes and offices is expected to grow after 1980 and to provide 15×10^{15} BTU each year after 1990. Finally, after 1985 the growing availability of solar energy and fusion energy, along with conservation measures, will fill the gap between existing sources and demand. This *forecast* is reasonably optimistic; unforseen events may render it obsolete. Nevertheless, the energy sources of the remainder of the twentieth century are primarily available to us now. It remains only for us to use them wisely.

Energy use in industrial societies is closely tied to man's desire for speed and rapid development. For example, transatlantic air travelers require an energy subsidy on the order of 20 million BTU while those content to travel more slowly by tramp steamer require only one-fortieth of the subsidy of the air traveler.

In the raising of beef using the highly intensive modern methods, it takes 14,300 kilocalories of energy subsidy to make a kilogram of beef.[25] That is about 84,000 kilocalories/kilogram of protein. Higher production from the land and from industry can be achieved only at the expense of increasing energy inputs.

We are also aware that our automobiles use proportionately more gasoline at higher speeds. Perhaps it is time to analyze the effect of ever greater haste and increased production per unit of time. Perhaps our society has reached the point of diminishing returns with respect to increased use of energy considering all the effects on our environment, our resources and our lives.

CHAPTER 19 REFERENCES

1. M.K. Hubbert, *Resources and Man*, W.H. Freeman and Co., San Fransicso, 1969: Chapter 8. Also in "The Energy Resources of the Earth," *Scientific American*, September 1971, pp. 31–42.

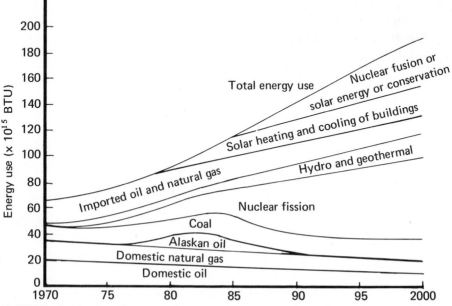

FIGURE 19-21   One forecast of the sources and use of energy in the United States during the period 1970-2000.

2. H.C. Hottel and J.B. Howard, *New Energy Technology*, The MIT Press, Cambridge, Mass., 1971.

3. D.P. Gregory, "Hydrogen: Transportable Storable Energy Medium," *Astronautics and Aeronautics*, August 1973, pp. 38–43.

4. "Certain Background Information for Consideration When Evaluating the National Energy Dilemma," Joint Committee On Atomic Energy of the United States Congress, United States Government Printing Office, Washington, D.C.

5. "The U.S.'s Oilman in Arabia," *Business Week*, Sept. 22, 1973, p.58.

6. P.J. Nowacki, "Long Term Planning in Poland," in *Growing Against Ourselves: The Energy Environment Tangle*, Lexington Books, Lexington, Mass, 1972.

7. J.H. Morris, "First Floating Nuclear Power Plant is Due in 1979, but Ecology Issues Muddy the Water," *Wall Street Journal*, May 24, 1972, pp. 31–32.

8. W.R. Cherry, "Harnessing Solar Energy: The Potential," *Astronautics and Aeronautics*, August 1973, pp. 30–36.

9. P.E. Glaser, "Solar Power Via Satellite," *Astronautics and Aeronautics*, August 1973, pp. 60–68.

10. J.F. Stacks, *Stripping*, Sierra Club, San Francisco, 1972.

11. A. Wolff, "Showdown at Four Corners," *Saturday Review*, June 3, 1972, pp. 29–41.

12. "World List of Nuclear Power Plants," *Nuclear News*, September 1973, pp. 53–62.

13. "An Assessment of Solar Energy as a National Energy Resource," *Report of the NSE/NASA Solar Energy Panel*, University of Maryland, Department of Mechanical Engineering, December 1972.

14. R.S. Greeley, "Energy, Resources and the Environment," *Mitre Corporation Report M73–61*, June 1973.

15. W. Lipman, "Congress Considers Penalizing Heavy Cars," *California Journal*, July 1973, p. 245.

16. "Aluminum Prosperity is Riddled with Troubles," *Business Week*, Sept. 1, 1973, pp. 54–55.

17. "Problems of Electrical Power in the Southwest," *Report of the United States Senate Committee on Interior and Insular Affairs*, Serial no. 92–24. 1972.

18. W.E. Winscke, K.C. Hoffman and F.J. Salzano, "Hydrogen: Its Future Role in the Nation's Energy Economy," *Science*, June 29, 1973, pp. 1325–1332.

19. D.P. Gregory, *op. cit.* (Reference 3)

20. L. Lessing, "The Coming Hydrogen Economy," *Fortune*, November 1972, pp. 138–146.

21. "Desalting—Answer to Water Crisis?" *U.S. News and World Report*, Sept. 24, 1973, p. 100.

22. "Enough Energy—If Resources are Allocated Right," *Business Week*, April 21, 1973, pp. 80–87.

23. W. Clark, "Interest in Wind is Picking Up as Fuels Dwindle," *Smithsonian*, October 1973, pp. 70–79.
24. B. Keenan, "The Energy Crisis and Its Meaning for American Culture, *The Christian Century*, July 18, 1973, pp. 756–759.
25. M. Slesser, "Energy Analysis in Policy Making," *New Scientist*, Nov. 1, 1973, pp. 328–333.
26. D.J. Rose, "Energy Policy in the United States, *Scientific American*, January 1974, pp. 20–29.

CHAPTER 19 EXERCISES

19–1. In the future it will be desirable to make a synthetic fuel from solar or geothermal energy that is storable, portable, cheaply transportable and made from abundant materials. Also, it must burn with little effect on the environment. Electricity from solar energy conversion devices could produce hydrogen by the process of electrolysis, for example. Pipelines can transport hydrogen underground relatively cheaply. It stores readily and it causes no pollution upon burning.[18, 19, 20] Hydrogen could be transported and stored more cheaply than electricity, and since it gives off only water vapor after combustion it is environmentally advantageous. One underground 36-inch diameter pipeline can carry the same as four overhead energy electric powerlines, thus vastly reducing the environmental impact of transporting energy. Examine the specific advantages of a hydrogen fuel system for heating, transportation, home lighting and industrial use.

19–2. The effort to convert salt water into fresh water is progressing significantly in the United States. Currently there are 330 desalting plants in operation and Congress has provided $27 million each year for several years for research and development of improved desalting processes. The largest American plant is a three million gallon-per-day plant in Fountain Valley, California which cost $26 million. Such plants supply fresh water at costs ranging from $.30 to $1.25 per 1,000 gallons. Another plant is planned for construction in Yuma, Arizona which would treat up to 100 million gallons per day.[21] Determine the energy needs of such desalting plants and examine the total energy requirements for desalting for 1980 if 400 plants treat 300 million gallons of water daily.

19–3. Electric appliances consume about 10% of the electric energy in the United States. The largest users of electric energy are refrigerators, ranges and air conditioners. Determine the efficiency of several models of these appliances and calculate the energy savings possible if devices of the highest efficiency were used in all homes.

19-4. Illumination of residences and commercial buildings accounts for 1.5% of American energy consumption. In some cases the level of illumination used in offices is twice that necessary. Also, many office buildings are lighted all day and night. In addition, incandescent lighting has only 40% of the lighting efficiency of fluorescent lighting. Develop an energy-consumption plan for your city and determine the savings of energy consumption for the city that would result from its adoption.

19-5. Electric utilities charge customers less as they use increasing amounts of electric energy. Thus, the more an industry or home uses, the less one pays per kilowatt-hour. This practice has long been justified on the basis of economies of scale, since each additional kilowatt-hour has cost less to generate and distribute and the savings have been passed on to the customer. This declining cost curve has stimulated increasing use of electricity. Develop a cost mechanism for increasing prices as use increases and examine the projected use of electricity under such a scheme.

19-6. Nuclear power plants have an efficiency of only 33%. The waste heat should be used effectively rather than be put into a river or lake. Design a new town near your home with a nuclear power plant sufficient for 25,000 people which uses its waste heat for activating a sewage treatment plant and heating the buildings.

19-7. The rise and fall of the tides provides intriguing prospects for electrical power generation in a few selected locations where tides are high and the land forms are such as to admit the possibility of damming of bays or other inlets. Explore the development of tidal power and the potential energy that may be realized.

19-8. Wind power has been used to propel ships and for pumping water. Windmills can be used for the generation of electric power.[23] One estimate states that by the year 2000, a major development in wind power could result in an annual yield of 1.5 trillion kilowatt-hours of electricity. This could be 10% or more of the electric energy consumed each year by the year 2000. In 1850, the use of wind power represented about 1.4 billion horsepower-hours (3.6×10^{12} BTU), but the availability of the steam engine cut down the use of the windmill in the following decades. Explore the potential for the use of windpower in the United States. Determine what means of storing energy you would suggest to account for windless days.

19-9. Operating buildings consumes 50% of all electricity produced in the United States. The new 110-story World Trade Center in lower Manhattan, with its extensive list of electrical requirements, consumes as much electricity each year as an entire city of 100,000 persons—a city like Schenectady, New York. Design new standards for heating, cooling and lighting for office buildings and commercial buildings.

Building designs run from three watts to over six watts per square foot of space. Determine the electric power specification for your school or office building. Three watts per square foot should, with proper design, provide sufficient lighting, heating and cooling.

19–10. In the winter of 1973–74, the United States experienced an oil shortage caused by production cutbacks and export cutbacks from the Arab countries. These cutbacks were imposed on the United States because the Arabs believed the United States was supporting Israel in the Arab-Israel conflict of October, 1973. While Arab oil accounts for only 5% of the United States total energy use, it does account for 10% of the oil used in the United States. Do you believe the United States is, and will be in the future, excessively dependent on imported petroleum? What alternative sources would you suggest? Can we expect to experience continuing gasoline shortages—illustrated by Figure E19–10?

FIGURE E19-10 Photo courtesy of David Flaherty and *Quest* magazine.

19–11. During the fuel shortage of 1973–74, several measures were suggested for reducing the dependence of the United States on imported fuel. The Senate Interior Committee considered: (1) Temperature restrictions in office and commercial buildings; (2) Limits on operating hours of commercial establishments and schools; (3) Restrictions on outdoor advertising; (4) A reduction in automobile speed limits to 50 miles per hour; (5) Institution of daylight savings time on a twelve-month basis. Discuss the wisdom of each measure and estimate the percentage of energy saving achieved by each approach.

19–12. The production of aluminum from bauxite ore consumes about 25 times more energy than the production of aluminum from recycled material. Currently the Aluminum Association has about 1,200 centers from which it recycles about two billion cans each year. A modest financial incentive of about ten cents a pound is provided. Examine the effects of recycling half of the aluminum produced each year on the use of energy in the United States. Consider the banning of aluminum cans or the imposition of a deposit charge on aluminum cans.

19–13. On November 25, 1973, President Nixon announced that he wished an imposition of a 50 m.p.h. speed limit for autos and a 55 m.p.h. limit for intercity buses and trucks effective January 1, 1974. Subsequently Congress required by law a speed limit of 55 m.p.h. for all vehicles. Explore the value of this approach in saving gasoline. Was it wise to set a different limit for buses and trucks? What is the benefit of a different speed limit?

20

ALASKAN OIL AND THE PIPELINE

IN 1968, OIL IN GREAT QUANTITIES was discovered under Alaska's bleak North Slope. The oil available in this area is estimated to be over 10 billion barrels. This oil can, in part, meet the energy needs of the United States during the last two decades of this century. However, the delivery of this oil to the United States markets has raised a controversey which involves environmental impact assessment, technology assessment and energy economics. The proposed Alaskan pipeline is a subject for considerable discussion. We examine this controversy as a case study of technology assessment, environmental impact reporting and energy economics.

PROSPECTING

The discovery leases to the Prudhoe Bay oil fields were granted in 1964. The search for oil on the northern slope of the Brooks mountain range at the north of Alaska had been going on for over a decade by 1964. The 1968 discovery of oil at Prudhoe Bay on the Arctic coast by Atlantic Richfield Company (ARCO) and Exxon Corporation set off a "black gold" rush. Derricks, pipe and drilling machines were shipped to the area. In 1969, in a lease auction that brought

Alaska $900 million, four other oil companies joined the original three and formed a consortium called Alyeska Pipeline Service Company. This oil lease brought almost 50% more than had ever been obtained in an auction previously.

THE PIPELINE

These companies then planned to tap the oil and transport it to the lower states. Transport by tanker or pipeline was examined. A tanker, the *S.S. Manhattan*, was brought to the waters of the Artic, but the ice pack precluded extensive all year use of tankers. Thus, it was decided to build a pipeline to deliver the oil to the refineries.

Conservative estimates placed the North Slope reserves at 10 billion barrels, and optimistic estimates increase that to 40 billion barrels.[1] At peak production, two million barrels of oil would pass through the projected pipeline each day.

In considering the Alaskan oil pipeline we are confronted with the issues of the environment, safety and economics. The pipeline is often viewed as a choice between energy and the environment. In fact, the "choice" involves many choices, singly and together quite complex, including technical, economic, procedural, legal, environmental, bureaucratic and intergovernmental political factors. However, the solution to this conflict must take into account the parties in the conflict, who are:

(1) The energy industries which accepted the original mandate to provide energy at the lowest cost

(2) A vocal group of awakened environmentalists who, in part, believe that the oil companies and the regulatory agencies are using the energy crisis as a justification to assault the environment with impunity and who demand equal status in the decision-making process

(3) The countries and states who fear encroachments into their jurisdictions

(4) The bureaucrats—Federal, state and local—each with a vested interest in his respective agency's mission

(5) The citizens who want cheap, plentiful energy with no environmental effects[2]

The Alaskan pipeline should be an important and contemporary example of a nation's attempt to place all such factors in perspective and decide what interaction of technology, energy use, environmental protection and national interest should be accommodated. Although public policy incorporates public opinion, private interests and international needs, policy is founded on how man wishes to live. The pipeline is one example of a technological development which illustrates man's attempt to harmonize his desires for a pristine natural environment with his dependence on energy available in the form of oil and other fuels. The benefits of technology are difficult to compromise with the effects of the technology. The Alaskan pipeline is no exception.

Authorization for the pipeline was passed early in 1974 over the objections of conservationists. Only history can tell how much of the decision to proceed was based upon throughtful analysis and how much was based upon rationalizations growing out of the fuel shortage at the time. Some have seen even the fuel shortage itself as part of diabolical machinations by the oil companies to force the issue of higher prices and new sources of supply.

CONSTRUCTION

The trans-Alaska pipeline will be the biggest construction project ever undertaken by private industry. The pipeline, 48 inches in diameter, would extend from Alaska's North Slope south 789 miles to the ice-free port of Valdez, as is shown in Figure 20–1. Initially five pump stations would move the oil through the line at a rate of 600,000 barrels per day. Later another seven stations would be added to boost capacity to two million barrels per day.

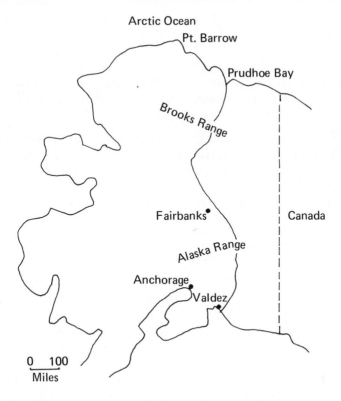

FIGURE 20-1   The proposed trans-Alaska pipeline route from Prudhoe Bay south to Valdez.

The pipe is rugged, with high ductility and weldability. It is able to withstand extreme temperatures.[3] The pipe has a wall thickness of one-half inch. The oil will be transported at a temperature of 145° F. How will this temperature affect the Alaskan wilderness?

PERMAFROST

Permafrost is ground, from solid rock to moist sand, that has remained below freezing temperature for two or more years. Permafrost may extend from a few feet to hundreds of feet below the surface. In the coldest regions of Alaska the permafrost is closest to the surface, covered only by a seasonal thaw layer and surface vegetation (tundra). It is believed that heat from the oil pipeline would cause no problem where the line is underlain by rock or gravel beds. But where ice and soils with high frozen moisture are present, the pipe would be in difficulty if placed directly on the soil. Therefore, the pipe will be elevated.

Other questions facing the pipeline engineers are:

(1) What areas along the route are stable and unstable permafrost?

(2) What if the line shuts down and the oil becomes gelatin-like at the low temperatures?

(3) What is the best technique for raising the line above ground?

(4) Will above-ground sections block the migration of caribou?

(5) Would terminal operation in Valdez harbor affect marine life in the area?

In order to assist in the solution of the problem of possible unstable permafrost, some 3,000 bore holes have been drilled, tested and analyzed along the pipeline route to determine where to elevate the line above the permafrost.

To examine the possibility of the oil turning to a gel after shutdown, a test was conducted during the 1969–70 winter near Prudhoe Bay using an eight-inch pipe. Calculations show that even a 21 day shutdown would not yield a gelatin-like oil. As a safety measure, however, a shutdown procedure will include the draining of oil from the line into 55,000 barrel relief tanks at pumping stations. The tanks will have oil heaters to send warm oil back into the line during start-up to aid in degelling of the oil.

The elevated portions of the pipeline will be laid out in a trapezoid zig-zag fashion to allow for contraction and expansion. The temperature range along parts of the line is from –65° F to +91° F during the year, so the expansion and contraction during the year is substantial. Above-ground pipeline sections will be laid on a thick berm of gravel and concrete to prevent permafrost degradation. Also, the above-ground pipeline will be insulated to maintain fluidity even after extended periods of shutdown. The pipeline will be buried in those areas where the permafrost is stable and where it is not subject to soil settlement caused by thawing. The pipeline will be anchored to resist earthquake forces and thermal forces from expansion and contraction. Insulation will maintain the temperature of the pipe itself about –20° F. Conventional burial techniques will be used for half the route. Depths will range from 18 inches to several feet depending upon terrain and soil. To protect against corrosion, below ground sections will be treated with an expoxy coating.

Early detection of leaks must be a central part of the pipeline system. An operations control center will monitor the pipeline continuously. A computer will provide four automatic alarms: pressure deviation, flow deviation, flow-balance deviation and line-volume balance. In the first three an alarm will ring when any one indicator deviates more than 1% from the present value. The data display then indicates the line section containing the leak. Pump stations can be shut down, isolating that section. The pipeline will also be patrolled by ground and aerial surveillance teams.

WILDLIFE

About 400,000 caribou live on the North Slope during the summer. They migrate south through the Brooks Mountain Range in the fall. To test the animals' reaction to a pipeline, a two-mile test section was erected in the path of the migration. overpasses and underpasses were built for this test line along its length. It appears that the animals will be willing to use these overpasses and underpasses during their migration.

PORT FACILITIES

The proposed terminal on the shore of Part Valdez will cover 800 acres and the facilities will include storage tanks, piping and valves, docks and tanker loading facilities, the operations control center, a ballast treatment center and other necessary buildings. All buildings will be built on bedrock, the best foundation for earthquake resistance. Furthermore, all structures will be of sufficient elevation to avoid tidal-wave damage. Initially 15 storage tanks, each holding 510,000 barrels, will be constructed; eventually there will be 44 storage tanks, each 250 feet in diameter and 62 feet high.

Tankers arriving at Valdez will unload ballast from their tanks into the ballast treatment plant which will remove oil contaminants.

A road must be built to provide access for construction along the pipeline route. Rock and gravel will be used for this road. An archaeologist must be with the construction on Federal lands so that if antiquities are discovered the construction may be halted or detoured. It is expected that archaelogical teams will be accompanying the construction teams because several ancient sites have been discovered already.

COST

The total cost of the trans-Alaska pipeline will be approximately $3 billion. It will cost another $1 billion for tankers to carry the oil to the west coast of the United States. The Prudhoe Bay unit of the Alaska oil wells will cover some 220,000 acres and about 150 wells will be needed to develop the field. An offshore drilling technique will be used for the wells. The wells will be spaced about 100 feet apart on gravel pads. The wells are drilled vertically through 2,000 feet of permafrost; it requires about 30 days to drill one of these holes.

The natural gas reserves of the Alaskan North Slope are estimated to be 26 trillion cubic feet. The United States uses approximately 25 trillion cubic feet each year, so the Alaska supply could provide a significant percentage of the gas source over a decade. Presently it is planned to construct a gas line through Canada and Alaska as shown in Figure 20–2. The 48-inch line would deliver about 3.5 billion cubic feet per day, or approximately 1.28 trillion cubic feet each year. This is about 5% of the annual American consumption. At that rate of use, it would last about 20 years. The length of a gas pipeline from Prudhoe Bay to Chicago via Edmonton would be about 2,400 miles. It could cost about $5 billion. It is estimated that gas production could begin about two years after oil production commenced. In the meantime, the gas that is produced as a result of the oil operations will be reinjected into the ground until needed.

The environmental considerations of such a significant project as the trans-Alaska pipeline are important. The legal and environmental actions of the past several years are illustrative of the interaction of environmental, economic, technological, political and legal aspects of a large technical development.

About a tenth of the United States lands are still in a wilderness condition and most of this wilderness is in Alaska. Thus, one immediate question is the value of the oil *versus* the necessity to invade the wilderness of Alaska with a 789-mile pipeline. Most of the Alaskan wilderness are United States Federal lands and State lands.

In 1969, The United States Department of the Interior was recommending the development of the pipeline. Also, the State of Alaska, with a state budget of $154 million in 1969, could envision the development of its lands and resources and new jobs and new tax sources. In addition, Alaska holds an additional 800,000 acres which it can subsequently lease to collect royalties. It was estimated that an income of over a million dollars a year would accrue to the state as a result of leases and royalties.

OBJECTIONS TO THE TAP

The oil consortium had supposed that it would get quick approval from the Interior Department and it would be in operation by 1972. But several environmental groups filed a lawsuit in 1970 and stopped the project. In early

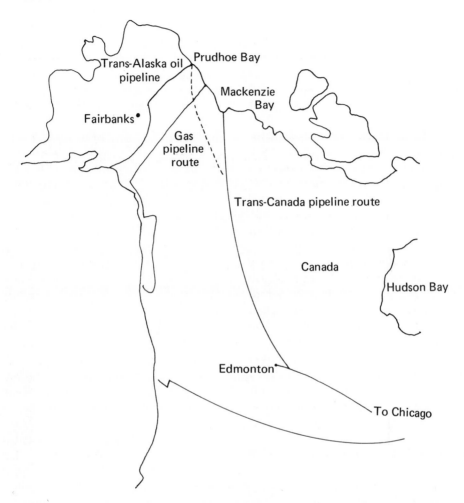

ϝIGURE 20-2   The Trans-Alaska oil pipeline and the proposed Trans-Canada pipeline
route are shown as well as a gas pipeline connecting from Prudhoe
Bay to Edmonton and on to Chicago.

1970 the President had signed the Environmental Policy Act of 1969, which
requires the analysis of environmental impacts of such projects as the pipeline. In
April 1970, a United States District Judge issued an injunction against the
pipeline and required an environmental-impact report considering such environmental factors such as the potential for oil spills, the degradation of the
permafrost and the effect on the Alaskan fishing industry. Would there be leaks
from the pipe? Would the 41-ship tanker fleet spill a significant amount of oil?

On March 20, 1972 an environmental impact report was released by the
Department of the Interior. The report was 3,500 pages long. As a result of this

report the judge removed his injunction, allowing the Department of the Interior to issue a permit for construction. The Wilderness Society, Friends of the Earth and the Environmental Defense Fund then issued a 1,300 page rebuttal. Several experts questioned portions of the report dealing with topics such as the quality of the pipe and the potential of landslides on the pipeline. On May 11, 1972 Secretary of the Interior Morton announced he would approve the pipeline in the near future.

Nevertheless, the environmentalists chronicled the issues of concern. Every year Alaska has more than 1,000 earthquakes, most of them slight. However, the pipeline would cross three faults where two dozen significant earthquakes have been recorded this century. The Alaskan earthquake of 1964, which cost $500 million in damage and carried a tidal wave that wiped out the old town of Valdez, had a magnitude of 8.5 on the Richter scale.[1]

Much of the ground along the pipeline route is unstable, and landslides could occur which would cause pipeline leaks and ruptures. Since the pipeline crosses 350 streams, there could be water pollution. In the event of a pipeline rupture, 14,000 barrels of oil would leak out during the time required for pump and valve shutdown. This oil spill might cover some three acres in winter. While the oil could be removed, it could severely damage the vegetation.

The Prudhoe Bay oil field would cover some 880 square miles and along with the line and road would cover less than 0.2% of the Alaskan land surface. It is a matter of judgment whether that is a significant incursion on the wilderness.

The tanker fleet could average between one and two collisions or groundings each year. In addition, accidental discharges from the tankers can occur. Some foresee, as a result, a significant number of barrels of oil washing up on the beaches and ruining fishing areas.

On February 8, 1973, as a result of litigation of several environmentalist groups, the United States Court of Appeals again stopped the issuance of a permit of construction on the basis of the Mineral Leasing Act of 1920 which limits the right of way across Federal lands to 25 feet on either side of the pipe. The oil companies found this width too narrow for their equipment and thus the pipeline was held up again. This obsolete law had been violated for years and the application of the law in this case caught the companies unaware. Subsequently, legislation was passed to permit pipeline construction.

A statement that perhaps sums up the environmentalists' opposition to the pipeline is the following issued by the Sierra Club in May, 1972: "In deciding to grant permits for an all-Alaskan route for an oil pipeline, Secretary Morton has chosen a short-term expedient which will confront us with the worst possible combination of long-range results."

Another objection to the Alaskan route comes from many observers who fear that the Alyeska Company wishes to use TAP and tankers so that oil not readily sold on the west coast of the United States can be sold to Japan at great profit.[5]

A CANADIAN ALTERNATIVE

One alternative proposal to the trans-Alaska pipeline (TAP) is a trans-Canada pipeline (TCP) which would go through Canada's Mackenzie River Valley to Edmonton and on from there to supply the midwestern United States. The proposed route of the TCP is shown in Figure 20–2 above. A Canadian pipeline offers some considerable advantages. It would deliver Alaskan oil to an area of the United States in need of increased supply, and it would avoid the use of super-tankers and their potential spills. Secretary of the Interior Morton approved the TAP because it was believed a large delay would occur while arrangements were made for the TCP. Also, it was estimated that the TCP would cost about twice as much as the Alaskan route and it was feared that the Canadian route would prevent the United States from having full control over its Alaskan oil. National security as a factor in preferring the TAP over the Canadian route often ignores the fact that the tankers are vulnerable to attack in the case of hostilities while the TCP pipeline would be less so. On the other hand, the Canadian government, at one point in the discussion, indicated that Canada would require 51% ownership of the pipeline. This deterred many American officials from accepting the TCP; they did not want another country to hold a majority interest over the flow of Alaskan oil to United States users.

It is also interesting to note that there is strong opposition to TCP in Canada on the basis of both environmental concerns and economic dependence on the United States.

LEGAL MANEUVERS

A series of legal activities was initiated by the plans for the Alaskan pipeline. The first suit was filed in April 1970 on behalf of the Alaskan villages claiming title to the land that TAP would cross eventually. In 1971 the Native Claims Act was passed by Congress settling the land claims and establishing procedures for payments for land use. Royalties up to $500 million will be paid over the life of the Alaskan oil fields.

As we noted earlier, the Wilderness Society, Friends of the Earth and the Environmental Defense Fund filed a suit against the Secretary of the Interior, the State of Alaska and the Alyeska Pipeline Service Company in April 1970. The suit was brought on the grounds that the Department of the Interior had not complied with the Environmental Policy Act of 1969 and that the width of the right of way had exceeded the statutory limit.

Another suit was filed in April 1971 by the Cordova District Fishing Union claiming that the Forest Service Permit which Alyeska got for its terminal operations at Valdez was invalid and the requirements of the National Environmental Policy Act were not met.

In his book *Defending the Environment*, attorney Joseph Sax points out: "In the ever expanding and elaborate procedures that the legislatures impose on administrative agencies, it is usually rather easy to find some procedural blunder or failing that can be called to the attention of a court."[4]

In mid-July 1973 the United States Senate engaged in seven days of debate on the trans-Alaskan pipeline (TAP) and approved its construction on July 17, 1973. The initial legislation, S.1081, gave the Secretary of the Interior the authority to grant a right of way across the Federal lands. Subsequently during the debate it was amended with very important additions. One amendment, offered by Senator Walter Mondale of Minnesota and Birch Bayh of Indiana on July 13, 1973, would have required an eight-month study of the Canadian route (TCP) by the National Academy of Sciences. This amendment was favored by the environmental groups and by midwesterners who wanted Alaskan oil to go to the midwestern United States. After completion of the study, the amendment said, Congress would decide the route to be authorized, a decision which would not be reviewable in court. In the end the amendment failed by a vote of 61–29.

The second major amendment was offered by Senators Mike Gravel and Ted Stevens of Alaska. It established a Congressional finding of fact that the the environmental impact statement issued by the Department of the Interior complied with the National Environmental Policy Act of 1969, thereby cutting off further court action. Many opposed this action on the grounds that it set a precedent for exemption from the NEPA. This amendment was passed on a vote of 49–48. On a motion to reconsider the vote, Vice-President Agnew's tie-breaking vote defeated it 50–49. The bill to approve the pipeline was said to provide the quickest way to start the oil flowing. The most controversial part of the bill states: "No right-of-way, permit, or other form of authorization which may be issued (for the pipeline) ... shall be subject to judicial review." All government actions taken to that point were to be considered legal.

The bill invested the Secretary of the Interior with the power to "issue, without further action under the National Environmental Policy Act of 1969 or any other law," whatever permits would be required, and it specifically approved the TAP route.

The bill contained a provision that the President must give advance approval of the export of any Alaskan oil to foreign nations, with Congress having the right to overturn such a decision.

On August 2, 1973 the United States House of Representatives passed a bill with provisions similar to the Senate bill. The House divided 222 to 198 in the critical vote on the issue. Final action then resulted from a resolution of the differences in a Conference Committee of the Congress. President Nixon signed the resulting bill into law on November 16, 1973.

During the discussion of the Senate bill, it was desired to know more of the Canadian view of the TCP possibility. The United States Department of State, after an inquiry of the Canadian government, stated, "The Departments of State and Interior are convinced that there is no Canadian alternative to the proposed Alaskan pipeline at this time." Furthermore, it said, "The Canadian Government has no strong current interest in the construction of a Mackenzie Valley oil pipeline." Part of the explanation was, "There appears to be increasing public concern in Canada for the environmental and native rights problems."[6]

Later it was revealed that the State Department may not have communicated the full intentions or attitudes of the Canadian government in this matter. Senator Walter Mondale labeled this as "misrepresentation" while others called it "an inadvertent slip-up." Mr. D. MacDonald, the Canadian Minister of Energy, Mines and Resources, remarked to the Canadian Parliament, "The view we take continues to be that the MacKenzie Valley would be the best route for bringing Prudhoe Bay oil to southern markets. However, in the case of a serious proposal, the government does feel an obligation to have an investigation of this alternative route to determine whether or not it is feasible and in general whether it would be in the national interest to promote it further."[7] Perhaps the delay of a study and the concern for pipeline ownership was an overriding factor in the Congressional decision.

Thus it appears that the trans-Alaska pipeline will be built during the 1970s, and oil may be flowing to the United States west coast by 1978. Oil imported from South America may be diverted from the United States west coast to the Gulf Coast to serve the needs of the midwest, and the Pacific Coast will be served by Alaskan oil. In his letter to the House of Representatives on August 1, 1973, Mr. Nixon stated, "As the House moves toward final consideration of legislation to remove the present legal impediment to construction of the proposed Trans-Alaska Pipeline, I want to share with you my view that prompt construction of this pipeline is of great importance to the national interest of the United States. The Department of the Interior has given very careful and detailed attention to the environmental issued posed by the Trans-Alaskan pipeline proposal. The Secretary of the Interior will assure that all necessary precautions are taken to protect the environment throughout construction and operation of the pipeline."

Thus, while all may not be assured of the quality of the environmental planning and monitoring of the pipeline, it will almost certainly be built. A large percentage of the American voters believe that the Alaskan oil is needed to meet the energy crisis. In a poll taken in California, of those aware of the proposal 64% favored construction of the TAP. The pipeline wil add 5,000 to 10,000 new jobs to Alaska's economy and will bring oil to an energy-hungry United States. Whether it was the right decision for the country, only time will prove.

SUMMARY

What we may have learned from the issue of the trans-Alaskan pipeline is that new institutional arrangements may be necessary to meet the complex technological systems of our future. A pipeline is not a purely technical problem. Although public policy must be formulated and carried out within boundaries set by limited funds and politically powerful interests, in the final analysis public opinion should set the limits of public policy. In the case of the pipeline, public opinion called for delivery of the oil. The need for a comprehensive, integrated program for energy use is clear. Responding on a piecemeal or *ad hoc* basis is not sufficient. A process which incorporates the legal, technical, industrial, economic and environmental needs is required. A balance between man and nature—and between technology and the environment—is necessary.

Time is short as we confront ever larger and more complex technological solutions to man's needs.

CHAPTER 20 REFERENCES

1. R. Sherrill, "The Trans-Alaskan Pipeline," *The Nation*, June 11, 1973, pp. 745–751.
2. J.A. Best, "New Institutional Arrangements to Resolve Power Plant Siting Conflicts: A Political Analysis," *Cornell University Energy Project Paper No. 72–4*, February 1972.
3. "Engineering the Trans-Alaska Pipeline," *Civil Engineering*, September 1972, pp. 79–83.
4. J.L. Sax, *Defending the Environment*, Alfred A. Knopf, Inc., New York, 1971.
5. C.J. Cicchetti, "The Wrong Route," *Environment*, June 1973, pp. 4–13.
6. "Oil Slip-up," *The New Republic*, August 11, 1973, pp. 11–12.
7. "Canadian Pipeline," *Congressional Record*, May 29, 1973, p. E3520.
8. R.A. Rice, "How to Reach that North Slope Oil: Some Alternatives and their Economics," *Technology Review*, June 1973, pp. 9–18.
9. B. Cooper, *Alaska: The Last Frontier*, William Morrow & Co., New York, 1973.
10. T.A. Morehouse, "Mythology on Alaska and Oil, *Technology Review*, November 1973, pp. 74–75.
11. T.M. Brown, "That Unstoppable Pipeline," *New York Times Magazine*, Oct. 14, 1973, pp. 34–35, 88–92.

CHAPTER 20 EXERCISES

20–1. Reconsider the potentially available methods of transporting the oil from the North Slope of Alaska. Which method do you favor and for which reasons?

20–2. Considering the balance between the environment of Alaska and the fuel needs of the 48 contiguous states, do you favor the TAP? Do you believe it will be environmentally safe? Discuss this balance and your evaluation of the TAP pipeline route.

20–3. The decision in Congress to proceed with the pipeline was taken under the pressure of the energy crisis of 1973 and the wish of some American citizens to proceed. Reevaluate this decision in the light of subsequent events and give several suggestions for how the political consideration of such a large engineering project can be improved.

20–4. It has been proposed that a two-track railway running from Prudhoe Bay through the MacKenzie River Valley to Edmonton and Chicago may be a viable alternative to a pipeline.[8] As we noted in Chapter 17 (see Table 17–2), pipelines are about 50% more energy-efficient than railroads. It is estimated that the cost of the railroad would be approximately equal to the total pipeline-and-tanker approach. The railroad could be used to ship equipment north as well as carry oil and gas south in tank cars. The railroad would help to open up the northern part of Alaska to settlers and commerce. Consider the advantages and disadvantages of a railroad for bringing the oil and gas of the North Slope to the midwestern United States.

20–5. Permafrost covers most of Alaska in one form or another. Most of the permafrost contains ice and frozen silt. Once thawed, this material begins an essentially irreversible process of erosion. Examine the potential effects of a hot-oil pipeline buried in this soil or lying on its surface.

20–6. Some studies have shown that the Caribou may refuse to use either ramps or underpasses to migrate by the pipeline.[10] Examine this problem and evaluate the importance of this difficulty.

20–7. There are many analogous considerations between the Alaskan pipeline and drilling for oil in the Santa Barbara channel. In early 1969, a disastrous blowout from an offshore well blackened the beaches of Santa Barbara. Recently, under the need for more domestic oil, there has been a call for resumption of drilling in the channel. Many Santa Barbara residents say they would rather have less energy than see oil platforms rise in their channel. Explore the benefits and potential costs of resuming offshore drilling. Would you recommend resumption of drilling to the Governor of California?

21

LIMITS TO GROWTH

CARDINAL NEWMAN WROTE IN 1864, "Growth is the only evidence of life." A decade ago the doctrine that economic growth was the touchstone for dealing with most social problems was virtually unchallenged. Europe and Japan basked in their economic miracles and every nation wished to emulate the economic growth of the United States. Now, however, the United States and Europe have re-evaluated this unfettered growth and considered the results as a possible crisis. The deteriorating environment, crowding in cities and basic attitude changes have resulted in challenges to the growth ethic. A concern with the continuing growth of population and the uses of energy and technology has led to a need for reevaluation of the mechanisms of growth.

MODELS FOR THE GROWTH PROCESS

A *model* is a qualitative or quantitative representation of a process showing the effects of those factors which are significant for the purposes being considered. Mathematical modeling of the dynamics of growth, assisted by a computer, has been used during the last decade to analyze the possible consequences of actions of public and economic policy and technological development. One aspect of modeling is the elucidation of the essence of a system, in this case the world system of dynamic behavior. When a computer is applied to this service we have a *computer model*. A computer model is a representation of a system of phenomenon in a mathematical or symbolic form suitable for demonstrating the behavior of the system or phenomenon.[1] A *computer simulation* uses a computer model and estimates of the actual conditions of the system being modeled to evaluate the situations of the investigator wants to understand. Computer simulations have been used to analyze the various approaches and tasks of the United States aerospace program. For example, a simulator was used to train astronauts for lunar landings.

In the past few years Professors Jay Forrester and Dennis Meadows have developed computer models of the dynamics of the world's resources, population and economics. In his book *World Dynamics*, Professor Forrester discusses his highly complex model of society.[2] The model attempts to interrelate population, capital investment, geographical space, natural resources, pollution and food production. The model is an attempt to represent concepts from economics, demography, agriculture and technology. According to Forrester, a model "makes a theory unambiguous," so that its assumptions can be criticized. When Forrester runs his model he obtains frightening conclusions, such as: "Rising pressures are necessary to hasten the day when population is stabilized ... . Civilization must be restrained rather than expanded."[2] This is clearly a call for limiting growth. A graph resulting from the computer simulation shows the quality of life declining with time as a result of the assumptions Forrester put into the computer. Forrester's model states that to bring the world system into equilibrium by 1980 would require the reduction of current pollution generation by 50%, the reduction of capital investment by 40%, the reduction of the birth rate by 30% and the reduction of food production and natural resource usage by 20% and 75% respectively. Mr. Forrester concludes, "A society with a high level of industrialization may be non-sustainable."

In the following article, Robert Boyd criticizes the World Dynamics model and analyzes the results of altering the model to include the benefits of technology toward the solution of some of the problems evidenced in the Forrester model.[3]*

WORLD DYNAMICS: A Note

Robert Boyd

In *World Dynamics* Forrester presents a simple simulation model of the world. This model consists of five first-order differential equations in five very highly aggregated state variables. For example, the variable NR is defined as "natural resources." He also includes some auxiliary variables that indicate performance; these he calls "quality of

*Reprinted from *Science*, Vol. 177, August 11, 1972, pp. 516–519. Copyright 1972 by the American Association for the Advancement of Science.

life." Forrester performs several simulations and, based on the behavior of the quality of life in these simulations, makes several quite provocative policy recommendations.

For most of the scientific community, modeling the world with only five state variables is a dubious project. Recommending policies based on such a simple, operationally vague model is doubly suspect. Forrester argues that his simulations are not at all dubious, and that since decisions are being made on the basis of simpler and fuzzier verbal models, decision-making can only be improved by using computer simulation. In this report I argue that this assertion is only true if the computer simulation is insensitive to different sets of plausible assumptions and I will show that the *World Dynamics* model does not demonstrate such insensitivity.

World Dynamics has been received very critically. Several reviews have taken Forrester to task for doing bad science. For example, Shubik wrote " ... the book is blatant and insensitive advocacy for unsubstantiated model building on a very large scale ... What is this book for? Its behavioral-scientific content is virtually zero ... None of his (Forrester's) book has any empirical content, yet the operation of all his models calls for large numerical inputs." These sorts of criticisms miss their mark; Forrester is not trying to do behavioral science. He has a less ambitious goal, to improve on the verbal and mental models that are now in use. And he would argue that computer models are always better than verbal or mental ones. His argument can be condensed into the following propositions:

1. People are quite good at perceiving relationships between various components of the world system.

2. People are not very good at deducing the whole system consequences of these individual relationships.

3. It is quite easy to use computer simulation to accurately deduce the consequences of any set of relationships. Therefore, verbal models or mental images are necessarily inferior to mathematical models.

4. Policy decisions are presently being based on nonmathematical models. Therefore policy can certainly be improved by reexpressing these vague models in a more exacting systems dynamics formalism.

It seems relevant to ask whether this conclusion is true. Does the simple translation of a verbal model into Dynamo improve its predictive ability?

Imagine, for a moment, one of Forrester's social dynamicists entering the political arena. There he would find decisions being based on an array of vague, hard to verify, mental models founded upon conflicting assumptions, and reaching different conclusions. As a first step the social dynamicist could re-express each of the different mental models as a different simulation of his model. But unless all these simulations show substantially the same results, the dynamicist has done little to improve the lot of the decision makers. Instead of being faced with an array of unverified mental models, he is confronted with an equal number of conflicting computer simulations. To be of use then, the social dynamicist would have to go to a more sophisticated, complicated model whose components can be independently verified in some way.

Let's apply this test to the model in *World Dynamics*. Roughly there exist two ideological poles in the controversy about the future of the world.

1. The Malthusian view. The adherents of this view argue that the earth is endowed with some fixed, finite amount of resources. We only have so much agricultural land, so much petroleum, and so on. Further, Malthusians argue that anything that makes life

better, be it better nutrition or increasing material standards of living, leaves birth rates constant and lowers death rates. New technology in this view can only temporarily alleviate shortages; population must inevitably overtake any increases in productivity. Eventually an equilibrium between high death rate and high birth rate is reached. This view is more completely expressed and defended in much of the conservation literature.

2. The technological-optimist view. The adherents of this view argue that there are no foreseeable limits on production of goods. Any particular scarcity will be eliminated by substitution technology. Further, the increasing stock of technology is seen to increase productivity and thus increase the standard of living. Increasing the standard of living is then supposed to produce lower birth rates. Eventually society is seen to reach an equilibrium between a low birth rate and a low death rate. A more complete explication can be found for example in Barnett and Morse or in Fredericksen.

Even a casual reading of *World Dynamics* shows that Forrester's model is fundamentally Malthusian. The quantity of natural resources is fixed, and the productivity of industry based upon them decreases with time. Agricultural productivity may be increased only by increasing capital investment in agriculture. The pollution output per unit of the material standard of living is irreducible. In several of his simulations Forrester invokes technology to, for example, reduce pollution output by some fixed fraction at some point in time, but this is hardly the constantly adjusting, powerful technology of the technological optimist. Birth rate in Forrester's model increases strongly with increasing food per capita. It decreases strongly with increasing pollution and crowding, and weakly with increasing standard of living.

To test the sensitivity of the *World Dynamics* model to changes in assumptions I have altered the model to conform to the technological optimist's view. This involved adding a new state variable, technology (T) and multipliers to express the effect of technology on the other state variables. Two birth rate multipliers were also altered. For the sake of consistency all these alterations are expressed in the systems dynamics idiom.

The results of the simulation are shown in Fig 21-1. They are exactly what a technological optimist would predict. Technology increases productivity, which, in turn, increases the standard of living. This eventually drives birth rates low enough that a "Utopian" equilibrium is reached. In *World Dynamics*, Forrester argues that social systems are counterintuitive, that they show behavior opposite to what one would expect. As an example he introduces birth control into his model and shows that it only worsens the eventual equilibrium. Forrester's version of birth control was added to my model and the results are shown in Fig 21-2. The results are again exactly what the technological optimist would expect. Birth control allows the world to go through the demographic transition at a lower population level and greatly enhances the eventual equilibrium.

Forrester's model fails the test. It is completely unable to resolve the technological optimist-Malthusian controversy. In fact the output of the model under each of the different sets of assumptions is the same as was reached without the use of a computer. Thus, the *World Dynamics* simulation is far from useful as a policy tool, and, even within his own framework, Forrester was unjustified in making such strong policy recommendations.

The great strength of Forrester's methodology is its ability to assimilate expert opinion easily. A model with a larger number of more disaggregated state variables would allow experts in various fields to involve their knowledge more usefully. Such a model

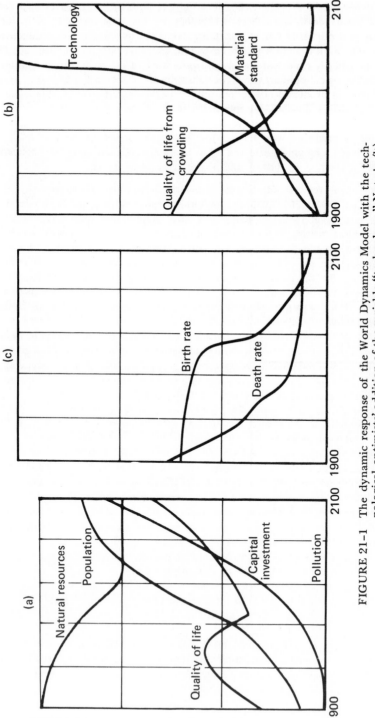

FIGURE 21-1 The dynamic response of the World Dynamics Model with the tech-nological optimists' addition of the variable "technology." Note in (b) that technology is increasing and allows an increasing standard of living.

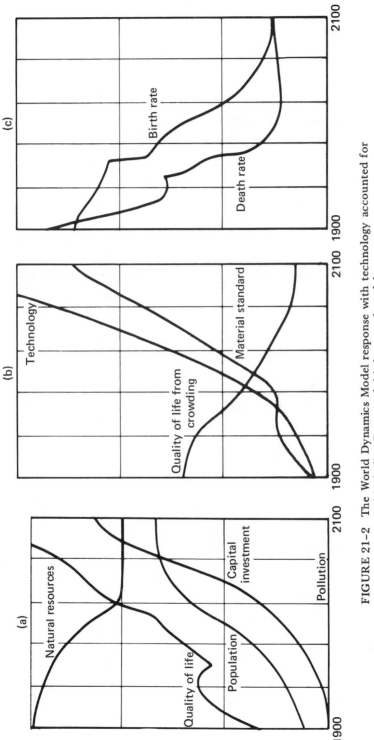

FIGURE 21-2 The World Dynamics Model response with technology accounted for and including Forrester's birth control model.

might be a practical policy tool. And in a way, such a model would be the good science that Forrester's reviewers were looking for.

In a more recent book, *Limits to Growth*, Professor Meadows uses Forrester's world model to investigate further the effects of various policies on population and natural resources. Meadows concludes, on the basis of his computer model, that the current exponential growth rates in population, industrialization, consumption of natural resources, food demand and pollution are rapidly approaching the planet's linear limit. Some sort of equilbrium must be reached soon in order to avoid disaster, in his view.

LIMITS TO THE MODEL'S EFFECTIVENESS

Computer models are constructed on the basis of assumptions and are limited in value as the prediction is extended forward in years. Also, all critical variables must be included and properly accounted for. The advantage of computer models are that system performance can be observed under all conceiveable conditions and the results of future response based on policy decisions can be observed. They permit experimentation when one cannot experiment with society or the world. Nevertheless, many critics believe that the Meadows model is inadequate at best and an "intellectual Rube Goldberg device" at worst. Using five variables to model the world leads to a very high level of aggregation and therefore lack of detail.

The basic policy alternative leading to survival, according to the model, is zero population growth and zero economic growth. Arresting economic growth appears to freeze the undeveloped countries into their state of poverty while permitting the developed countries to remain affluent. Also, this policy ignores the possible beneficial effect of technology which often enables man to replace exhausted resources with new materials and processes. The use of technology to replace depleted resources and overcome the effects of pollution is largely overlooked in the *Limits to Growth* model.

In addition, the *Limits to Growth* model may be quite sensitive to a change in its numbers. One recent study found some altered responses with a change in one of the model parameters.[5] In addition, revised economic policies can yield a stable, prosperous world response from the *Limits* model, according to this study.[5] In this case it is suggested that (1) the economics of the agricultural sector be modified by discouraging the use of capital-intensive farming above the level where soil depletion becomes serious and diverting the capital to land development; and (2) the demand for food and good be moderated.

IN FAVOR OF GROWTH

In a recent book, *Retreat from Riches—Affluence and Its Enemies*, Professors Parsell and Ross outline the benefits of economic growth and attempt to refute the no-growth argument. They argue that economic growth is the only way to cure America's environmental and social ills. The incentive to pollute the atmosphere and waters can be overcome by charging the industry or individual for pollution. Also, in their view, growth provides new jobs and overcomes poverty; it must be encouraged.

In another recent book, *The Logarithmic Century*, Ralph Lapp states his view that many of the exponential growth curves of the past will level off during the twentieth century. As we noted in Chapter 18, the finite reserves of fossil fuels indicate that our reserves of oil and gas may be essentially exhausted over the next fifty years. Clearly, the rapid growth of other variables must slow down or they will equal the finite limit of the earth or the population. As an example of an exponential growth which eventually slowed, recall the rapid growth of the number of students attending college during the 1960s. This growth eventually stopped as the percentage of the college-age group attending college reached 50%. If this rapid growth continued unchecked, an extended curve might imply that all the citizens of the United States would be attending college eventually.

In California and Oregon there has been a massive assault against unchecked growth during the past several years.[8] A no-growth campaign of sizable proportions has emerged, perhaps partly traceable to common recognition of writings such as *World Dynamics* and *Limits to Growth*. Crowding, pollution, overpopulation, and the unchecked use of natural resources have led some people to attempt to counter this growth with legislation and economic actions. Many cities in California have recently elected city councils on the platform of no-growth for the city. The Coastline Initiative passed in California in 1972 establishes the Coastal Zone Preservation Act, which may severely restrict the development of the California coastline. The revolt against the construction of new freeways is another example of an attempt to slow growth and land-use. Perhaps the people of California and Oregon have recognized the limitations of the finite earth and wish to preserve the qualities of their land and environment. Slow growth toward equilibrium is desirable, for while there may be a disagreement concerning the carrying capacity of the earth, no one would dispute the fact that there is some maximum population that the earth can support at a reasonable and desirable level of affluence.

In the following article, Jørgen Randers and Donella Meadows, members of the *Limits to Growth* modeling team, discuss the limits to growth implied by the computer-model results.[9] In this article Meadows and Randers discuss the finite amount of arable land and the constraints on resource use and pollution absorption. Following the basic behavior of the world model, the authors note that a collapse of industrial output and population may be the result by the year 2050. Alternative policies and the policy of stabilizing population and industrial investment after 1980 are proposed in the article.*

*"The Carrying Capacity of Our Global Environment: A Look at the Ethical Alternatives." Reprinted by permission of the authors. Copyright © 1972 by Jørgen Randers. Note: Figures have been re-numbered for use in this text.

THE CARRYING CAPACITY OF OUR
GLOBAL ENVIRONMENT:
A LOOK AT THE ETHICAL ALTERNATIVES

Jørgen Randers and Donella Meadows

The main thesis of this reading is very simple: because our environment—the earth—is finite, growth of human population and industrialization cannot continue indefinitely. The consequences of this simple and obvious fact pose an unprecedented challenge to mankind, for in a limited world we cannot maximize everything for everyone. In the near future we will have to decide just what ethical basis we should use for making the trade-offs which will be necessary in a world with finite limits.

THE ENVIRONMENT IS FINITE

It should be quite unnecessary to point out that the global environment is finite. However, most considerations of the world's future tend to lose sight of this finitude. Because it is not generally recognized how very close we are to the physical limitations which define the carrying capacity of our globe, it will be worthwhile to spend some time discussing these limitations.

Agricultural Land

The quantity most obviously limited on our earth is arable land. There are about 3.2 billion hectares of land suitable for agriculture. Approximately half of this land is under cultivation today. Immense investments of capital will be required to settle, clear, irrigate, or fertilize the remaining half before it can produce food. The capital costs are so high that the United Nations Food and Agriculture Organization, which is seeking desperately to stimulate greater food production, has concluded that in order to expand food output it must rely on more intensive use of currently cultivated land, not on new land development.

If we were to decide to incur the costs necessary to cultivate all possible arable land and to produce as much food as possible, how many people could we expect to feed? A graph indicating the increasing need for agricultural land caused by the growing world population is shown in Figure 21–3. The lower curve in Figure 21–3 shows the amount of land *needed* to feed the growing world population, assuming that the present world average of 0.4 hectares per person is sufficient. (If we wanted to feed everyone at U.S. standards, we would require 0.9 hectares per person.) From 1650 to 1970 world population grew from .5 billion to 3.6 billion; the land that has been needed to feed these people at the present world standard is depicted with a heavy line. The lighter line indicates future need for land, assuming that the population continues to grow at the projected 2.1 percent per year after 1970. The upper curve indicates the actual amount of arable land *available*. This line slopes downward because each additional person requires a certain amount of land (0.08 hectares assumed here) for housing, roads, waste disposal, power lines, and other uses which essentially "pave" land and make it unusable for farming. The graph in Figure 21–3 tells us that, even with the optimistic assumption that we can utilize all possible land, we will still face a serious land shortage before the year 2000 if the population continues to grow at 2.1 percent per year.

Figure 21–3 also illustrates two other very important facts about exponential growth within a limited space. First, it shows how the condition of mankind can change within a few years from great abundance to great scarcity. The human race has had an

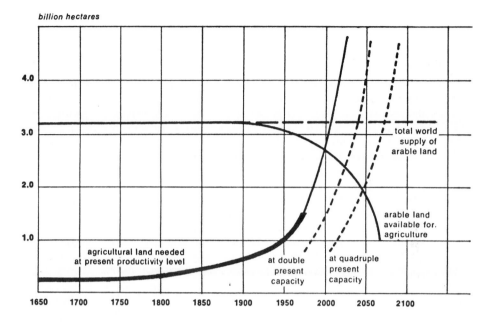

FIGURE 21–3   Total world supply of arable land is about 3.2 billion hectares. About 0.4 hectares per person of arable land are needed at present productivity. The curve of land needed thus reflects the population growth curve. The light line after 1970 shows the projected need for land, assuming that world population continues to grow at its present rate. Arable land available decreases because arable land is removed for urban-industrial use as population grows. The dotted curves show land needed if present productivity is doubled or quadrupled.

overwhelming excess of arable land for all of its history, and now, within thirty years, or one population-doubling period, mankind will be forced to deal with a sudden and serious shortage. A second lesson to be learned from Figure 21–3 is that the exact numerical assumptions made about the limits of the earth are essentially unimportant when one faces the inexorable progress of exponential growth. For example, one might assume that *no* arable land is taken for cities, roads, or other nonagricultural uses. Then the amount of available land would be constant, as the horizontal dashed line shows, and the point at which the two curves cross can be delayed, but by only about ten years. Or one can suppose that the productivity of the land will be doubled or even quadrupled through advances in agricultural technology, the effect of which is shown by the dotted lines in Figure 21–3. Each doubling of productivity gains us just one population-doubling period, or about thirty years.

Some people look to the sea to provide the extra food we will need as our population grows. But the total world fish catch in 1969 represented only a few percent of the world's caloric requirements, and the total catch decreased in 1970. The 1970 catch was the first decreased total since World War II, and it occurred in spite of increasing investment and technological developments in the fishing industry. Most experts agree that the world's fish banks have been overexploited and that future prospects for output from the sea indicate further decline, not increase. The seas thus cannot make up for the constraints limited land imposes on growth.

Heat Release

Man faces further obvious constraints on natural resources, such as fresh water, metals, and fuels. Indications are that resources such as these will be in short supply even at higher prices within the next forty years, if present growth continues. However, it is frequently argued that by mining low grade ores and desalting sea water we can alleviate resource shortages, which may indeed be so, assuming we can satisfy the enormous demands for energy such operations would present. Whether or not these arguments are based on reasonable assumptions is one of the topics Preston Cloud discusses in "Mineral Resources in Fact and Fancy."

A consideration of the energy that will be necessary to meet man's growing needs leads to a more subtle but much more fundamental physical limitation imposed by our environment. Even if one assumes that we find the means to *generate* the needed energy—for instance, through controlled fusion—we still face the fundamental thermodynamic fact that virtually all energy unleashed finally ends up as heat in the environment. An everyday example is the energy originally stored in the gas of a car. A significant part of this energy is immediately released as heat through the engine and the radiator, because the engine is necessarily inefficient in converting the potential energy in the gas to the useful mechanical motion of the wheels. But the point is that even the energy resulting in *useful* mechanical motion finally generates heat, in the brakes as they slow the wheels, in the tires as they turn on the road, in the road, and ultimately in the air surrounding the road and the auto. The dissipation of energy into heat is characteristic of all energy-using processes. On a scale larger than autos, there is the vast heat release created by the process of condensation of water vapor to form distilled water in a desalination plant—another example of energy usage adding heat to the environment.

The final fate of energy expended is easily confused with what is commonly called "thermal pollution," namely the waste heat produced locally at power plants, for example, in the generation of electricity, but energy dissipation and thermal pollution are not the same. Waste heat is given off by generating processes, which, in obedience to physical laws, cannot be completely efficient, and the consequent "thermal pollution" heats the environment, of course. But once again, the point is that even the *useful* energy output from the power plant finally ends up as heat, regardless of whether the energy was generated by burning coal or oil, or by nuclear reactions, and regardless of what the energy is being used for. It is theoretically impossible to avoid heat release if one wants to consume energy. No technical gadgetry or scientific breakthrough will circumvent it.

The heat released from all of mankind's energy-using activities will begin to have worldwide climatic effects when the released amount reaches some appreciable fraction of the energy normally absorbed from the sun. Experts disagree on exactly what the fraction is. They do agree, however, that if man wants to avoid major unpredictable changes in the climate he must recognize the fundamental limit to the amount of energy that can be consumed on the surface of the earth.

If worldwide energy consumption increases at four percent per year for another 130 years, man will then be releasing heat amounting to one percent of the incoming solar radiation—enough to increase the temperature of the atmosphere by an estimated $\frac{3}{4}$ degree centigrade. That may sound like an unimpressive figure, but on a worldwide basis it may lead to climatic upheavals, such as increased melting of the polar ice caps. Local weather perturbations may come much sooner. In just thirty years it is estimated that in the Los Angeles Basin heat released through energy consumption will be eighteen percent of the normal incident solar energy of that area.

Pollution Absorption

A third limitation to population and industrial growth is the earth's finite capacity for absorbing pollution. As we've noted, until quite recently the environment was considered essentially infinite: it seemed impossible that the use of soap for one's laundry or pesticides for one's roses could affect the workings of the world ecosystem. But clearly man's activities *do* affect the world ecosystem. Lake Erie, one of the world's largest fresh-water lakes, has been brought to the verge of ecologic death, CO_2 has increased in the atmosphere all around the world, and the sale of swordfish in the U.S. has been prohibited because of its mercury content. It is becoming abundantly clear that our environment is able to absorb and degrade only a limited amount of emissions and wastes every year. When we exceed this absorptive capacity, we not only cause pollutants to accumulate in nature faster than they can become degraded, but we also run the risk of destroying the natural degradation processes themselves, thus decreasing the future absorptive capacity. This general principle can be described with very real examples. Discharging a small amount of waste into a pond will only slightly lower the water quality, because the pond's microorganisms manage to degrade the pollution as it occurs. A constant higher discharge rate will result in a constant water quality that is lower. The absorptive capacity of the pond is exceeded, however, if one increases the discharge rate to the point where the absorbing microorganisms die because of oxygen depletion or accumulation of toxic wastes. When that happens, continued constant discharge to the pond will simply build up, continually reducing the water quality. This, essentially, is what has nearly "killed" Lake Erie, by no means a small pond.

Thus man is beginning to realize that absorptive capacity—far from being a good in unlimited supply—is an extremely valuable, scarce resource, which in fact limits the total possible pollution arising from human activity.

THE PRESENT GLOBAL TREND: GROWTH

Growth in a Finite World

Having acknowledged the existence of purely physical limitations of our earth (and we have described here only a few of the biological and physical limits that exist), we may now ask whether mankind's present behavior takes into account their existence.

On a global scale man is presently experiencing an exponential growth in population and in what we will call capital—buildings, roads, cars, power plants, machinery, and ships. Some inevitable consequences of this growth are the exponentially increasing demands for food and energy and also the exponentially increasing additions of pollution to the environment.

Because we know that there are upper limits to the supply of food and energy the earth can provide and limits also to the amount of pollution that can be absorbed by the environment, it seems obvious that the material growth that brings us toward these limits cannot continue indefinitely. In fact, matters are most urgent, since indications are that we will surpass several of these constraints within the next few generations if current growth continues. The growth must stop.

How? Are there mechanisms in the world system as it is currently organized that will

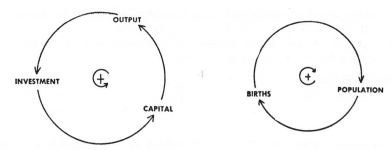

FIGURE 21-4   The positive feedback loops governing the growth in population and
               capital.

bring about smoothly the necessary shift from present physical growth trends to other
trends consistent with the world's finite capabilities? Or will the transition be sudden and
stressful?

These are the questions the Systems Dynamics Group of the A. P. Sloan School of
Management at the Massachusetts Institute of Technology set out to illuminate when, in
the fall of 1970, it embarked on an effort to devise a model for mathematical simulation of
population and capital growth in the world system.

The World Model

The world model is a set of assumptions relating world population, industry, pollution,
agriculture, and natural resources. Explicitly, the model represents the growth of
population and industry as a function of the biological, political, economic, physical, and
social factors that influence this growth. The model also recognizes that population and
industrial growth in turn provide feedback to adjust each of those biological, political,
economic, physical, or social factors.

The exponential growth of population and capital is inextricably linked with most
global problems—unemployment, starvation, disease, pollution, the threat of warfare, and
resource shortages. All are influenced importantly by the dynamic interaction of
population and capital. No attempt to understand man's long-term options can succeed
unless it is firmly based on an understanding of the relationships between these two forces
and the ultimate limits to their growth. We can briefly outline some of these relationships
as they are reflected in the assumptions on which the world model is based.

Population and births are represented in our model by a positive feedback loop. If
there are more people, there will be more births each year, and more births result in more
people (as Figure 21-4 shows). Wherever there is a dominant positive feedback loop of
this form, exponential growth will be observed. Capital and investment constitute another
positive feedback loop, also shown in Figure 21-4. Capital produces output of goods and
services. If all else is equal, greater output results in a larger investment and thus in more
capital. The interactions among population and capital determine the rate at which each
grows. These interactions take many forms, as are shown in Figure 21-5.

Let us examine material output. As output is diverted from investment, the growth of
capital decreases. Output may be diverted to services, to agricultural capital such as
fertilizer, tractors, or irrigation ditches, and to consumption. As output is diverted to
services, these increase: health and education improve, average lifetime becomes greater,

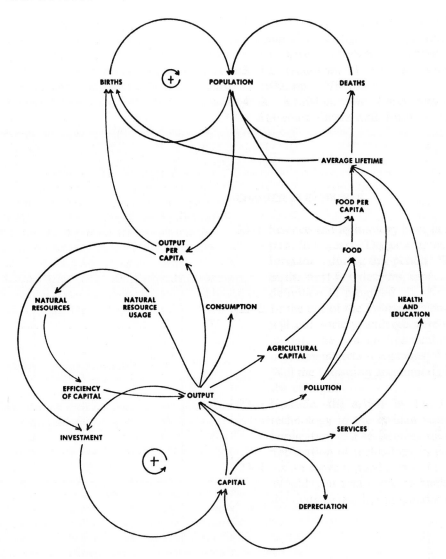

FIGURE 21-5 Basic interactions between population growth and capital accumulation.

deaths decrease, and population grows. Similarly, output diverted into agricultural capital
results ultimately in more food and a higher average lifetime. Primarily what determines
how much output is reinvested and how much is diverted is per capita output. Where
output per capita is low, most of it must be diverted to consumption, services, and food,
allocations which reduce the rate of accumulation of the capital base and, at the same
time, stimulate the growth of population. Population can increase much more easily than
capital in societies that have maintained traditional technologies. Hence, output per capita
remains low in these countries and they find it very difficult to achieve economic growth.

When output is diverted to consumption it subtracts capital from the system and does not generate future growth directly. Production of material goods consumes resources, and output diverted to consumption through industrial manufacture leads to the depletion of natural resources. As natural resources decline, lower-grade ores must be mined and raw materials must be transported longer distances. Since more capital must be allocated to obtaining resources, the overall production efficiency of capital decreases, and more capital is needed to produce a given amount of output.

Output per capita is the single force with a potential for slowing the population explosion in the model. As output per capita increases, the family size thought to be most desirable declines and birth control efficiency increases. The birth rate then goes down and the population growth rate decreases. That this is a typical trend can be seen in Figure 21–6, where birth rate is plotted against per capita product. The influence of this trend is accelerated somewhat by the fact that as death rates decline there is a further decrease in desired family size, for a large number of the world's parents bear children primarily as a source of support during old age. If there is a high mortality rate, parents must bear three

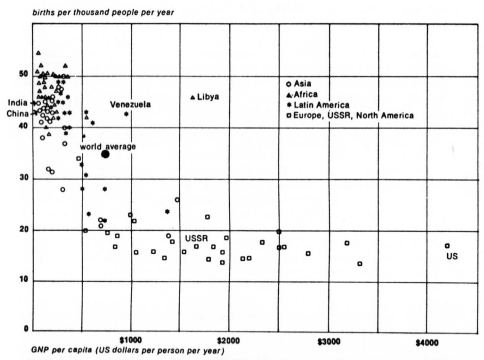

births per thousand people per year

FIGURE 21–6 Birth rates in the world's nations show a regular downward trend as GNP *per capita* increases. More than one-half of the world's people are represented in the upper left-hand corner of the graph, where GNP *per capita* is less than $500 per person per year. The two major exceptions to the trend, Venezuela and Libya, are oil-exporting nations, where the rise in income is quite recent and income distribution is highly unequal. [Source: *Population Program Assistance* (Washington, D.C.: Government Printing Office, 1970).]

or four sons to insure that one will live. Thus, as the perceived death rate decreases, so too does the birth rate decline. Material output has one additional impact: it leads to the generation of pollution. Pollution may decrease food production and also decrease the average lifetime.

The simple set of interactions shown in the feedback loops of Figure 21–5 contain the roots of most important global problems. But though these constitute the essential structure of the world model, a much more accurate description of the assumptions is needed before the model can be used for formal analysis of the consequences of the world assumptions. Figure 21–7 shows a diagram of the degree of detail included in the final world model. Four main advantages result from collecting assumptions about the world into a formal computer model of this sort. First, the assumed interrelations are listed explicitly, and they are readily available for criticism and improvement by those with knowledge in any specific problem area. Second, it is possible, with the aid of a computer, to follow the way a set of assumptions about the world system will behave as a function of time. Third, it is possible through such simulation to test the effect of some change in the basic assumptions on which the model rests, and hence, one may investigate which interrelations are critical to the behavior of the system (and thus deserve close study) and which are not. And fourth, the model permits one to study the effects of policies introduced to improve the behavior of the system. Use of the computer speeds up the calculations necessary to consider all the interconnections between variables, mathematical work which would otherwise be very tedious and time consuming.

Given the assumptions about the behavior of the world system outlined above and statistics about world limits and realities, the computer produces diagrams such as those in Figures 21–8, 21–9, and 21–10 showing differing effects of different policies with respect to natural resource usage, pollution control, and allocation of capital investment.

The computer plots the change in eight quantities over time: population (the total number of people), industrial output per capita (measured as the dollar equivalent of output per person per year), food per capita (measured as kilogram-grain equivalent of food per person per year), pollution (measured in relation to the 1970 amount), and nonrenewable resources (measured in relation to the reserves that existed in 1900). The crude birth rate (births per 1000 persons per year) is plotted as B, the crude death rate (deaths per 1000 persons per year) as D, and services per capita (dollar equivalent of services per person per year) is plotted as S.

Each of these variables is plotted on a different vertical scale. The vertical scales have deliberately been omitted and though the horizontal time scale runs through a two hundred year period from 1900 to 2100, these dates have also been made somewhat symbolic in order to emphasize the general behavior modes of the computer results instead of numerical values, which are not precisely but only approximately known. The scales are, however, exactly equal in all the computer runs presented here, so results of different runs may be easily compared. The computer results show only the consequences of the assumptions made in constructing the model. Because the simulation runs are not exact numerical predictions of the future, providing instead qualitative projections of possible future trends, the precise timing of events is less significant than the changes in behavior among simulations, and the years shown at the bottom of each figure are given only as approximate reference points. As mentioned earlier, when we attempt to study the consequences of exponential growth toward the limits of the earth, precise numerical values have relatively little effect in altering the time over which such growth can take place.

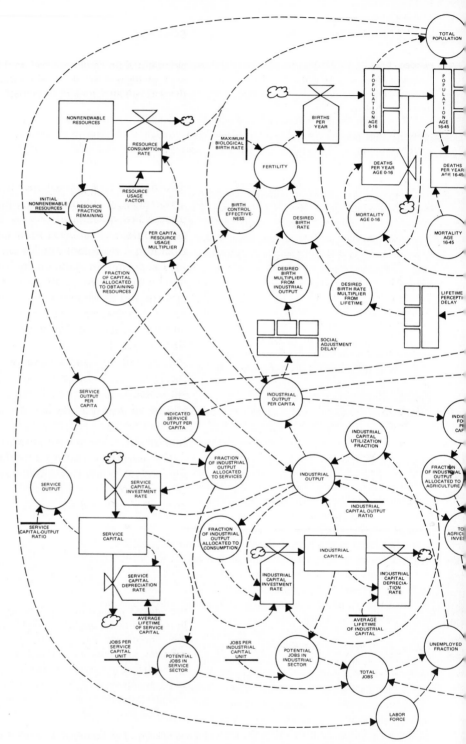

FIGURE 21-7 The entire world model is represented here by a flow diagram, in formal system dynamics notation. Levels, or physical quantities that can be measured directly, are indicated by rectangles, rates that influence those levels by valves, and auxiliary variables that influence the

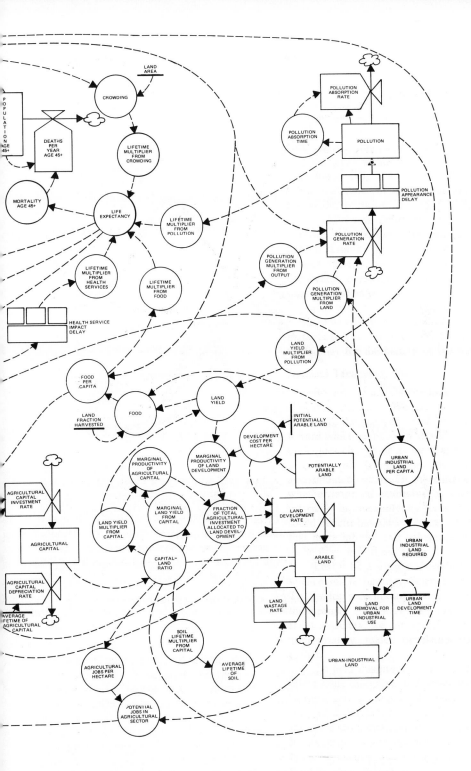

rate equations by circles. Time delays are indicated by sections within
rectangles. Real flows are shown by solid arrows; causal relationships
by broken arrows. Clouds represent sources or sinks that are not
important to the model behavior.

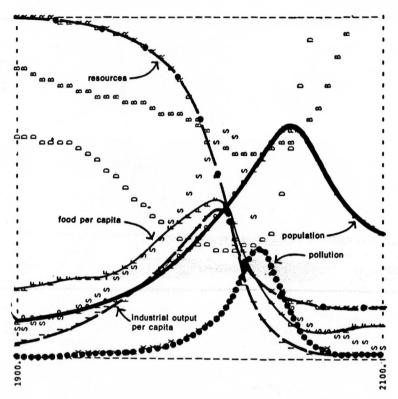

FIGURE 21-8   The "standard" world model run assumes no major change in the
physical, economic, or social relationships that have historically governed the development of the world system. All variables plotted here
follow historical values from 1900 to 1970. Food availability, industrial output, and population grow exponentially until the rapidly
diminishing resource base forces a slowdown in industrial growth. But
because of natural delays in the system, both population and pollution
continue to increase for some time after the peak of industrialization.
What finally halts population growth is a rise in the death rate due to
decreased food availability and decreased medical services.

Continued Growth Leads to Collapse

The computer simulations show that only if physical growth halts will the earth be able to
support comfortable human life. But they show something else too: the transition from
growth to nongrowth must be made smoothly. The transition will have to be engineered by
man—not by nature. Furthermore, there are no currently existent social mechanisms that
will bring physical growth to a smooth and orderly end when the maximum growth that
can be supported by the finite global environment is reached. Continued societal emphasis
on physical growth will overshoot environmental limitations. A basic redirection of
society's goals—a value change away from the current ideal of maximizing material
growth—seems to be needed to achieve a smooth end to the current growth.

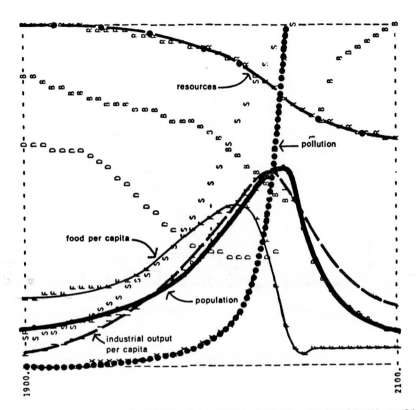

FIGURE 21-9   As means of avoiding the resource depletion depicted in Figure 21-8, resource depletion in the world model system is eliminated by two assumptions: first, that "unlimited" nuclear power will double the resource reserves that can be exploited, and, second, that nuclear energy will make extensive programs of recycling and substitution possible. If these changes are the *only* ones introduced in the system, growth is still stopped, by rising pollution.

What will happen in the absence of societal value changes? The physical growth will stop, of course, even without a change through values oriented toward educed growth. But in that event society will not experience an orderly transition to some feasible final state. Rather the human population will most likely overshoot the physical limitations of the earth, and then be forced into a traumatic decline back to some level of population and industrialization that can be supported by the physical environment—an environment which by then will be sorely depleted. For once we exceed any natural constraint, tremendous natural pressures will develop to halt growth. If it happens that the pollution we create exceeds the environmental absorptive capacity, the pressures will take the form of increases in death rates due to impurities in food, water, and air, decreases in crops and fish catches due to reductions in plant and animal life, and a significant reduction in the effectiveness of investment, due to the high costs of controlling the extreme pollution which will then exist in all input factors. These involuntary pressures will mount until

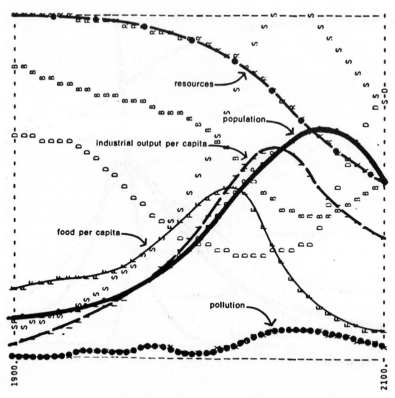

FIGURE 21-10 A further technological improvement is added to the world model to
avoid the resource depletion and pollution problems of previous
model runs. Here we assume that in 1975 pollution generation per
unit of industrial and agricultural output can be reduced to one-
fourth of its 1970 value. Resource policies are the same as those in
Figure 21-9. The changes allow population and industry to grow
until the limit of arable land is reached. Then food *per capita*
declines, and industrial growth is also slowed as capital is diverted to
food production.

population and industrialization finally start to decline, and the pressures will cease only
when the world returns to levels of population and industrialization which are consistent
with the supporting capacity of the physical environment.

   If man attempts to continue growth by removing one kind of pressure—for instance,
by increasing food output with fertilizers and high-yield grains—he alleviates the situation
only until he encounters the next constraint. If he manages to remove that constraint, he
will soon reach another one. World limitations are analogous to a room with infinitely
many ceilings in which man is forever growing taller: one does not *solve* the problem by
removing the first ceiling upon contact, nor by removing the second, third, etc. The only
solution to growth in a finite space is to halt the growth.

THE ETHICAL BASIS FOR ACTION

The Short-Term Objective Function

We face the fact that continuation of current material growth-practices will inevitably lead to some sort of collapse, with a subsequent decrease in the cultural and economic options of the human race. Inasmuch as this is clearly undesirable, one is naturally led to ask: What shall we do?

It is important to realize that an answer to this question is completely dependent on a choice of criteria for what is "good." If we do not know what we want to obtain, if we don't know our "objective function," it is meaningless to try to decide what to do in a given situation. If one's objective is to maximize the benefits of the people alive today, one's actions will be quite different from what they will be if the goal is to maximize the benefits of all people who are going to live on our planet over the next 200 years.

At least in principle (and it is clear that this is far from being realized) present human behavior is guided by the general idea that all people alive *today* are equally important and that man's objective function is to maximize today's total benefits for all of these people. People have decided in the Western democracies that this objective is best served by letting each individual be free to pursue his own interest. It is assumed, very simply, that if every citizen and institution in the society acts to maximize his own position in the short term, society as a whole will benefit.

This acceptance of the "invisible hand" of classical economic theory has, however, introduced a strong emphasis on short-term benefits in present societies and a disregard for the future. When an action will bring both benefits and costs over time, it is considered acceptable business practice to appraise the net *present* value and to discount the *future* implications in determining whether an action is profitable—and hence whether it should be taken. The result of this procedure is that essentially zero value is assigned to anything happening more than twenty years from now. In other words, plans are acted on even though their cost to society twenty years hence will be enormous—just because the benefits are larger than the costs in the short run (for example, over the next decade).

If we continue to adhere strictly to the objective of maximizing the short-term rewards for the present generation, then we disregard longterm environmental trade-offs. We simply continue what we have practiced in the past, maximizing the current benefits and neglecting any possible future costs. The question about use or nonuse of DDT, for example is easily resolved for the present generation. The fact that 1.3 billion people today can live in safety from malaria thanks to DDT outweighs the costs—for instance, in the form of inedible fish—inflicted upon future generations through continued use of the chemical, if the short-term objective is man's only concern.

What about the value of human life in the short term? The short-term objective function supports the currently accepted belief that the value of each additional human being is infinite. The severe restrictions each additional human will impose on the choices and perhaps even on the lives of future generations by virtue of his consumption of nonrenewable natural resources and his contribution to the destruction of the life support system of the earth—this is completely neglected when the short-term objective function is used to judge the value of another human being.

Thus we see that adherence to the short-term objective function avoids very simply all trade-offs of current benefits for future costs. Of course, man is left with the more usual trade-offs that affect people alive today—for instance, the choice between denying the firm upstream freedom to dump waste in the river and denying those who live downstream pure drinking water. But conflicts within the short term are not our concern here, because we *do* have mechanisms in our society to resolve conflicts between people alive today.

We do *not* have, however, mechanisms or even moral guidelines for resolving conflicts between the population of the present and the people of the future. This at the same time the world model indicates that the present preoccupation with what seems pleasant and profitable in the short run fuels the growth which will ultimately cause the world's human population to overshoot some physical constraint, forcing us—but especially our descendants—into a period of abrupt and traumatic change.

The Long-Term Objective Function

It is, however, possible to change the objective function, in the same way that, for Western man, the Judeo-Christian religious tradition turned the objectives of man away from selfish gratification and toward consideration of the welfare of all people living at the same point in time. We could, for example, adopt as our cardinal philosophy the rule that no man or institution in our society may take any action that decreases the economic and social options of those who will live on the planet over the next 100 years. Perhaps only organized religion has the moral force to bring about acceptance of such a rule, but perhaps it could result also from an enlightened, widespread program of public education.

So basically there is only one ethical question in the impending global crisis. Should we continue to let our actions be guided by the short-term objective function, or should we adopt a long-term perspective? In other words, what time horizon should we use when comparing the costs and benefits of current actions?

The moral and ethical leaders of the world's societies should, we believe, adopt the goal of increasing the time horizon that forms the context within which all the activities of mankind are set—that is, they should urge acceptance of the long-term objective function that maximizes the benefits for those living today, subject, of course, to the constraint that it should in no way decrease the economic and social options of those who will inherit this globe, our children and grandchildren. This goal is of course not completely foreign to contemporary society. People in general feel responsibility for the lives of all offspring entering the world within their lifespan, and the long-term objective function seems to be the value implicit in the actions of environmental conservationists. However, ultimately it must be present in *all* of our activities—as it is said to have been for the native tribes of Sierra Leone, where nothing could be done to the jungle which would leave it unfit for the use of *any* future generation.

GLOBAL EQUILIBRIUM: A DESIRABLE POSSIBILITY

A Lasting Solution

Assuming that we accept the long-term objective function as the guideline for our actions, what can we do about the approaching collision between our growing societies and the physical limits of the earth? As soon as we are committed to the creation of a long-term, viable world system, our most important task will be to avoid the trauma caused if we actually exceed any of the earth's physical limitations—food production capability, pollution absorption capacity, or resource supply. This can only be done through a deliberate decision to stop physical growth. We must engineer a smooth transition to nongrowth—a "global equilibrium" or steady state in accord with the earth's physical limits. Our generation must halt the growth by developing and utilizing legal, economic, or religious pressures, social substitutes for those pressures that would otherwise be exerted by nature to halt physical growth.

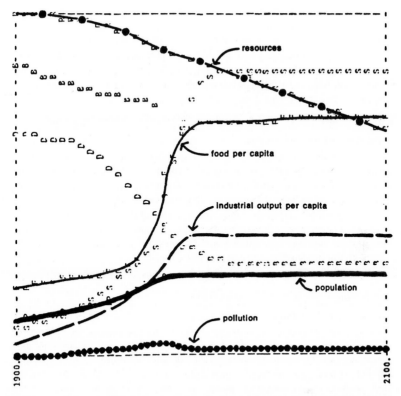

FIGURE 21-11   An equilibrium state has been achieved that is sustainable far into the future. Technological policies include resource recycling, pollution control devices, increased lifetime of all forms of capital, and methods to restore eroded and infertile soil. Value changes include increased emphasis on food and services rather than on industrial production. Births are set equal to deaths and industrial capital investment equal to capital depreciation. Equilibrium value of industrial output *per capita* is three times the 1970 world average.

By starting now we may still be able to *choose* the set of pressures we prefer to employ in stopping population and capital growth. We cannot avoid pressures. As we continue growing, nature will supply counterforces—forces that *will* pressure us until growth stops. However, a deliberate choice to live with those counter-forces which are least objectionable is likely to leave many more of our fundamental, long-term objectives intact than will a refusal to recognize the limits nature can enforce through the blind and random action of such natural forces as starvation or social breakdown.

The first requirement for a viable steady state is a constant level of population and capital—that is, the number of people and physical objects must remain constant. A second requirement is equally important. Since we want to create a system capable of existing for a long time, the state of global equilibrium must be characterized by minimal consumption of nonrenewable materials and by minimal production of nondegradable waste. In this way we can maximize the time before resources are depleted and avoid a critical load on the environment.

A computer simulation of the achievement of equilibrium is depicted in Figure 21-11. Many different possible paths to global equilibrium exist, however, and the choice depends on society's objectives. For instance: should the world support many people at a low material standard of living or a few people at a higher standard? Should the societal objective be exotic, fancy foods or just the basic daily ration of calories, protein, and vitamins?

In an equilibrium phase of human civilization, science and technology will be busy developing ways of constructing products which last very long, do not emit pollution, and can be easily recycled. Competition among individual firms may very well continue, the main difference being that as the total market for material goods will no longer expand, emphasis will shift to repair and maintenance and away from new production.

Although global equilibrium implies nongrowth of all *physical* activities, this need not be the case for cultural activities. Freed from preoccupation with material goods, people may throw their energy into development of the arts and sciences, into the enjoyment of unspoiled nature, and into meaningful interactions with their fellow man, a goal that can be achieved if the production of services flourishes.

The Distribution of Wealth and Responsibility

Stopping the population explosion is becoming more readily accepted as an important task to be accomplished as fast as possible, but what about stopping physical growth? Can the world's wealthy nations really suggest a deliberate restriction of material production, leaving the poor of the world in their present miserable situation?

Striving towards global equilibrium does *not* imply "freezing" the present configuration of the world's rich and poor nations and peoples. What must finally stop is *overall* material growth, but that does not preclude redistribution of the world's existing material wealth. One possibility is for the developed nations deliberately to stop their growth, possibly even to let themselves "shrink" somewhat, while the developing world is allowed (and maybe helped) to grow materially to an acceptable, but not unbounded state. Thus initially it will be the developed world which has to take the lead in the path toward equilibrium; however, the developing world will have serious responsibilities in attempting to stop its rapidly growing populations.

Many people believe that the goal of maximizing economic growth must be clung to simply because we are still so *very* far from having attained the utopia where everything is plentiful for everyone. However, we must remember the conclusion that a continued reliance on short-term objectives and continued physical growth only makes it certain that there will be no acceptable future—for *any* country. In other words: such a utopia does not exist, and striving toward it is futile.

Also it should be made quite clear that material growth as we have experienced it over the last century in *no* way has resulted in increased equality among the world's people. To the contrary, growth in its present form simply widens the gap between the rich and poor, as indicated by the data in Figure 21-12.

An end to overall physical growth, however, might very well ultimately lead to a more equitable distribution of wealth throughout the world—because no one would accept material inequalities in the present under the (false) pretense that they would be removed through future growth. Of course the state of global equilibrium will also have its

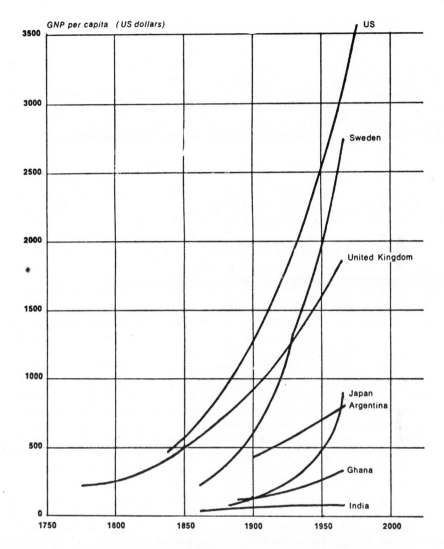

FIGURE 21-12 The economic growth of individual nations indicates that differences in exponential growth rates are widening the economic gap between rich and poor countries. (Source: Simon Kuznets, *Economic Growth of Nations*, Cambridge, Mass.: Harvard University Press, 1971).

problems, mainly political and ethical. As Herman Daly notes in the introduction to this volume:

> For several reasons the important issue of the steady state will be distribution, not production. The problem of relative shares can no longer be avoided by appeals to growth. The argument that everyone should be happy as long as his absolute share of the wealth increases, regardless of his relative share, will no longer be available. . . .

The steady state would make fewer demands on our environmental resources, but much greater demands on our moral resources.

But these political problems have solutions, and society is certainly more likely to find those solutions in an equilibrium state than in a collapsing one.

Stopping the overall physical growth on our planet should not be construed as an attempt by the rich countries to divert attention from economic development to the protection of "their" environment. Rather, global equilibrium is a necessity if mankind wants to have an equitable future on what is altogether a small, fragile planet.

The Golden Age

The presence of global equilibrium could permit the development of an unprecedented golden age for humanity. Freedom from the pressures of providing for ever-increasing numbers of people would make it possible to put substantial effort into the self-realization and development of the individual. Instead of struggling merely to keep people alive, we could employ more of our energy in developing human culture, in increasing the quality of life for the individual far above the present subsistence. The few periods of equilibrium in the past—for example, the 300 years of Japan's classical period—often witnessed a profound flowering of the arts.

The freedom from ever-increasing capital—i.e., from more concrete, cars, dams, and skyscrapers—would make it possible even for our great-grandchildren to enjoy solitude and silence. The desirable aspects of the steady state were first realized more than a century ago. John Stuart Mill wrote in 1857:

It is scarcely necessary to remark that a stationary condition of capital and population implies no stationary state of human improvement. There would be as much scope as ever for all kinds of mental culture, and moral and social progress; as much room for improving the Art of Living and much more likelihood of its being improved, when minds cease to be engrossed by the art of getting on. Even the industrial arts might be as earnestly and as successfully cultivated, with this sole difference, that instead of serving no purpose but the increase of wealth, industrial improvements would produce their legitimate effect, that of abridging labor.

This, then, is the state of global equilibrium, which seems to be the logical consequence of the adoption of the long-term objective function. The changes needed during the transition from growth to global equilibrium are tremendous, and the time is very short. But the results seem worth striving for, and the first step must be to increase our time horizon—to accept the long-term objective function.

In a recent book, *Models of Doom*, a University of Sussex, England team criticizes the *World Dynamics* and the *Limits to Growth* model. In this book Professor Harvey Simmons notes that the conclusions resulting from using the model are not surprising, but rather must lead us to reexamine our political processes, priorities and values. Excerpts from this article follow:[10]*

*Reprinted from "System Dynamics and Technocracy" by Harvey Simmons. From *Models of Doom: A Critique of the Limits to Growth* Edited by H.S.D. Cole, Christopher Freeman, Marie Jahoda, and K.L.R. Pavitt. Universe Books, 1973.

If we turn to the world dynamics model, much the same point can be made. If one starts, as Forrester does, with fixed assumptions about exponential growth, it is apparent that the world is headed for imminent disaster from overcrowding, pollution or exhaustion of natural resources. Moreover, the notion that the only way to stabilise the system is to limit industrialisation, given the premise that pollution and exhaustion of resources are both made directly dependent on industrialisation, is not surprising. In fact there is nothing at all surprising in Forrester's conclusions given his assumptions. The model has only to be stood on its head for the solution to appear.

The difficulties inevitably encountered in trying to analyse and then design policies for real complex social systems play a large role in Forrester's books. He spends some time describing what he considers to be the characteristics of complex social systems so that in the end we have a picture of their apparently innate qualities. Indeed, after a while, they seem to take on not only certain characteristic qualities but even a personality. Forrester, at one point, refers to the "traps" set by the character of social systems for "intuitively obvious solutions to social problems". These are: "first, an attempt to relieve one set of symptoms may only create a new mode of system behaviour that also has unpleasant consequences. Second, the attempt to produce short-term improvement often sets the stage for long-term degradation. Third, the local goals of a part of a system often conflict with the objectives of the larger system. Fourth, people are often led to intervene at points in a system where little leverage exists and where effort and money have but light effect". But could it not be misleading to speak of "traps" set by social systems as if they were sentient beings motivated by evil considerations? Forrester does speak of the "diabolical" nature of social systems, but this tendency to reify social systems could hide the fact that it is not necessarily the system which is tricking the observer but rather the observer who is either making mistakes or is having difficulty in trying to deal with a very complex phenomenon. Most of the "traps" mentioned by Forrester occur not because of the peculiar nature of systems but because our political values and preferences structure the way in which we deal with the system. Given different political values, these "traps" might disappear. The problem is not only a theoretical one of "system dynamics" but more a practical one of politics.

• • •

One of the more interesting aspects of the Forrester/Meadows argument is its popularity. Surely one might have expected people to be sceptical about such an apocalyptic thesis and to have asked for a much more serious examination of the data and the premises on which the model is built before reaching any conclusions about its soundness. Yet this has not been the case. Why is it that a group of computer engineers and systems analysts who have frankly admitted the tentative nature of their findings, have found such a ready audience? ("It (the world model) perhaps contains the tenth part of 1% of what we should know about the real world if we want to make long-term statements ... ") There are a number of reasons for this, some of them may be more tentative than others, but all of them may tell us something about the intellectual mood of our era.

For some time now, especially in the USA, but increasingly in Western Europe, there has grown the belief that we are in the midst of a profound environmental crisis. People are now concerned about issues that, a few years ago, would scarcely have provoked a raised eyebrow except from the most concerned citizens. We have already seen that there have been earlier periods of "environmental pessimism". This new pessimism, however, is connected with the fact that people are sceptical of the idea that the Western capitalist cornucopia will eternally overflow with wealth that must eventually trickle down to even the most deprived elements of the population. Many people are firmly convinced that, on the contrary, the overflow will eventually clog the sewers and pollute lakes and rivers. In a recent lecture, Robert Heilbroner has illuminated this change in attitude with a particularly apt analogy. Suppose, he says, that if in 1930 "someone had been told that the United States GNP in the 1970's would surpass a trillion dollars effectively doubling the real *per capita* income within the life span of the majority of the population then alive—I am sure he would have felt safe in predicting an era of unprecedented social peace and good will, particularly if he had also been informed that the distribution of income in 1970 would not be significantly less unequal than in the 1930's ... Yet ... social harmony has not resulted".

This disillusion with the healing properties of capitalism has combined with a generalised and widespread feeling of despair at the apparent breakdown of certain societal values. Concern over rising crime rates, decline in the environment, changing family structure and relationships have all contributed to a general feeling of unease. It is as if society was on the verge of, or perhaps even in the first phase of, some kind of vast social or cultural revolution whose exact nature has not yet been understood but of which people see premonitions all around them. Strangely enough, these feelings of pessimism and despair are felt even in societies where it is probably easier to argue that, in fact, things have been getting better and better all the time. For everyone who criticises the cost to the environment of industrial pollution, there is someone else who will use the rising standard of living in the west as proof that for certain costs, i.e., pollution, there are even greater benefits. What we are witnessing then, and what the widespread acceptance of the pessimistic argument indicates, is a gradual change in our cultural attitudes.

Values are now changing and the notion that quality of life (however vaguely defined) is more important than such things as increased material goods is the cause of the unresolvable argument between those whose hierarchies rank cars and clean air at different levels of goodness. The pessimistic ecologist's case fits well with this change in outlook since it confirms the suspicion so many people have of the negative effects of unrestrained growth. Moreover, unlike past arguments for caution in utilising environmental resources, the pessimists argue that the result of carelessness will not be merely a loss of certain amenities in life but disaster on a global scale.

There is another apparent reason for the widespread acceptance of the pessimist's case, and this has to do with the anti-political overtones of its message. We have seen how Forrester in particular expresses pessimism about the ability of politicians to solve the major problems posed by the world dynamics model. And although Meadows is, characteristically, more cautious than Forrester, he is also sceptical of political systems which are unable, supposedly for electoral reasons, to plan for more than five years ahead.

Once again, it should be emphasised that it is not necessarily the political process that is irrational; rather, politicians may be responsive to an electorate and to interest groups that do not share the values of its critics. The irrationality that certain industrialists or scientists in particular sometimes perceive as inherent in the process of policy-making is sometimes nothing more than their unexpressed dissatisfaction with the goals and values of the polity at a given moment of time. Given a certain amount of consensus, however, there is no reason why even the most democratic political system could not formulate long-term measures extending from 5 to 50 years—and indeed, many of the decisions being made now, to construct atomic power plants, to construct dams, to build nursery schools, or to use natural gas as opposed to oil—all have longterm implications. Thus, the suspicion the pessimists and their supporters have of the political process is based to a great extent on disagreement over what should be the primary goals pursued by the political system. After all, not everyone believes that industrial growth should be stopped. What evidence is available suggests that, although they may well be concerned to devote greater resources to combat some types of pollution, the great majority of people in all countries do not share this view. It is natural, however, that the discontent many people feel about the apparent unwillingness of politicians to press for limitations on pollution, etc, fits in well with the abstract nature of the system dynamics approach which attributes to the political system itself defects which are integral to any situation, where conflict over values must be resolved by reaching political concensus. Moreover, belief that the world dynamics model is somehow above politics lends to it a spurious aura of clarity and precision which, in turn, leads to the conclusion that it is only the obdurate and inexplicable irrationality of politicians that prevents men of good will from rapidly developing a programme to achieve the ends so clearly indicated by the world model.

Having abjured any particular political or ideological viewpoint, carefully dissociating themselves from politicians (despite the rather important positions some of its members hold in semi-public bureaucracies) it is quite clear how anxious they are to appear to be above politics. They are open to the most radical kinds of proposals as long as the techniques used to arrive at them are well within contemporary canons of what constitutes an acceptable scientific-technological approach, and they are supporting a number of research programmes of model-building outside MIT and the USA. It is precisely because the world dynamics model says nothing about the means necessary to achieve the stable system that seems indicated by its conclusions, that it appeals to the "intellectual technologists" of the Club of Rome as well as to ordinary citizens. That this apolitical attitude is also dictated by their desire to appeal to decisionmakers from diverse political systems and holding sometimes contradictory values does not vitiate the fact that, as have been noted above, certain political choices are logically inherent in the world dynamies model and in the notion of reaching a stable state.

There is a further reason why the MIT model may have such an immediate and intense appeal for certain elements of the population in the west. As we have seen, those who support the notion of a limit to growth and who share the model's pessimism almost always articulate a pessimistic view about the current state of society. Now it is generally agreed that one of the major causes of civil and criminal conflict today is that some people feel society has not given them their due, that financial, cultural, social or prestige rewards have not been allocated in a just fashion. The almost universal belief in progress implied that in one way or another everyone would benefit from the increased productivity of advanced capitalist society. But even though in absolute terms the standard of living has risen for almost everybody in the west, conflict today often arises because deprived groups are irritated by the gap between what they think they ought to have and what they actually receive. Moreover, expectations of what is their due are constantly raised by the vast array of goods that are produced in capitalist societies.

In the stable state, however, all this would change and people would have to adjust to a system where growth was limited and the infinite accumulation of goods would finally end. What kind of society would this be? A world-wide radical egalitarian levelling of incomes and property could be taken as a necessary implication of a stationary state and Meadows himself sometimes appears to adopt this view.

· · ·

It is now apparent that the vision offered by some of the ecodoomsters is not at all one of world disaster and conflict. Despite the moralistic tone injected into Meadows' book by the quotation from Dr. Daly, the real thrust of the message is clear. Once the stable state has been achieved all the anxieties, frustrations and conflict engendered by the frantic endeavour to accumulate goods will have disappeared. Man will be able to realise the spiritual side of his nature. Rivers and lakes will be clean, population controlled, urban problems resolved. The golden age would begin.

In common with other chiliasts, the new scientific chiliasts are utopians at heart. Like the great prophet of world salvation through world breakdown, Karl Marx, their apocalyptic visions of the immediate future are tempered by the glittering image of utopia barely discernible through the fire and brimstone that rages in the historical foreground. This is not to denigrate the beliefs of the Forrester/Meadows school in any sense; rather, it is to suggest that they too, despite the surface appearance of scientific neutrality and objectivity, bring us a message which can only be fully understood in the context of their own beliefs, values, assumptions and goals.

Professor Dennis Meadows and his colleagues respond to Dr. Simmons's critique and those that accompany it in the book *Models of Doom*. In the following three paragraphs, Professor Meadows succinctly summarizes his view.[9] The question remains a difference of opinion between what may be called the technological optimists and those who, with Malthus, fear we have strained our planet beyond its potential to accommocate its burdens. Professor Meadows's summary follows.*

Let us go on from this false analysis of a misunderstood difference to the real difference in "concept of man" that seems to be dividing the world into camps of "optimists" and "pessimists." One possible concept of man, the one that is held by the Sussex group, is that *Homo sapians* is a very special creature whose unique brain gives him not only the capability but the right to exploit for his own short-term purposes all other creatures and all resources the world has to offer. This is an age-old concept of man, one firmly rooted in Judeo-Christian tradition and newly strengthened by stunning technical achievements in the last few centuries.

Not only ingenuity but, increasingly, understanding not luck but systematic investigation, are turning the tables on nature, making her subservient to man.

According to this belief, man is essentially omnipotent, he can develop at no cost a technology or a social change to overcome any obstacle, and such developments will occur instantly upon the perception of the obstacle. Mankind's social, economic, political, and technical institutions operate flexibly and without error, and the best response to any apparent problem is to encourage these institutions to do more of whatever they have done in the past.

The opposite concept of man is also an ancient one, but it is more closely related to the Eastern religions than to the Western ones. It assumes that man is one species with all other species embedded in the intricate web of natural processes that sustains and

*Reprinted from "A Response to Sussex" by D.H. Meadows et al from *Models of Doom: A Critique of the Limits to Growth*, edited by H.S.D. Cole, Christopter Freeman, Marie Johoda, and K.L.R. Pavitt, Universe Books, 1973.

constrains all forms of life. It acknowledges that man is one of the more successful species, in terms of competitiveness, but that his very success is leading him to destroy and simplify the natural sustaining web, about which he understands very little. Subscribers to this view feel that human institutions are ponderous and short-sighted, adaptive only after very long delays, and likely to attack complex issues with simplistic and self-centered solutions. They would also point out that much of human technology and "progress" has been attained only at the expense of natural beauty, human dignity, and social integrity, and that those who have suffered the greatest loss of these amenities have also had the least benefit from the economic "progress." People who share this concept of man, as we do, would also question strongly whether technology and material growth, which seem to have caused many problems, should be looked to as the sources of solution of these same problems in the future. Technological optimists invariably label this view of the fallibility of man as "pessimistic"; Malthusians would simply call it "humble."

We see no objective way of resolving these very different views of man and his role in the world. It seems to be possible for either side to look at the same world and find support for its view. Technological optimists see only rising life expectancies, more comfortable lives, the advance of human knowledge, and improved wheat strains. Malthusians see only rising populations, destruction of the land, extinct species, urban deterioration, and increasing gaps between the rich and the poor. They would say that Malthus was correct both in his own time and today in his observation that

the pressure arising from the difficulty of procuring subsistence is not to be considered as a remote one which will be felt only when the earth refuses to produce any more, but as one which actually exists at present over the greatest part of the globe.

In some excellent and meaningful way, while the *World Dyamics* and *Limits to Growth* models may not be definitive, they are important. For they represent modern, computer-oriented man's attempt to foresee his future and to control it. Man's beliefs, values and goals inevitably lead to his model of the world and his political solution to yield his goals.

Whether man can control his technology for his benefit and avoid an ultimate "collapse" is unanswered as yet. Nevertheless, the debate between the technological optimist and the Malthusian pessimist continues. A computer model is a modern form of the issue. It may well be an improved way to confront the issue because it is duplicable and can be tested by many parties to the dispute.

Very likely, the growth dilemma will be resolved by a number of significant modifications in the present system. For example, while technology will continue to produce new products to replace the old, the new products must not be or make a threat to the environment. The recycling of products and the wiser use of energy must follow soon. Technology can have beneficial effects on energy use if solar energy is used more for heating and cooling. Transportation problems and air pollution can be ameliorated with wise technology—if there is time. The advocates of zero-growth question whether there is sufficient time or knowledge to solve our problems with technology. As the authors of *Limits to Growth* point out, it would take 70 years for the population of the United States to stabilize even if all Americans began immediately to limit their families to the so-called replacement rate (average) of 2.1 children per couple. Thus it is clear we must choose our available options early and wisely. Computer models will be one of the aids we will need to make an early, judicious and acceptable choice.

CHAPTER 21 REFERENCES

1. R.C. Dorf, *Introduction to Computers and Computer Science*, Boyd and Fraser Publishing Company, San Francisco, 1972, Chapter 13.
2. J.W. Forrester, *World Dynamics*, Wright-Allen Publishing Company, Cambridge, Mass., 1971.
3. R. Boyd, "World Dynamics: A Note," *Science*, August 11, 1972, pp. 516–519.
4. D.H. Meadows, D.L. Meadows, J. Randers and W.W. Behrens, *The Limits to Growth*, Universe Books, New York, 1972.
5. T.J. Boyle, "Hope for the Technological Solution, *Nature*, Sept. 21, 1973, pp. 127–128.
6. P. Passell and L. Ross, *The Retreat from Riches*, Viking Press, New York, ʾ1973.
7. R.E. Lapp, *The Logarithmic Century*, Prentice-Hall, Inc., Englewood Cliffs, N.J., 1973.
8. T. Harris, "Californians Are Saying 'No' to Growth In a Spreading Revolt that Makes Strange Allies," *California Journal*, July 1973, pp. 224–229.
9. J. Randers and D. Meadows, "The Carrying Capacity of Our Global Environment: A Look at the Ethical Alternatives," *Sloan Management Review*, Winter 1972. (Also in *Toward a Steady-State Economy*, H.E. Daly, Editor, Freeman, 1973.)
10. H.S.D. Cole *et. al. Models of Doom*, Universe Books, New York, 1973.
11. C. Starr and R. Rudman, "Parameters of Technological Growth," *Science*, Oct. 26, 1973, pp. 358–364.
12. R.G. Ridker, "To Grow or Not to Grow: That's Not the Relevant Question," *Science*, Dec. 28, 1973, pp. 1315–1318.
13. H.E. Daly, *Toward a Steady State Economy*, W.H. Freeman and Co., San Francisco, 1973.
14. K.E.F. Watt, *The Titanic Effect*, Sinaver Associates, Inc., Publishers, Stamford, Connecticut, 1974.

CHAPTER 21 EXERCISES

21-1. Many cities and towns are establishing limits for their population by means of limiting new construction of homes and apartments. Do you believe it is wise and fair for a town to do so? Discuss the effects on the nearby towns and cities.

21-2. There is ultimately a limit to the total population of the earth. What other factors of the quality of life are ultimately limited?

21-3. If the effect of technology is incorporated in a model of the world, do you believe there are no limits to growth? Can technology overcome pollution and provide new substitutes for resources as they are exhausted?

21-4. Professor Simmons states in his article that the "traps" mentioned by Forrester and Meadows may disappear if the political values of our system are altered. Consider the effect of our value systems on the limits to growth.

21-5. The key effect of technological growth on the limits to growth of pollution, resources, capital and population is major. The parameters of technological growth have been identified as societal resources and societal expectations.[11] Technological growth itself usually follows an exponential path. For example, the performance curve for digital computers is shown in Figure E21-5. This exponential growth of technology fits the view of the technological optimist. Discuss the technological growth of the United States and Europe and compare it with technological growth in less developed nations. Is technology sufficient to overcome an exponential growth of the population of a nation? What about the ultimate limits of overcrowding? Can technology overcome this problem?

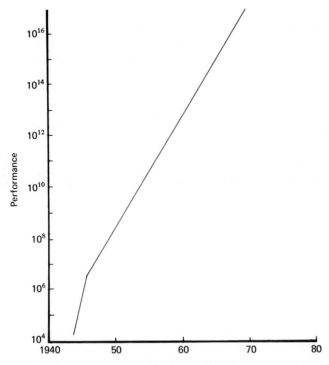

FIGURE E21-5 Computer performance from 1944 to 1973. Performance is defined as the ratio of capacity in bits to add time in seconds.

22

TECHNOLOGY, THE STATE AND THE MILITARY

TECHNOLOGY AND PUBLIC POLICY are closedly related phenomena. Since the early civilizations, the nations have looked to technology to serve their needs, both social and military. The great pyramids were the result of governmental policy decisions to satisfy the pharaohs' desires. Classical civilizations emphasized public works, military technology, transportation and communication. The Greek and Roman governments built aqueducts to furnish good water for drinking and bathing. In Renaissance Italy, Leonardo da Vinci was one of many great men who worked as a military and civil engineer for the rulers of Milan and Romagna.

The application of technology to the needs of the nation vary from agriculture in the land-grant colleges to research and development at the large industrial firms. The National Aeronautics and Space Administration (NASA) is an example of a large Federal agency with responsibility for technological developments. Other United States Federal agencies with responsibility for technology are the Bureau of Standards, the National Science Foundation and the Environmental Protection Agency.

APPLIED TECHNOLOGY IN HISTORY

The role of the engineer in the development of military weapons and equipment has been extensive since the times of the Greeks and Romans. The first extensive military engineering developments date from the time of the ascendance of Alexandria around 300B.C. Alexander the Great encouraged the development of mechanized warfare. Improvements of the catapults and ballistics for hurling missiles was accomplished.

In the eighth century the stirrup used for riding horses reached the Frankish empire. Earlier a horseman had to cling to his steed by pressure of his knees and cavalry was used primarily for bowmen. With the addition of stirrups, the rider could wield a lance and shield while maintaining himself on the horse. When combined with a saddle, stirrups made a single organism of horse and rider and revolutionized eighth century battle.[1] Mounted warriors became the basis on which Charlemagne enlarged the Frankish empire at the beginning of the ninth century.

Modern warfare developed with the industrial revolution. The techniques for drilling gun barrels were used for boring steel for pistons and cylinders for the steam engine. Military technology moved into its modern form following the French Revolution in 1789. Conscription, economic mobilization, propaganda and internal security measures transformed France after 1793. The education of military engineers was initiated at the École Polytechnique in France in 1794 and at the United States Military Academy at West Point a short time later.

Some of the technical achievements of the nineteenth century were in the improvements of small firearms. During this period the percussion cap and the breech-loading rifle were invented. Rifling in the muzzle of firearms improved the accuracy of the weapon. Rapid-fire guns such as the Gatling gun were invented during the last half of the ninteenth century. Also, artillery weapons using breech loading and rifled barrels were used extensively during the late ninteenth century.

Wide-ranging improvements in weapons, transport and communications coincided with the development of mass-production techniques. Thus, with mass-produced military equipment and the mobilization of a nation, mankind moved into the present century prepared for total warfare—warfare based on the machine.

World War I incorporated the efforts of engineers and scientists in the war effort on both sides. For example, the United States Naval Research Laboratory employed civilian engineers and scientists in problems of the Navy. The Germans used submarine warfare to attack the United States and British shipping and Navy effectively. Big Bertha, an enormous gun, shelled Paris from 76 miles away, and many other technological accomplishments were used by the combatants.

MILITARY TECHNOLOGY SINCE 1920

Annual appropriations for military research were about $4 million a year during the 1920s. In 1935, this was raised to $9 million to permit new research into aircraft. As war broke out in Europe in 1939, the United States formed the

National Defense Research Committee with Vannevar Bush, Frank Jewett, James Conant (President of Harvard University) and Karl Compton as members. With Bush providing extensive leadership, contracts were established with universities and industrial firms to carry out military research and development. By 1941, the Office of Scientific Research and Development was founded with Bush as its director. OSRD marshalled the engineers and scientists of the United States in the service of the military and in many ways this established the present tradition of industry and the universities serving the military needs of the nation. Altogether, more than $3 billion passed through the OSRD and the work accomplished was respected by those in the Congress and the Executive branch. This close relationship between the Federal government and American technology was established during the years of World War II. In addition to OSRD, the Naval Research Laboratory, the Army Signal Corps and other agencies of the Federal government distinguished themselves in technology during this period. The contract for research and development with universities and industrial firms became an established mode of military development. Following World War II, the United States government established the Office of Naval Research and the Atomic Energy Commission to carry on the development of weapons for the country. The development of nuclear weapons, initiated during the early 1940s, was continued by the A.E.C., resulting in the hydrogen bomb and other devices. World War II brought into use the modern airplane, tanks, submarines and the ballistic missile in the form of the German U-2.

The merger of the ballistic missile and the atomic bomb has brought a cold war based on poised, massive retalitory strike forces. The technology of warfare has led mankind to the ability to annihilate his enemy with long-range ballistic missiles. This balance of terror against the hope for arms control of nuclear weapons is a product of the atomic age and modern technology. In many ways, the consequences of technology are immense in the application of military technology.

The grave drawbacks of a strictly technological approach to national security are clear to those who have lived for the past 20 years in a balance of terror. Technology has given us the power to exterminate ourselves. It will be fitting (if it is indeed possible) if technology, through arms-control surveillance, can provide a safe outlet for an excessive military potential.

The defense budget of the United States remained below $1 billion until the late 1930s. However, the military budgets grew during World War II and remained high during the 1950s and 1960s in order to meet the challenge of the cold war. In 1969, the United States had a military budget of $78 billion. However, Richard Kaufman estimates that the true cost of military expenditures in 1969 were 42% of all Federal spending.[2] In this calculation he includes military assistance funds, veterans's benefits, the A.E.C. and that portion of the interest on the national debt attributable to past military spending. Yet one must also account for all the man-years of work, the land space consumed by military bases, and the resources used by the military, to say nothing of human resources and lives. The United States military holds a land area of about 30 million acres valued at $40 billion. A modern fighter plane costs over $10 million each and a like amount to maintain over its life. In 1968 and 1969 the amount the Massachusetts Institute of Technology received military funds for research and development equal to $119 million, representing about half of M.I.T.'s total budget.

In 1961 the defense budget was $48.1 billion compared to over $90 billion today. In a farewell address by President Dwight Eisenhower, delivered in Janurary 1961, the President warned of a military-industrial complex. President Eisenhower said:

A vital element in keeping the peace is our military establishment. Our arms must be mighty, ready for instant action, so that no potential aggressor may be tempted to risk his own destruction.

Our military organization today bears little relation to that known by any of my predecessors in peacetime, or indeed by the fighting men of World War II or Korea.

Until the latest of our world conflicts, the United States had no armaments industry. American makers of plowshares could, with time and as required, make swords as well. But now we can no longer risk emergency improvisation of national defense; we have been compelled to create a permanent armaments industry of vast proportions. Added to this, three and a half million men and women are directly engaged in the defense establishment. We annually spend on military security more than the net income of all United States corporations.

This conjunction of an immense military establishment and a large arms industry is new in the American experience. The total influence-economic, political, even spiritual-is felt in every city, every State house, every office of the Federal Government. We recognize the imperative need for this development. Yet we must not fail to comprehend its grave implications. Our toil, resources and livelihood are all involved; so is the very structure of our society.

In the councils of government, we must guard against the acquisition of unwarranted influence, whether sought or unsought, by the military-industrial complex. The potential for the disastrous rise of misplaced power exists and will persist.

We must never let the weight of this combination endanger our liberties or democratic processes. We should take nothing for granted. Only an alert and knowledgeable citizenry can compel the proper meshing of the huge industrial and military machinery of defense with our peaceful methods and goals, so that security and liberty may prosper together.

TECHNOLOGY AND THE STATE

The newer applications of technology to problems of the cities and states are often the responsibilities of state and local governments. If the Federal investigation of such problems as pollution, traffic congestion and housing construction is to be useful, the ability of the state and local governments to apply new solutions involving new technology is of critical importance. Yet public policy for science and technology at the state and local level is often inadequate. Many states have, however, recently developed Science and Technology Advisory Boards and new agencies to utilize and control the new technologies available to them.

The professional societies for engineers will, in the future, join further in the effort to develop public policy regarding technology and its effects. Examples of influential societies are the National Academy of Engineering and the National Society of Professional Engineers. Eight engineering professional societies, representing 440,000 scientists and engineers, have opened offices in the District of Columbia. They include The Institute of Electrical and Electronics Engineers (IEEE), The American Society of Civil Engineers (ASCE), The American Institute of Aeronautics and Astronautics (AIAA), The American Society of Mechanical Engineers (ASME), The American Society of Engineering Education (ASEE), The National Society of Professional Engineering (NSPE) and The American Chemical Society (ACS).

These societies can contribute to a wise public policy by attempting to inform the public about new developments and by completing objective studies of alternative technologies.

INDIVIDUAL ENGINEERS AND THE STATE

The involvement of the individual engineer in public policy is also of great importance. Matters such as air pollution, transportation and energy production are technical in origin and yet critical to the social wellbeing of a nation. The engineer, while attempting to excel in his engineering efforts, must also attempt to be a person in public affairs. An engineer is one who best utilizes, with a definite purpose, the scientific, technical and material resources of his nation and the world. The analyses of the engineer must include the economic, social and political effects of alternative developments if they are to constitute a complete approach. In the best sense, good engineering leads to good public policy. There is no public policy that is entirely technical, nor is there any that is purely political. Good public policy incorporates all the aspects of the sciences, engineering and politics.

Since the time of Bacon, leaders of science and engineering have considered themselves to be in the vanguard of political progress. Currently, the assumption that science and engineering contribute to the human welfare through its Federal and state institutions is under question. Technological alternatives can no longer be separated from the potential for undesirable consequences. Engineers must play a rational and effective role in politics and adopt a strategy that is more modest in its hopes for the perfectability of mankind and the inevitability of progress.[3] In the belief that technology does not shape public policy, engineers will then be free to devote their own analytic and synthetic skills to the formulation and criticism of policies which may control technology in the public interest.

CHAPTER 22 REFERENCES

1. L. White, *Medieval Technology and Social Change*, Oxford University Press, New York, 1966.
2. R.F. Kaufman, *The War Profiteers*, Bobb-Merrill, New York, 1971.
3. D.K. Price, "Purists and Politicians," *Science*, Jan. 3, 1969, pp. 25–31.
4. L. Mumford, *The Myth of the Machine: The Pentagon of Power*, Harcourt, Brace, Javanovich, Inc., New York, 1970.
5. R. Gilette, "Smart Bombs: Air Warfare Undergoes a Reluctant Revolution," *Science*, June 9, 1972, pp. 1108–1109.
6. M.L. Harvey, L. Goure and V. Prokofieff, *Science and Technology as an Instrument of Soviet Policy*, University of Miami Press, Miani, 1973.
7. R. Filette, "Military R and D: Hard Lessons of Electronic War," *Science*, Nov. 9, 1973.
8. A. Chayes and J.B. Wiesner, *ABM: An Evaluation of the Design to Deploy an Antiballistic Missile System*, Harper and Row, New York, 1969.
9. H. York, *Race to Oblivion*, Simon and Schuster, New York, 1970.

CHAPTER 22 EXERCISES

22–1. Technology transfer is the process by which a technology is applied to a purpose other than the one for which it was intended. Many technologies developed for military needs have been transferred to the civilian sector. One example is the communications sattelite developed as a result of military communications needs and transferred to civilian needs at a later date. What other technological developments were originally developed for the military and have resulted in new or improved civilian products?

22–2. By the end of World War II technology had completely changed the nature of warfare. Previously, many leaders thought war was a cruel and destructive, but rational, tool of international policy. The dropping of the first atomic bomb on Hiroshima in 1945 made it clear that major war can no longer be considered normal or rational. Total war had become a war without a victor or vanquished. What is the role of the military in a world living on the edge of technological annihilation?

22-3. In his recent book *The Myth of the Machine*, Lewis Mumford states, "Though the power system can be adequately represented by abstractions, the concrete form of the Pentagon in Washington serves even better than its Soviet counterpart, the Kremlin, as a symbol of the absurdity of totalitarian absolutism."[4] What is the size of the Pentagon and how many people work there? Have the military machines of the superpowers gorwn beyond man's ability to comprehend and control them?

22-4. The use of modern technologies makes the actions of individuals and organizations affect the well-being of others to a greater degree than in the past. Hence there is greater need for the ᴿegulation and control of private activities by government agencies. Also, technology-based affluence has increased social interdependence in favor of collective goods and services. The provision of public goods and services and the regulation of private activities are made in the public sector. Discuss the rise in number and size of the Federal and state government agencies.

Has modern technology led to greater and more extensive government?

22-5. The United States Air Force used laser-guided bombs during the last years of the Vietman war. The laser-guided bombs are called "smart bombs" since they work with a guidance system in contrast to regular bombs. These smart bombs permit a high degree of accuracy and are capable of hitting trucks and other small targets.[5] A small kit of equipment, costing about $3,500, is installed on a regular bomb. This involves mounting a laser light sensor on the bomb's nose and some movable steering vanes on its body. An attacking aircraft aims a laser beam at the intended target and the bomb homes in on the target. The laswer-guided bomb is a current example of technological warfare. The Polaris submarine and flying radar stations are other examples.

How can a nation avoid continually striving to maintain its place in the military technology race? What is the cost to most modern nations for new military equipment as a percentage of their annual GNP?

22-6. In the United States more than one thousand billion dollars has been devoted to military expenditures since World War II. The nations of the world presently spend about $200 billion a year on war or preparations for wars. Among the poorest two-thirds of the world's people, military expenditures exceed expenditures for all forms of education and health services. Coupled with military technology is the technology of communication and its offshoot propaganda. Examine ways in which the communications media in our society had influenced and is influencing the citizens' attitudes toward Federal expenditures on military technology.

22-7. President Eisenhower's Farewell Address of January 1961 pointed out the danger of public policy becoming a captive of a technological elite. A portion of his speech follows:

Akin to, and largely responsible for the sweeping changes in our industrial-military posture, has been the technological revolution during recent decades.

In this revolution, research has become central; it also becomes more formalized, complex, and costly. A steadily increasing share is conducted for, by, or at the direction of, the Federal government.

Today, the solitary inventor, tinkering in his shop, has been overshadowed by task forces of scientists in laboratories and testing fields. In the same fashion, the free university, historically the fountainhead of free ideas and scientific discovery, has experienced a revolution in the conduct of research. Partly because of the huge costs involved, a government contract becomes virtually a substitute for intellectual curiosity. For every old blackboard there are now hundreds of new electronic computers.

The prospect of domination of the nation's scholars by Federal employment, project allocations, and the power of money is ever present-and is gravely to be regarded.

Yet, in holding scientific research and discovery in respect, as we should, we must also be alert to the equal and opposite danger that public policy could itself become the captive of a scientific-technological elite.

It is the task of statesmanship to mold, to balance, and to integrate these and other forces, new and old, within the principles of our democratic system-ever aiming toward the supreme goals of our free society.

Has public policy become a captive of a technological elite during the past ten years? What safeguards would you suggest?

22–8. Science and technology have been seen as instruments of the cold war between the super-powers. In a recent book it is stated, "Science and technology are not merely instruments for healthy, peaceful competition and cooperation, at best designed to demonstrate the superiority of Soviet social-economic organization, but (are) essential means for enabling the Soviet Union to achieve supremacy on a global scale and gain its ultimate objective, i.e., the destruction of capitalism."[6]

In this book the Soviet Union is seen using technology primarily for military purposes. Also, the authors state that 80% of all research and development expenditures are for military purposes. Examine this viewpoint and attempt to determine what role technology plays in the military stance of Russia.

22–9. The war in the middle east in the fall of 1973 was a proving ground for Soviet and American military technology. By the time of the October 22nd truce, something on the order of $4 billion worth of military technological equipment lay wrecked on both sides of the Suez canal and along the Golan Heights.[7] The extensive use of Soviet missiles by Egypt and Syria demonstrated their effectiveness against fighter planes and tanks. The Soviet SAM (surface-to-air missile) changed the future of the military tank. The Israelis are said to have lost 500 tanks to SAM's. The United States is currently developing new electronic countermeasures for SAM's and also developing a new tank (which has cost over $2 billion to develop so far) to be ready by the end of 1977. What are the new paths for military technologies, and is there any possibility that nations may reduce their investment in military research and development?

22–10. During the past several years there has been a great debate in the United States concerning the necessity for developing an antiballastic missile system (ABM).[8,9] The ABM system, as proposed, is a highly sophisticated system of radars, computers and missiles with the mission of destroying attacking ballistic missiles approaching the United States. Evaluate the technology of such a system and its contribution to the arms race. Examine the cost of such a system and estimate its benefits. Do you support the deployment of an ABM system?

23

TECHNOLOGY AND THE FUTURE

TECHNOLOGY HAS HAD A MASSIVE INFLUENCE on modern society and will continue to influence society over the coming decades. During the past decade, a great debate has ensued concerning the possible catastrophes emanating from new technologies and the control of technology. The possible catastrophes —nuclear or biological warfare, ecological disaster, population collapse, or massive outbreaks of famine—seem to be more distant or capable of being controlled. Nevertheless, as a result of the debate, a decade of Earth Days, a Stockholm Conference on the World Environment and anti-Vietnam peace marches, one may surmise that science and technology have undergone a transformation during this period. The criticisms and attacks on technology during the last decade have undermined the naive faith many formerly held in the complete efficacy of technology. The naive analysis that even if most of our problems stem from technology (which is an exaggeration of the first order), then only more technology will ever cure them, has proven too frail a view even for the ardent technocrat.

On the other hand, blanket rejection of all things scientific and technological is equally indefensible today. One new view emerges: that the new way of life, after the energy crisis and technological crisis of the 1970s, will involve a new technology, different from the narrower technology of the past decade. Many sources and indications suggest a new technology incorporating or facilitating the values of democracy, compassion and humanity. The new climate is perhaps best evidenced by the literature and the words used, such as alternative technology, people's technology, ecological technology and "soft" technology.[1] Perhaps the critics, in league with the engineers, are transforming the old technologies slowly along new directions toward new forms of progress. To generalize, this movement strives to put men before machines, people before government, practice before theory, smallness before bigness, organic materials before synthetic ones, craftsmanship before expertise and quality before quantity. We find emerging a new technology designed not so much to dominate nature as to mesh with it.

ANTICIPATION

One of the features of the new technology is how it deals with possible side effects of heretofore unforeseen social consequences. Previously, the tendency was to estimate the negative effects and to determine whether they were smaller or larger than the positive effects. We used only such concepts as safe average levels, or acceptable general levels of radiation, or ecological damage. In the future we shall include the effects on individuals and communities as well as the effects averaged over the state or nation.

The call for a new or alternative technology is based on the thesis that it is the form of contemporary technology which is primarily at fault, and not the existence of technology itself. The view held is that a new technology must be developed so that it is not inherently polluting. For example, there is need to develop the use of solar and wind power rather than to continue to increase the use of fossil fuels. Alternative technologies should be non-polluting, cheap and labor-intensive, non-exploitive of natural resources, compatible with local cultures, functional in a non-centralist sense and non-alienating. Perhaps there would be a strong interrelationship between technology and a culture so that there might be a Japanese technology as there is Japanese culture.

In an era of alternative technologies, sewage would be returned to the land and used as natural fertilizer. Food might be grown locally using new methods that bypass the high energy intensity of contemporary agriculture. Fish farming in village ponds may be a new "soft" technology that provides high-protein foods at low energy cost. The new technology will tend to use renewable resources rather than nonrenewable resources.

The political system of a nation may be said to go through three major eras in the relationship of development to environment. They can be characterized as: (1) the Political Economy of Expansion and Resource Development; (2) the Political Economy of Environmental confrontation; and (3) the Political Economy of Ecology.[2] Most developed nations are moving from the resources development phase to an era of environmental confrontation. The developing nations are still moving into the period of resource development. Moving from phase 1 to 2 depends upon the citizens' recognition of environmental problems and the achievement of a level of affluence and surplus which permits paying for a clean environment and a higher "quality of life."

The characteristics of the three eras are summarized in Table 23–1. The objectives of a nation during the growth phase are:

(1) Greater variety, diversity and available life styles in the society

(2) Increased rates of economic growth and rising living standards

(3) Larger tax bases and increased government revenue

(4) Greater national security

(5) Protection for unique geological or archeological areas such as national parks and monuments.

During this era, the politics of expansion constitute a positive non-zero sum game. Everyone pursues his own objectives at minimal perceived costs to others.

When the society reaches an acceptable level of affluence and its perception of the costs of technology are such that new concern for the environment is evidenced, then a nation enters an era of environment concern and confrontation. New objectives emerge, which may include:

(1) Concern for the capacity of the environment to absorb or regenerate renewable resources (water, air, soil, etc.)

(2) Attention to the health and aesthetic costs of pollution rather than concern for the conservation of "scarce" non-renewable resources

(3) Protection from "external costs" for individuals and groups so affected; imposition of these social costs on polluters who previously did not have to calculate such costs in their operation.

Then the politics of this era become a zero-sum game. There are winners and losers and compromise is necessary between environmental effects and social demands. Then the thrust of government becomes regulative in the environmental sector (at least).

The third phase is the more hypothetical era of ecology. In this era, ecological concerns are of pre-emminent value. The severity of the environmental problem pushes key national leaders to forge comprehensive, integrated solutions. The utility of communal solutions rises. As a result, the politics for the nation becomes a positive non-zero sum situation. At this point technology is used effectively to overcome the adverse effects of resource-use and industrial development. There is in the third phase a heavy emphasis on recycling of materials and the full use of energy at its level of entropy.

Whether in Phase Two or Three, the engineer will no longer be content to sit behind his desk or drawing board, or white-coated in his laboratory, busily designing new devices without considering whether or not they are also for society's benefit. The engineer will be required to document not only the technological value of his development, but also the social benefit. the engineer will need to assist in overcoming the appalling global flight from the country to the city. This world-wide trend needs to be reversed in the name of ecology and

Table 23-1

Summary Table of the Three Eras of Economic Development and Environment.

Area of Comparison	Era	Political Economy of Expansion and Resource Development	Political Economy of Environmental Confrontation	Political Economy of Ecology
Major Political Objectives		Economic Growth, Resource Exploitation, Unique Area Preservation	Continued Expansion and Environmental Quality (Pollution)	Crisis, Survival Environmental Quality (Pollution Resources)
Locus of Policymaking		Legislative Committees Low Level Administrative Agencies	Entire Legislature High Level Administrative Agencies	Administrative Branch
Nature of Policymaking System		Logrolling, Pork Barrel Ad Hoc, Remedial Segmented and Sectoral Decentralized Local Initiative and Solutions	Bargaining, Compromise Remedial National Solutions More Centralized Group Conflict	Administrative Decision Making, Centralized Comprehensive and Integrated Routinized, Cybernated Problem Shed Approach
Dominant Political Force		Self Interested Economic Groups (Industry, Mining, Agriculture) Traditional Geographical and Political Areas	Self Interested Economic Groups Environmentalist Groups	Environmentalist Groups (Balance Shifted by Elite Intervention)
Type of Government Policy		Distributive Specific to Single, Isolated Areas of Activity	Regulative Sectoral but More Unified	Redistributive and Regulative Unified, Regional and International Solution
Type of Political Game		Positive Non-Zero Sum	Zero Sum	Positive Non-Zero Sum as Result of Threatened Negative Non-Zero Sum Situation

Level of Conflict	Minimal	Significant	Minimal, More Cooperation
Amount of Coercion	Minimal	Some, Working Through Individual Conduct	Moderate Amount, Working Through Individual and Environment of Conduct
Techniques for Achieving Policy Objectives	Subsidies, Tax Credits, Depreciation Allowances, Depletion Allowances, Nuisance Laws	Standards, Criteria	Major Commitment of R&D Funds, Large Scale Technological Projects; Project Forms of Environmental Administration; Effluent Fees, Pollution Taxes with Minimum Standards
Specific Interests Benefitting from Government Activities	Extraction and Manufacturing Industries; Municipal Sewage Systems	Environmental Interests; Extraction and Manufacturing Industries; Municipal Sewage Systems	Environmental Interests
Type of Technological Base	Minimum	Pollution Control Systems	Recycling Systems; Revolutionary New Propulsion and Energy Producing Technologies
Relationship to Energy Needs and Environmental Problems	Unlimited Expansion of Capacity; Policy in Terms of Specific Sectors (Oil, Coal, etc.)	Movement Toward an "Energy Policy"; Increasing Government Regulation	Conversion to Revolutionary Technologies; Limits on Growth and Siting; Combined Energy and Environment Policy

society. also, the engineer must assist in the real development of the Third World. This development must be beneficial and compatible with existing social norms and cultures, and it must make maximum use of local resources. The ecological crisis must be resolved in favor of a reintegration of man with the natural systems of which he is a part. Finally, the engineer must assist in placing technology in its proper perspective—surely different from its overreaching position today with regard to other craft and creative activity.

One day the landscape may be dotted with solar houses on farms using intensive but ecologically sound practices. The food production would be dependent upon the integration of many species of animals and plants. Perhaps the community of the future will free men from excessive dependence on external services and living in new community. At present, in the United States about 50% of the people live on 1% of the land and more move to cities each day. New communities are needed. But this is only the bare beginning.

The use of technology to solve the problems of our society in order to make a new future is discussed in the following article. In this article Dr. J.E. Goldman discusses the need for new technologies to satisfy the needs of tomorrow. While in the past we invested our technical resources in the military and space activities, we need now to invest in new civilian technological development, according to the author.*

Toward a National Technology Policy

J.E. Goldman

As we look back over the past decade, we have good reason to be uncomfortable with our national investment in technology. Lack of support is not the cause; in fact, there has been no diminution in the magnitude of technology expenditures over these 10 years. Nor is the magnitude of the total national effort the critical issue now. My own reasons for concern are twofold:

First, it seems to me we have given far too little support to the creation of new technologies, including too little toward the support of science itself. Rather, we have spent our dollars—and the energies of our technical people—on the exploitation of given features of the old technologies, principally through scale-up and increasing reliability.

Second, our technical priorities have held relatively stable, despite the fact that the world about us is undergoing enormous change. During this past decade, to cite one specific fact, our national government has invested some 3 billion to 4 billion man-years in research and development (R & D) programs; less than 1 percent of this enormous investment has gone toward R & D support relating to such critical problems as housing, crime, the urban environment, and ground transportation.

As we correct this imbalance and begin now to channel a larger part of our technical effort toward programs of social significance, we must remember one of the important lessons of our recent past: the history of the space program, it seems to me, is a lesson in

*J.E. Goldman, "Toward a National Technology Policy," *Science*, Vol. 177, Sept. 22, 1972, pp. 1078–1080. Copyright 1972 by the American Association for the Advancement of Science.

the mastery of the institutional techniques necessary to bring together the segments of the intellectual, industrial, and technological community needed to fulfill goals in a timely fashion. If we choose to ignore this, if we set aside the space program's experience as nothing more than a $20-billion waste, then by our irrationality we will endanger the ultimate achievement of such important societal objectives as better housing and the renewal of our inner cities.

Managing the Technology-Creating Process

The formula that worked for the National Aeronautics and Space Administration is clear and straightforward. It is a four-step process:

1) Identify clear-cut goals.

2) Institutionalize the mechanisms for achieving these goals.

3) Engage all segments of society whose talents and resources are needed to fulfill these goals.

4) Create a market to receive the new output (and new technology).

Consider applying the formula to an area where new social needs have arisen. As a test case, let us try it in the field of transportation. To begin with, clear-cut goals do not exist and have never existed. Even the supersonic transport was smuggled into the "system"—not because of its relevance as a component of our long-range transportation programs, but for extraneous rationales about national prestige, trade balance deficits, and aerospace unemployment. For urban mass transport and rail transport in general, the planning process has been fragmented and, consequently, ineffectual. We boast of our great mobility, and yet our current planning effort for the future of this mobility is not nearly sufficient to the task before us.

If we can agree that the optimization of multimodal transportation is a desirable goal for our nation, we can then identify and spell out our transportation goals. Having defined them, we must then step up to the need to institutionalize the mechanisms for evolving the technologies that could help meet those major goals. What we may need is something akin to an Office of Naval Research (ONR) for transportation or, alternatively, something patterned after but broader in outlook than the old National Advisory Committee for Aeronautics that played so important a role in making possible the growth of our civilian air transport industry by providing technical support both through its considerable inhouse capabilities and by contracting for outside technical expertise.

It is a fact that often the most creative talents for a field under development lie outside the original institutional scope of the group charged with its development. Such talents may reside in other industries, universities, at the research institutes, or wherever. A mechanism has to be found for involving them and for attracting their input into the evolving system. The Commerce Technical Advisory Board report (1967) "Innovation, its Environment and Management" speaks eloquently on this point.

The military, and later the space program managers, found the way. Looking at one of today's specific problems, we may ask why a bright theoretical physicist with clever ideas on transportation systems analysis should have to scrounge around to find $10,000 to support his exciting researches. This man, and others like him, are the people who can see how the world could be, rather than how it is. Such people should be made a part of the technical input apparatus, much as the U.S. Navy Department, in the days of ONR, came to have at its disposal the expertise that it needed in undersea warfare. The wisdom and management skill of ONR made that possible. And that is the third step of our process: Engage all segments of the society that can contribute to the achievement of the goals, with special efforts toward coupling the academic with the industrial community, and both with the government.

Finally, all this will come to naught if there is not an established market for the product. In our example, transportation, I think it no accident that this field is one of the hard currency producers in which the United States is threatened by foreign suppliers. Our government may have to create the market for new transportation products and services that will, in turn, stimulate the creation of technologies to fill that market.

Defining Goals for Technology

The mechanisms for assessing the needs, defining the goals, and creating the markets necessitate some high order systems analyses that put into perspective the cost-benefit relationships: the implications of environmental effects (air pollution, noise), economics, and raw material conservation, for instance. The private sector left to its own devices—at least in this field—has not performed. Nor is it likely to perform this function in an objective manner. If it had, we would have a suitable rail system, good urban mass transit, and electric cars available right now. But the stimuli have been lacking. If we had had a Rand Corporation for the Department of Transportation, and a billion dollars a year for procurement of new, alternative transportation systems (developed in parallel, as weapons systems were), we might not have a transportation problem today. It seems almost incredible to realize that our $50-billion federal highway program has gone forward without such national systems analysis—with the result that our total transport network is overbalanced in favor of the automobile, to the detriment of other transport modes. And I think the analogs in housing, urban development, health care, crime control, energy, water resources, and such, are obvious.

In the absence of this kind of mechanism, we will see the perpetuation of misplaced emphasis and lack of balance and lack of choice in the efforts and resources our society elects to expend on its technical choices: in energy, the dominance of oil and gas; in urban transportation, the dominance of the car; in health care, the dominance of the classical hospital; in housing, the dominance of single-family suburban dwelling; and the ancient Roman sewerage for waste disposal.

The first element of the first step, then, in the articulation of a set of national goals for technology, is to make specific the uses toward which we wish to apply that technology. This is a role suited to the federal government; in fact, the federal government is in the best position to give wide credibility and gain consensus and acceptance for the set of goals in the diverse set of social areas. For example, no group outside the federal government could have set up the landing of a man on the moon as a goal for U.S. technology in the 1960's. This is not to say that the goal-setting role should be the exclusive province of the federal government; rather, it is to point out its responsibility for leadership. Nor do I say that goal-setting will be easy. Establishing goals in the civil areas, each fraught with powerful pre-existing vested interests, is likely to be more difficult than it was in the new field of space exploration. Hence all the more reason for beginning to exercise judiciously the federal leadership responsibility today, as a principal means for bringing together the analytical, evaluative, and synthesis skills of the widest possible range of institutions.

Adapting to New Conditions

The technology story has another side that I have not yet touched upon; it also bears some analysis. It is observed by many, most recently by the Haggerty panel of the President's Science Advisory Council, that our economy is moving steadily toward dominance by the service sector. Where once we were predominantly an agricultural economy, and subsequently an industrial economy, now the production of services exceeds the production of physical goods: the values of services are now a larger component of our gross national product than the values originating from manufactured goods. For a country that established its position of world leadership largely through its wealth of natural resources, plus the initiative to exploit those resources efficiently through advances in manufacturing technique and in productivity, this is indeed a major change. All indications point to further dominance by the service sector.

This remarkable evolution has important implications for technology. It tells me that the impact of technology on the economy in the next decades will not come primarily from technology's function of enhancing manufacturing productivity. Rather, I believe technology will be directed toward the improvement of services, or, more broadly expressed, improving the quality of life. Perhaps this is an even more demanding area for application of technology.

New Values Added

Concomitantly, we must ask ourselves whether such a redirection of technological goals will enable us to maintain a position of international preeminence. I believe it will. By utilizing technology to perform more efficiently the services made possible by the artifiacts of the preceding technology, we not only raise the "quality of life" within our own borders, but we also add new value to the services we already export to the rest of the world. In the past we have found the same pattern both fulfilling and enriching. We gave to the world an agricultural revolution; we have contributed dramatically to health care, pharmaceutical technology, and synthetic materials, in each field reducing man's dependence on material resources. As sources of new value-added functions that we can export to the rest of the world we might look to transportation, pollution abatement, energy, health, and education. The higher the technology content of the products or services we can evolve, the more likely we are to maintain some measure of advantage in international economics. In applying our technology and innovation we must recognize the economic, social, and political trends; in fact, technology should help us lead those trends, whenever we are perceptive enough to see them coming.

Federal Technology Policy

To set the current thinking about technology policy in perspective, we should examine the federal government's response to the new social goals. As a measure of that response, I shall take the distribution of the federal R & D budget. An analysis of this area was completed a few months ago by Paul Shapiro of the Sloan School, as part of a summer study led by J. Herbert Hollomon of Massachusetts Institute of Technology. That study shows that, from 1961 to now, the federal government supported (i) 2 million man-years of defense R & D, (ii) about 1 million man-years of space R & D, and (iii) about 175,000 man-years of noncivilian nuclear R & D.

In contrast: (i) the total of all housing, urban social, and crime research that the federal government has ever funded is less than 13,000 man-years; (ii) the total R & D sponsored by the federal government for nonaviation transportation is of the order of 10,000 man-years; (iii) since 1969, 53,000 man-years of R & D have been expended by the federal government for environmental improvement.

During the current fiscal year, federal R & D expenditures are supporting the equivalent of 230,000 scientists and engineers for noncivilian purposes and about 84,000 scientists and engineers for civilian purposes, a ratio of about 3 to 1. In 1961, that ratio was 6 to 1, also with 230,000 scientists supported for noncivilian purposes. Could there be hope for reaching parity someday?

But, even within the federally supported civilian R & D area, we must point out that there has not been a balance of funding among all fields. Health is the largest single component here: of the $4.2-billion civilian R & D in 1970, only $2.4 billion were for non-health purposes. Of a $15.2-billion total, less than $2 billion of federal funds was allocated for R & D for the total of the principal remaining civilian purposes: education, housing, nonaviation transportation, urban social problems, crime control, agriculture, natural resource development, basic research via the National Science Foundation, and civilian nuclear power. Distressing as these numbers may be, it should at least be said that non-health, non-aviation civilian R & D has grown at an average rate of 12 percent per year since 1961.

However, with less than $2 billion per year expended for R & D in these civilian areas, most of which are not subject to the supply-demand-profit relationships of classical markets, it is not surprising that we are not receiving the shot in the arm required to couple technology for the benefit of the public/civilian sector.

National Technology Policy

But let me underscore that national technology policy is more than federal R & D allocation practices. We must encourage all segments of our society—the academic, the industrial, the governmental, and the public at large—periodically to redefine the goals toward which technology should be applied, and to reassess objectives as the environment undergoes change or as we change it.

The initial condition, then, for developing a technology policy must be a re-assessment of goals. In the emergence of national goals during the past two decades, whether imposed externally by such cataclysmic events as the cold war, or sputnik, or internally more calmly, by goals commissions or deliberate private efforts, technology for social benefit has always come out at the small end of the horn. What emerges is a feeling that more intense and coherent social forces will be needed to foster the translation of technology into these areas.

Lest it be a source of confusion, we must remember the salient difference between science and technology. Science is both a means and an end. Technology is only a means. To develop a technology policy, we must identify the ends for which technology will be the means. Technology has no meaning in the abstract, only in relation to specified goals. If, as a society, we can specify these goals, technology can be applied to achieve them, appropriately guided or channeled according to the timetested processes mentioned earlier.

To help this process operate efficiently, the technical community at large must also effect some discipline and brush away some of the old polemics: the schism between the scientist and the engineer for one. Created by some artificially imposed pecking order, this socalled distinction has tended to impose some sort of special favor on the scientist during the last quarter century. But it served no useful purpose then, and we certainly do not need it now, especially in the new social environment. We need the respective contributions of both the scientist and the engineer; they must work together under conditions that allow both to reach their most creative levels. John Gardner put it in its proper perspective when he wrote: "A society that scorns excellence in plumbing because plumbing is a humble profession and exalts mediocrity in philosophy because philosophy is thought to be a noble profession—such society is doomed to failure. Neither its pipes nor its theories will hold water."

THE ENGINEER'S PERSPECTIVE

The role of the future engineer is wide, and it requires men and women with broad educations. Engineers will take new leading roles in industry, commerce, transportation, banking and other areas of our society. Leadership will require new approaches and new insights. The engineer must continually maintain his ability and current knowledge in order to avoid solving tomorrow's problems with yesterday's methods. Ethics and integrity will increasingly become measures of an engineer's professionalism. Now the engineer is often a member of a large firm or bureaucratic organization. As new organizations emerge in our society, the engineer will, one hopes, be freed to serve society increasingly as an individual rather than a component of a large organization. The engineer must be prepared to act as the interface or link among technology, economics, organization and politics. He must resynthesize the fragments caused by the specialization of other men. Such an engineer will not just be interested in costs or plans, but in the final results. He will not be excused when "human factors" defeat his technical accomplishments. Instead, his reputation will rest on how he blends plans and politics, needs and desires, machines and people. The engineer must be a designer, a synthesizer and a doer. He must concern himself with human organizations and institutions; ethics and efficiency; wisdom and leadership.

The need for a new society, the new engineer, and new forms of technology is clear. Kenneth Keniston discusses the need for a new society beyond the old technology in *The Uncommitted*. In the following paragraphs Keniston's thesis is succinctly stated.[4]*

With the age-old goal of universal prosperity within sight, we must question whether the methods—the technological values and virtues, the instrumental goals of our affluent society—that helped us approach this goal will serve to take us beyond it. For most of us, the urgency has gone outof the quest for prosperity . . . .

We are approaching, I believe, a new turning point in American society. Despite our growing affluence, despite the triumphant march of technology, despite the inundation of our society with innovations, something is clearly wrong. All the signs are present: our mid-century malaise, increasingly shrill cries to "rededicate" ourselves to outworn ideologies which can no longer inspire our commitment, a loss in the sense of social power, and all of the attitudes, feelings, and outlooks I have here called the "new alienation." The vision of an affluent society no longer excites us; and so too, we are losing our implicit faith in the ancillary beliefs of technology. In nations where affluence is still a distant dream, the situation is different: in Peru or Nigeria, in Thailand or Samoa, the struggle to attain some small freedom from suffocating poverty is still a compelling struggle. Nor is the achievement of affluence complete even in America: the spate of recent books on the "forgotten fifth" of the nation, on our "invisible poor" eloquently documents the distance we must still travel. Yet these same books, with their appeal for affluence for *all*, indirectly attest to the triumph of technology. Who, a century ago, would have complained that *only* 80 per cent of the people were prosperous? And who would have dared insist that *all* might be well fed, well housed, well educated, and well leisured?

Thus, paradoxically, at the very moment when affluence is within our reach, we have grown discontented, confused, and aimless. The "new alienation" is a symptom and an expression of our current crisis. The individual and social roots of our modern alienation, I have tried to suggest, are complex and interrelated; yet if there is any one crucial factor at the center of this alienation, it is the growing bankruptcy of technological values and visions. If we are to move toward a society that is less alienating, that releases rather than imprisons the energies of the dissident, that is truly worthy of dedication, devotion, idealism, and commitment, we must transcend our outworn visions of technological abundance, seeking new values beyond technology.

The values, requirements, demands, and virtues of technology are not in themselves bad; but their unquestioned supremacy is a human and social misfortune. Judgments of skill, competence, and effectiveness have replaced usefulness, beauty, and relevance to human needs as criteria of worth; instrumental values have replaced final purposes; and cognitive skills have replaced virtuous character as standards of human value. Nor is the technological process necessarily destructive: it only becomes destructive when men and women serve it, rather than vice versa. The human problems in our society stem not from the fact of technology, but from the supreme place we assign it in our lives. We grant it this place largely because in the recent American past higher productivity *was* a prerequisite to a decent life for our people. But this is no longer true, and increasingly technology dominates by default—because it is *there*, and countervailing values, goals, and purposes are not. The dominance of technology therefore springs ultimately from the failure of positive values in our society, and from a collective failure of imagination in the West as a whole. To ascribe causal primacy to technology itself—to make economics the motor of society—is a mistake. Equally important is our willingness to *allow* it to be the motor, and this willingness is ultimately a matter of ideology and social myth.

The primacy of technology leads to the dissociation and subordination of all that is not technological, of all that we can call the affective and expressive sides of life . . . .

Furthermore, the supremacy of technological values means that our society has no honored place for those who do not possess the virtues of its values . . . .

What our society lacks, then, is a vision of itself and of man that transcends technology. It exacts a heavy human toll not because technology exists, but because we allow technology to reign. It alienates so many not simply because they do not share its wealth, but because its wealth includes few deeply human purposes. It is a society that too often discourages human wholeness and integrity, too frequently divides men from the best parts of themselves, too rarely provides objects worthy of commitment. In all these ways, it exacts a heavy human toll.

Toward a More Human Society

If we are to seek values beyond technology, purposes beyond affluence, visions of the good life beyond material prosperity, where are these values, purposes, and visions to be found? . . . The turning point at which we stand today requires a similar translation of already existing dreams of human fulfillment and social diversity into the concrete goals of individuals and of our society. The values we need are deeply rooted in our own tradition: we must merely begin to take them seriously.

• • •

For those who do not want materially and are not oppressed politically, the quest for fulfillment beyond material goods becomes possible and urgent. ... Today, in America, and increasingly in other technological nations, ... we can now begin to imagine realistically that a whole society might commit itself to the attainment of the greatest possible fulfillment for its members.

To be sure, by the quantitative and reductionistic standards of our technological era, goals like "human wholeness," "personal integration," "the full development of human potentials" are inevitably vague and imprecise. They point to the quality of individual life, rather than to quantitatively measurable entities. ...

Just as there are from birth many distinct individuals, each with his own unique genetic and environmental potential, there must remain many paths to fulfillment. ...

The re-emergence of humane values over purely technological values may be the cornerstone of the last decades of the twentieth century. As we seek new visions in new uses of technology in its proper perspective, we shall seek a more human society. While the new goals, such as the quality of life, are inevitably vague and difficult to measure, they are not less important than quantifiable goals.

The Scarecrow and the Tin Woodsman are talking on their way to Oz.

"I don't know enough," replied the Scarecrow cheerfully. "My head is stuffed with straw, you know, and that is why I am going to Oz to ask him for some brains."

"Oh, I see," said the Tin Woodman. "But, after all, brains are not the best things in the world."

"Have you any?" inquired the Scarecrow.

"No, my head is quite empty," answered the Woodman, "but once I had brains, and a heart also; so, having tried them both, I should much rather have a heart."

We can dream that in the future our society will have both brains and a heart; both knowledge and compassion; both machines and ecological wholeness; both technology and new humanity. This may be the new dream of man concerning his new technology.

CHAPTER 23 REFERENCES

1. R. Clarke, "Technology for An Alternative Society," *New Scientist*, Jan. 11, 1973, pp. 66–68.

2. D. Schooler, Jr., "The Politics of Environment and National Development," from *Growing Against Ourselves: The Energy-Environment Tangle*, edited by S.L. Kwee and J.S.R. Mullender, Lexington Books, Lexington, Mass., 1972.
3. J.E. Goldman, "Toward a National Technology Policy," *Science*, Sept. 22, 1972, pp. 1078–1080.
4. K. Keniston, *The Uncommitted: Alienated Youth in American Society*, Harcourt, Brace and Jovanovich, New York, 1965, Chapters 13 and 14.

CHAPTER 23 EXERCISES

23–1. Science and technology have been subjected to extreme criticism during the past five years. Do you agree that technology has undergone a transformation during this period? What new values do you believe will emerge in the next few decades, especially concerning technology? Is there a new definition of progress?

23–2. In the era of the Political Economy of Environmental Confrontation, how will our society undergo new transformation? Are there ways to accommodate the new environmental demands without some severe difficulties occuring in certain segments of industry and to certain groups of engineers? Will the transition accommodate the need to bring the economy along with the changes?

23–3. Consider the article by Dr. Goldman and discuss the applicability of technology in the civilian sector. Discuss the added difficulties that the application of a technology incurs in the civilian sector compared with the application of technology in the military or aerospace sectors.

23–4. As we move toward a more humane society, what values must be stressed? Would you retain the technologically-oriented values of efficiency, productivity and other pragmatic values?

INDEX